PROGRESS IN
MATHEMATICS

Volume 9

Algebra and Geometry

PROGRESS IN MATHEMATICS

Translations of Itogi Nauki – Seriya Matematika

1968: Volume 1 – Mathematical Analysis
Volume 2 – Mathematical Analysis

1969: Volume 3 – Probability Theory, Mathematical Statistics, and Theoretical Cybernetics
Volume 4 – Mathematical Analysis
Volume 5 – Algebra

1970: Volume 6 – Topology and Geometry
Volume 7 – Probability Theory, Mathematical Statistics, and Theoretical Cybernetics
Volume 8 – Mathematical Analysis

1971: Volume 9 – Algebra and Geometry

In preparation:

Volume 10 – Mathematical Analysis
Volume 11 – Probability Theory, Mathematical Statistics, and Theoretical Cybernetics
Volume 12 – Algebra and Geometry

PROGRESS IN MATHEMATICS

Volume 9

Algebra and Geometry

Edited by

R. V. Gamkrelidze

V. A. Steklov Mathematics Institute
Academy of Sciences of the USSR, Moscow

Translated from Russian by Nasli H. Choksy

SPRINGER SCIENCE+BUSINESS MEDIA, LLC • 1971

The original Russian text was published for the All-Union Institute of Scientific and Technical Information in Moscow in 1969 as a volume of *Itogi Nauki – Seriya Matematika*

Library of Congress Catalog Card Number 67-27902
SBN 306-39209-7

The present translation is published under an agreement with Mezhdunarodnaya Kniga, the Soviet book export agency

ISBN 978-1-4684-3308-1 ISBN 978-1-4684-3306-7 (eBook)
DOI 10.1007/978-1-4684-3306-7

Originally published by Plenum Press, New York
Softcover reprint of the hardcover 1st edition 1971

United Kingdom edition published by Plenum Press, London
A Division of Plenum Publishing Company, Ltd.
Donington House, 30 Norfolk Street, London W.C.2, England

Preface

This volume contains five review articles, two in the Algebra part and three in the Geometry part, surveying the fields of categories and class field theory, in the Algebra part, and of Finsler spaces, structures on differentiable manifolds, and packing, covering, etc., in the Geometry part. The literature covered is primarily that published in 1964–1967.

Contents

ALGEBRA

GEOMETRY

ALGEBRA

Categories

M. S. Tsalenko and E. G. Shul'geifer

§1. Introduction

The present survey includes papers reviewed in the "Mathematics" section of Referativnyi Zhurnal from October 1962 to October 1967. During the period indicated there was a sharp increase in the number of publications on various aspects of category theory: the previous survey [17] comprised about 40 papers, while now there are more than two hundred. The authors have done their best to pinpoint the directions of research which in their opinion are becoming consolidated and to illuminate in more detail the fundamental papers in each direction; at the same time, certain papers are mentioned only in the Literature Cited section. On the other hand, the extensive group of papers by Ehresmann [70-107] on categories of structured structures is almost not at all reflected in this survey for two main reasons: first, the too-free use of the concept of class has not received wide support in the literature; second, the large number of awkward definitions makes it impossible to present Ehresmann's results within the scope of this article; the interested reader can refer to Ehresmann's book [95]. In the section on homological algebra we have included papers which are of general categoric nature because papers on homologies in concrete, Abelian or non-Abelian categories are not essentially of concern to category theory.

The authors have attempted to unify the terminology at least to some degree and have therefore departed, in a number of cases, from the terminology adopted in the papers cited and in their reviews. In view of the deviation from the notation of the previous survey and of the article [14] by Kurosh, Livshits, and Shul'geifer, we present certain notation used in this survey:

Categories	$\mathfrak{A}, \mathfrak{B}, \mathfrak{K}, \mathfrak{L}, \mathfrak{M}, \mathfrak{N}, \ldots$
Category of sets	$\mathfrak{S}$
Objects	A, B, C, D, X, Y, . . .
Functors	F, G, R, S, T, V, . . .
Morphisms	$\alpha, \beta, \gamma, \ldots$
Null object	0
Null morphism from object A into object B	0_{AB} or 0
Identity morphism	1_{AB} or 1.

The product of morphisms and of functors is written from left to right; in particular, if $F:\mathfrak{K}\rightarrow\mathfrak{L}$ and $G:\mathfrak{L}\rightarrow\mathfrak{M}$ are two unary functors, then FG(A) = G(F(A)) for any object $A\in\mathfrak{K}$. For each category $\mathfrak{K}$ the bifunctor from $\mathfrak{K}$ into $\mathfrak{S}$ which associates the set of morphisms from A into B with each pair of objects $A, B\in\mathfrak{K}$ is denoted by H(A, B) or $H_{\mathfrak{K}}(A, B)$.

§ 2. Foundations of the Theory of Categories

The use of the concept of class in the definition of a category leads to difficulties with the set-theoretic foundations of mathematics. First of all it turns out that the category of sets $\mathfrak{S}$ is not a universal category since not every category is concrete, i.e., not every category can be isomorphically embedded in category $\mathfrak{S}$ (Isbell [138], Freyd [116], Drbohlav [61]). Further, the "structures," connected with the natural concepts of subobject and factor object of an object of a category, sometimes turn out to be meaningless. Finally, such important examples as the categories of functors and their natural transformations are not always admissible for consideration. The papers by MacLane [172], Sonner [217], and Isbell [138] are devoted to surmounting the set-theoretic difficulties connected with the use of classes of objects and morphisms in the theory of categories.

As a set-theoretic system of axioms Sonner suggested a system in which a universal set U containing X is assumed to exist for each set X. A set U is called universal if the following conditions are fulfilled:

U1. If the sets $X_\alpha \in U$ and if the set Ω of indices $\alpha \in \Omega$ belongs to U, then $\bigcup_{\alpha\in\Omega} X_\alpha \in U$.

U2. If $x \in U$, then the set consisting of just the one element x also belongs to U.

U3. If $x, y \in U$, then $\{x, y\} \in U$.

U4. If $x \in X \in U$, then $x \in U$.

U5. If $X \in U$, then the set of subsets of set X belongs to U.

Within the framework of such a theory of sets it is reasonable to consider categories, objects, and morphisms which belong to a fixed universal set. Under such an approach such transformations as categories of functors also become admissible. Also see Gabriel [118].

Another approach was proposed by MacLane [172]. We recall that a category $\mathfrak{K}$ is called small if its objects make up a set. Now let $\mathfrak{A}$ be an Abelian category. A full subcategory $\mathfrak{A}'$ of category $\mathfrak{A}$ is called left-dense if together with the morphism $\alpha \in \mathfrak{A}'$ its kernel and cokernel are contained in $\mathfrak{A}$ and if for any morphism $\alpha: A \to B \in \mathfrak{A}'$ and for any epimorphism $\eta: C \to B \in \mathfrak{A}$ we can find an epimorphism $\eta': D \to A \in \mathfrak{A}'$ and a morphism $\alpha': D \to C$ such that $\eta'\alpha = \alpha'\eta$. The category $\mathfrak{A}$ is called left-bounded if any set of its objects is contained in some small left-dense subcategory $\mathfrak{A}'$. Right-bounded categories are defined in dual fashion. It turns out that the category ${}_R\mathfrak{M}$ of all left modules over a ring R is left- and right-bounded. The structure of derived functors suggested by Buchsbaum turns out to be correct in left-bounded categories.

Isbell [138] stated conditions on a category under which the consideration of the category of functors becomes admissible.

We also note the isolation of locally small categories, i.e., categories in which the subobjects and factor objects of any object form a set; if the set consists of only subobjects (factor objects), then we call the category locally left-small (right-small).

In a long series of papers [70-107] Ehresmann, in general, relinquished the requirement that for any objects A and B of category $\mathfrak{K}$ the collection of morphisms from A into B comprises a set, and admitted an extraordinarily broad operation with classes: the direct products of classes, classes of classes, etc. However, such a rejection of all restrictions has not yet received wide support.

§ 3. Fundamentals of the Theory of Categories

1. Three monographs on category theory appeared during the period being considered: the books by Freyd [116], Ehresmann [95], and Mitchell [184]; the fundamental concepts of category theory were also presented in MacLane's monograph [19]. The articles by Braconnier [53], Roux [206], Sonner [217], Vidav [227], and Kauffman [147] were devoted to the exposition and popularization of category theory fundamentals. Tondeur's article [223] is a good survey of articles on category theory.

2. Simultaneously with these there occurred a refinement of the definitions of some of the fundamental concepts, a clarification of their properties, and a further building of the tools of category theory. In particular, there was a continuation of the study of special types of monomorphisms and epimorphisms of a category, as well as different investigations connected with the refinement of the concepts of subobject and of the image of a morphism. This was mainly in connection with the following circumstances. 1) Monomorphisms and epimorphisms are not, respectively, one-to-one mappings and "onto" mappings in all the sufficiently well-known concrete categories; therefore, there arose the need to pick out only such monomorphisms and epimorphisms in the category setting. 2) Not in every category does the subobject of an object A, defined as the collection of equivalent pairs (U, σ) consisting of an object U and a monomorphism $\sigma : U \to A$, correspond to the natural understanding of a subobject of an object in a given category, even if the monomorphism σ is a one-to-one mapping. 3) The desire to define the image of an arbitrary morphism $\alpha : A \to B$ as a special form of the subobject (U, σ) of object B such that the morphism α is representable as the product $\alpha = \nu\sigma$, where ν is some special form of epimorphism, and such a representation of morphism α would be defined uniquely to within equivalence. Semadeni [211, 212] and Drbohlav [62] defined a bicategory as a category $\mathfrak{K}$, considered together with two fixed subcategories $\mathfrak{E}$ and $\mathfrak{M}$, satisfying the following conditions: a) $\mathfrak{E}$ and $\mathfrak{M}$ consist, respectively, of certain epimorphisms and certain monomorphisms of category $\mathfrak{K}$ (the epimorphisms of $\mathfrak{E}$ and the monomorphisms of $\mathfrak{M}$ are called, respectively, admissible epimorphisms and admissible monomorphisms of bicategory $\mathfrak{K}$); b) the intersection $\mathfrak{E}\cap\mathfrak{M}$ consists of all isomorphisms (invertible morphisms) of category $\mathfrak{K}$; c) each morphism α of category $\mathfrak{K}$ is uniquely representable, to within equivalence, as the product $\alpha = \nu\mu$, where $\nu\in\mathfrak{E}$ and $\mu\in\mathfrak{M}$. When we need to stress that a category $\mathfrak{K}$ is being considered as a bicategory we shall write: $\mathfrak{K} = (\mathfrak{K}, \mathfrak{E}, \mathfrak{M})$. A category $\mathfrak{K}$ cannot be allotted the structure of a bicategory uniquely. For example, Semadeni [212] showed that the structure of a bicategory can be defined on the category of locally compact Abelian groups in two ways, by choosing as the admissible epimorphisms and monomorphisms continuous homomorphisms on everywhere dense subgroups and, in the one case, bicontinuous one-to-one "into" homomorphisms and, in the other, continuous "onto" homomorphisms and continuous one-to-one "into" homomorphisms. Also see Pupier [194].

Drbohlav [60] showed that in the categories of semigroups and of associative rings not every epimorphism is an "onto" mapping and at the same time in the category of all one-type universal algebras every epimorphism is an "onto" mapping. Burgess devoted his paper [56] to the study from this same point of view of the epimorphisms and monomorphisms in certain concrete categories. Semadeni [212] proposed the following definition of a subobject of object A of a concrete category $\mathfrak{K}$. The subset $B\subseteq A$ is a subobject of object A if: 1) B is an object of category $\mathfrak{K}$; 2) the natural embedding $\iota: B \to A$ is a morphism of category $\mathfrak{K}$; 3) if $\varphi: C \to A$ is a morphism of category $\mathfrak{K}$ and $\varphi(C)\subseteq B$, then $\varphi: C \to B$ also is a morphism of category $\mathfrak{K}$; 4) for any element $a\in A\setminus B$ there exist an object D and morphisms $\varphi, \psi: A \to D$ such that $\varphi(b) = \psi(b)$ for all $b\in B$ and $\varphi(a) \neq \psi(a)$. The proposed definition of a subobject coincides with the natural definition of a subobject in such concrete categories as, for example, the category of groups, of topological spaces, and some others, but not in all concrete categories. A purely categorical characterization of the closed subsets of topological spaces has been presented by Trnková [24] and Hušek [135]. Hušek delineated monomorphisms satisfying the following condition: if the monomorphism μ is representable as the product $\mu = \nu_1\mu_1$ of a bimorphism ν_1 and a monomorphism μ_1, then ν_1 is an isomorphism. These same monomorphisms were studied by Isbell [139] who called them extremal monomorphisms. For each object A of category $\mathfrak{K}$ Isbell considered the so-called fundamental contravariant functor $H^A = H(\cdot, A)$ of category $\mathfrak{K}$ into the category of sets, which associates with every object B the set of morphisms $H^A(B) = H(B, A)$, and considered the subfunctors of the functor H^A, i.e., contravariant functors I of category $\mathfrak{K}$ into the category of sets such that the relations $I(B) \subseteq H^A(B)$ and $\alpha \cdot I(C) \subseteq I(B)$ are satisfied for every object B and every morphism $\alpha : B \to C$. If $\sigma : U \to A$ is some monomorphism, then denoting by $I_\sigma(B)$ the subset of the set of morphisms H(B, A) represented as $\beta\sigma$, we obtain an example of a subfunctor I_σ of the functor H^A. Such a subfunctor I_σ is called a representable monomorphism σ. A subfunctor I of the functor H^A is called polar if for any object $B \in \mathfrak{K}$ and any morphism $\alpha \in H^A(B) \setminus I(B)$ there exist morphisms $\varphi, \psi : A \to C$ such that $\beta\varphi = \beta\psi$ for every $B \in I(B)$ and $\alpha\varphi \neq \alpha\psi$. A monomorphism $\mu: U \to A$, representing a polar subfunctor is called a polar monomorphism. The monomorphism $\mu: U \to A$ is polar if and only if for any morphism $\gamma: V \to A$ satisfying the condition $\gamma\varphi = \gamma\psi$ each time that $\mu\varphi = \mu\psi$, there exists a morphism $\gamma' : V \to U$ such that $\gamma = \gamma'\mu$. Polar monomorphisms are extremal monomorphisms, and they possess many of the properties of normal monomorphisms. An extremal subobject and a polar subobject of an object are naturally defined. Isbell analyzed a locally left-small category $\mathfrak{K}$, in which for every family of objects there exists their direct product and every polar subfunctor of any fundamental contravariant functor is representable. In the category $\mathfrak{K}$ considered the intersection of any set of extremal subobjects of any object exists and is an extremal subobject. Furthermore, the product of two extremal monomorphisms is an extremal monomorphism. The class of extremal monomorphisms is in reality the smallest class of monomorphisms closed relative to intersection and multiplication and containing all polar monomorphisms. The category $\mathfrak{K}$ is a bicategory relative to the class of all epimorphisms and the class of all extremal monomorphisms. A somewhat different approach to the concepts of subobject and factor object of an object of a category was outlined by Salvioli and Bosch [210].

Special classes of monomorphisms and epimorphisms have also been considered by Semadeni [213] and by Jurchescu and Lascu [145].

3. Several papers have taken up the question of the possible introduction of the concept of a free object in category theory. Semadeni [211, 212] called an integral object U a fundamental free object of category $\mathfrak{K}$, if U is the retract of any other integral object U' and if any integral object which is a retract of object U is isomorphic to object U. (An object A is called a retract of object B if there exist morphisms $\sigma : A \to B$ and $\pi : B \to A$ such that $\sigma\pi = 1_A$.) The free product of any set of copies of a fundamental free object U is called a free object. Conditions were derived under which in the bicategory $\mathfrak{K}$ for every object A there exists an admissible epimorphism $\nu : P \to A$, where P is some free object.

Another approach to the concept of a free object was projected by MacLane [19], Mersch [182, 183], Frei [114], and Tondeur [223]. Let $F : \mathfrak{K} \to \mathfrak{L}$ be a functor from category $\mathfrak{K}$ into category $\mathfrak{L}$. An object $A \in \mathfrak{K}$ is called free relative to functor F with a generating object $M \in \mathfrak{L}$, if there exists a morphism $\lambda : M \to F(A)$ such that for any morphism $\gamma : M \to F(B)$ there exists an identity morphism $\gamma' : A \, \sigma \, B$ of category $\mathfrak{K}$, such that $\gamma = \lambda F(\gamma')$. Each object of category $\mathfrak{L}$ is a generating object for some free object of category $\mathfrak{K}$ relative to functor F if and only if the functor F possesses a left-adjoint functor.

4. The study of the properties of direct and free products and of closely related operations has continued. Lecouturier [163] has shown that the direct product of a finite number of epimorphisms, in a category $\mathfrak{K}$ satisfying the Hoffmann conditions [132], is an epimorphism. Furthermore, if in a category $\mathfrak{K}$ the free product of a finite number of objects always coincides with the direct product of those same objects, then category $\mathfrak{K}$ is Abelian. Tsalenko [29] defined the proper product of objects of an arbitrary category $\mathfrak{K}$ with null morphisms in the following way: an object A is the proper product of a family of objects A_i, $i \in I$, if there exist morphisms $\sigma_i : A_i \to A$ and $\pi_i : A \to A_i$ such that $\sigma_i \pi_j = 0$ when $i \neq j$, $\sigma_i \pi_i = 1_{A_i}$, and $\alpha = \beta$ follows from $\sigma_i \alpha = \sigma_i \beta$ for all $i \in I$. The free product, of course, is a special case of the proper product. Dual to the proper product there is defined the special subdirect sum of objects, a special case of which, dual to the free product, is the direct product of objects. Under the condition that direct and free products exist in a category $\mathfrak{K}$ the object A is the proper product of objects A_i, $i \in I$, if and only if the canonical morphism $\prod^*_{i\in I} A_i \to \prod^{\times}_{i\in I} A_i$ of a free product into a direct product is representable in the form $\prod^*_{i\in I} A_i \overset{\nu}{\to} A \overset{\sigma}{\to} \prod^{\times}_{i\in I} A_i$ and here ν is an epimorphism. The proper product satisfies both the postulate of Mal'tsev and the postulate of MacLane. We remark here, however, that under somewhat more stringent restrictions on category $\mathfrak{K}$ Shul'geifer [37] has shown that each object A of category $\mathfrak{K}$, possessing a system of ideals (A_i, σ_i), $i \in I$, such that $\bigcup_{i\in I}(A_i, \sigma_i) = (A, 1_A)$ and $(A_i, \sigma_i) \cap (\bigcup_{j\neq i}(A_j, \sigma_j)) = (0, 0)$ for every $i \in I$ (the union and the intersection are taken in the complete structure of the ideals of object A), is a special subdirect sum of objects A_i with embeddings σ_i, $i \in I$, if and only if the complete structure of the ideals of any object $A \in \mathfrak{K}$ is completely Dedekind, i.e., for any two systems (X_i, τ_i) and (Y_i, λ_i), $i \in I$, of ideals of object A, satisfying the condition $(X_i, \tau_i) \leq (Y_j, \lambda_j)$ when $i \neq j$, there is satisfied the relation

$$(\bigcup_{i\in I}(X_i, \tau_i)) \cap (\bigcap_{i\in I}(Y_i, \lambda_i)) = \bigcup_{i\in I}((X_i, \tau_i) \cap (Y_i, \lambda_i)).$$

The proper product of objects A_i, $i \in I$, simultaneously also being the special subdirect sum of those same objects, is called the discrete direct product of objects A_i. Under the condition that normal images exist in category $\mathfrak{K}$ as well as direct and free products, Livshits, Tsalenko, and Shul'geifer [18] have shown that the discrete direct product is a covariant functor of its own factors. Discrete direct products were considered also by Wyler [232] who, among other results, derived certain conditions for the coincidence of the discrete direct product of a pair of objects with the direct product of the same objects.

5. A number of papers have appeared with various generalizations of the fundamental concepts of category theory. Hoehnke [131] generalized the concept of a null system of morphisms. Lavendhomme [159] introduced the concepts of monomorphism, epimorphism, kernel and cokernel of a morphism relative to a two-sided ideal $\mathfrak{a}$ of category $\mathfrak{K}$. Here the two-sided ideal $\mathfrak{a}$ of category $\mathfrak{K}$ is defined naturally as the class of morphisms of category $\mathfrak{K}$ such that if the morphism $\alpha \in \mathfrak{a}$, then $\alpha\beta$, $\gamma\alpha \in \mathfrak{a}$ for any morphisms β and γ of category $\mathfrak{K}$, for which the products indicated have meaning.

The concepts of the kernel and cokernel (or, in other words, the left and right equalizers) of a system of morphisms $\gamma_i : A \to B$, $i \in I$, are important generalizations of the concepts of the kernel and cokernel of a morphism. The kernel $\mathrm{Ker}(\gamma_i, i \in I)$ of the system of morphisms $\gamma_i : A \to B$, $i \in I$, is defined as the morphism $\mu : K \to A$ such that $\mu\gamma_i = \mu\gamma_j$ for any $i, j \in I$, and if for some morphism $\varphi : C \to A$ there is fulfilled the condition $\varphi\gamma_i = \varphi\gamma_j$ for any $i, j \in I$, then there exists an identity morphism $\varphi' : C \to K$ such that $\varphi = \varphi'\mu$. The cokernel of a system of morphisms is defined in dual manner. It is evident that the kernel of a system of morphisms is a special case of the inverse limit. Freyd [116] and Mitchell [184] called a category $\mathfrak{K}$ (finitely) left-complete if for every (finite) family of objects in $\mathfrak{K}$ there exists their direct product and there exists the kernel of any pair of morphisms. A right-complete category is defined in dual manner. A category which is simultaneously left- and right-complete is called complete. Maranda [176], as well as Eckmann and Hilton [67] and Mitchell [184], proved that in order for every functor F from any small category $\mathfrak{L}$ into a category $\mathfrak{K}$ to possess an inverse limit, it is necessary and sufficient that category $\mathfrak{K}$ be left-complete. Here we note, however, that Maranda also proved that the functor $F : \mathfrak{K}_1 \to \mathfrak{K}_2$ commutes with the inverse limits if and only if it commutes with the direct products and with the kernels of pairs of morphisms. As Gray [123], Tondeur [223], and Mitchell [184] have shown, the existence of a left-adjoint functor for the functor F is a sufficient condition for the functor $F : \mathfrak{K}_1 \to \mathfrak{K}_2$ to commute with the inverse limits.

6. An essential tool by means of which the various structure diagrams in category theory are constructed, are the so-called couniversal and universal squares [or, left and right intersections of pairs of morphisms with a common end (beginning), or, pull-back and push-out]. The couniversal square for a pair of morphisms $\alpha : A \to C$ and $\beta : B \to C$ is nothing else but their inverse limit, i.e., the commutative diagram

$$\begin{array}{ccc} D & \xrightarrow{\beta'} & A \\ {\scriptstyle\alpha'}\downarrow & & \downarrow{\scriptstyle\alpha} \\ B & \xrightarrow[\beta]{} & C \end{array} \qquad (1)$$

such that for any two morphisms $\varphi: E \to A$ and $\psi: E \to B$ satisfying the condition $\varphi\alpha = \varphi\beta$ there exists an identity morphism $\gamma: E \to D$ such that $\varphi = \gamma\beta'$ and $\psi = \gamma\alpha'$. The universal square is defined in dual fashion. In the case when α and β are monomorphisms the structure of the couniversal square is the structure of the intersection of the subobjects (A, α) and (B, β) of object C. If α and β are null morphisms, then $D = A \times B$ is the direct product of objects A and B. The complete pre-images of subobjects relative to some morphism are constructed with the aid of the couniversal squares. In the category of groups the structure of a universal square for a pair of natural embeddings of a subgroup into a group is the structure of the free product of two groups with a combined subgroup. The investigations of the various properties of universal and couniversal squares can be found in Hilton and Ledermann [128], Gabriel [118], Pupier [194], Mersch [181], Tondeur [223], Freyd [116], Mitchell [184], and MacLane [19]. Here we note also that Trnkova [224] has proven the existence, for any two small categories $\mathfrak{K}_1$ and $\mathfrak{K}_2$ with some common subcategory $\mathfrak{L}$ of the free product of categories $\mathfrak{K}_1$ and $\mathfrak{K}_2$ with a combined subcategory $\mathfrak{L}$, i.e., the existence, in the category of small categories of a universal square for two small categories $\mathfrak{K}_1$ and $\mathfrak{K}_2$ and the natural embeddings of subcategory $\mathfrak{L}$ into categories $\mathfrak{K}_1$ and $\mathfrak{K}_2$.

7. Researches devoted to total and partial summability of morphisms have continued. Livshits [15] considered a category $\mathfrak{K}$ with null morphisms and with a direct product for each family of objects. A family of morphisms $\alpha: A \to B$, $i \in I$, is said to be summable if there exist a direct product $C = \prod\limits_{i \in I}^{\times} C_i(\pi_i, \sigma_i)$ and morphisms $\beta_i: A \to C_i$, $\gamma_i: C_i \to B$, $\beta: A \to C$, $\gamma: C \to B$ such that $\alpha_i = \beta_i\gamma_i$, $\beta_i = \beta\pi_i$, $\gamma_i = \sigma_i\gamma$. Necessary and sufficient conditions were derived for the partial summation of morphisms thus introduced to satisfy all five Kurosh conditions [12], and here, for the direct product of any family of objects to coincide with the direct product of these same objects in the sense of Kurosh's paper already cited. Isbell [140] considered a category $\mathfrak{K}$ with null morphisms and with free and direct products for every finite family of objects, as well as the functors $\times$ and $*$, respectively, of a direct and a free product of a pair of objects of category $\mathfrak{K}$. The total or partial n-ary operation Q, defined on the sets H(A, B), is called natural if $Q(\beta\alpha_1\gamma, \ldots, \beta_n\alpha\gamma) = \beta Q(\alpha_1, \ldots, \alpha_n)\gamma$ every time one of the sides of the equality has meaning. A one-to-one correspondence was established between all natural transformations of the functor × into the functor * and all total natural binary operations over the morphisms of category $\mathfrak{K}$, The natural transformation of the functor × into the functor *, corresponding to a commutative and associative natural binary operation +, satisfying the condition $\alpha + 0 = \alpha$ for any morphism α and for the null morphism 0, was pointed out. If category $\mathfrak{K}$ is, furthermore, a bicategory $\mathfrak{K} = (\mathfrak{K}, \mathfrak{E}, \mathfrak{M})$ and if $A * B \overset{\gamma^\times}{\to} A + B \overset{\gamma_\times}{\to} A \times B$ is the canonical expansion of the canonical morphism $\gamma : A*B \to A \times B$ in bicategory $\mathfrak{K}$ then $A + B$ is the discrete direct product of objects A and B. It was shown in what way the partial summation of a pair of morphisms can be introduced with the aid of the discrete direct product, and conditions were derived under which this summation will be natural. Lecouturier [163] also has considered the partial summation of morphisms.

8. Freyd [116] and Mitchell [184] have shown that an Abelian category can be defined with the aid of the following conditions: 1) the null object exists; 2) for each pair of objects their direct and free products exist; 3) every morphism possesses a kernel and a cokernel; 4) every monomorphism is the kernel of some morphism and every epimorphism is the cokernel of some morphism. Gabriel [118] and Mitchell [185] considered a right-complete Abelian category $\mathfrak{K}$. In such an Abelian category $\mathfrak{K}$, as follows, for example, from Maranda's results [176] mentioned in **5**, for every functor $F : \mathfrak{D} \to \mathfrak{K}$ from a small category $\mathfrak{D}$ into category $\mathfrak{K}$ there exists the direct limit $\varinjlim F$. The taking of the direct limit is a functor from the Abelian category $H(\mathfrak{D}, \mathfrak{K})$ of functors and their natural transformations from category $\mathfrak{K}$ into category $\mathfrak{D}$ in the Abelian category $\mathfrak{K}$. In order for any small category $\mathfrak{D}$ of the indicated functor $\varinjlim : H(\mathfrak{D}, \mathfrak{K}) \to \mathfrak{K}$ to be exact, it is necessary and sufficient that for each upward directed sequence of subobjects (U_i, σ_i), $i \in I$, and for any subobject (V, λ) of an arbitrary object $A \in \mathfrak{K}$ there is satisfied the equality $(V, \lambda) \cap (\bigcup_{i \in I} (U_i, \sigma_i))$ $\bigcup_{i \in I} ((U_i, \sigma_i) \cap (V, \lambda))$. When the last condition is fulfilled we say that $\mathfrak{K}$ is an Abelian category with exact direct limits. In the papers mentioned it is also proven that in an Abelian category $\mathfrak{K}$ with a set of generating objects and with exact direct limits, the injective envelope (see [17]) exists for every object. In the cited paper by Gabriel it is proved that every Abelian category $\mathfrak{K}$ is defined uniquely to within equivalence by any complete subcategory J of the injective objects of category $\mathfrak{K}$, such that for every object $A \in \mathfrak{K}$ there exists at least one monomorphism $\sigma : A \to I$, where I is some injective object from J.

9. In certain papers, mainly as auxiliary results, lemmas are proved of the type of the lemma on five morphisms, the nine lemma,* the lemma on isomorphism, and the lemma on connecting morphism and its various generalizations. For Abelian categories these lemmas are contained in the papers of Puppe [195] and Mersch [181]. For various non-Abelian categories all or some of the lemmas indicated have been proved by Leicht [167], Livshits, Tsalenko and Shul'geifer [18], Bănică and Popescu [45], Wyler [231], and Lecouturier [163]. Here we also note that in Leicht's paper cited it is also proved that in a quasi-exact category, i.e., in a locally small category in which every morphism decomposes into the product of a normal epimorphism and a normal monomorphism and which possesses a kernel and a cokernel, the structure of the ideals of an arbitrary object is Dedekind.

§ 4. Embeddings of Categories

Several papers have been devoted to the possibility of embedding a given category $\mathfrak{K}$ as a subcategory into some category $\mathfrak{K}'$ satisfying some additional conditions. Shul'geifer [35, 36] derived a sufficient condition under which category $\mathfrak{K}$ can be embedded (preserving normal monomorphisms) as a complete subcategory into a category in which every normal subobject is an ideal of the object. This condition is fulfilled, for example, when in category $\mathfrak{K}$ every morphism possesses a kernel and the collection of all ideals of any object forms a set. The same paper presents conditions for a similar embedding of category $\mathfrak{K}$ into a category in which every morphism possesses a kernel or into a category in which the collection of all ideals of every object forms a complete structure.

Tsalenko [30] has shown that every category $\mathfrak{K}$ can be embedded as a complete subcategory into a category $\bar{\mathfrak{K}}$, in which a

*Translator's note: This is more customarily called the 3×3 lemma.

free product exists for any family of objects. The properties of a category $\mathfrak{K}$, which can be carried over to the category $\bar{\mathfrak{K}}$ are also indicated. Dually, an arbitrary category $\mathfrak{K}$ can be embedded as a complete subcategory into a category in which a direct product exists for any family of objects. It was proved also that any category can be embedded as a complete subcategory into a category in which free and direct products exist for any family of objects, whose cardinality does not exceed an arbitrarily chosen fixed cardinal number.

Hasse and Michler [126] and Hoehnke [130] examined the possibility of embedding a category as a subcategory into a category in which all morphisms or some preassigned part of the morphisms of the original category are invertible morphisms. Roux [208] stated that for any category $\mathfrak{K}$ there exists an embedding functor $F:\mathfrak{K}\to\mathfrak{K}'$ of the category $\mathfrak{K}$ as a complete subcategory into the category $\mathfrak{K}'$ with inverse limits, such that for any functor $G:\mathfrak{K}\to\mathfrak{L}$ of category $\mathfrak{K}$ into some category $\mathfrak{L}$ with inverse limits there exists a unique, to within equivalence, functor $G':\mathfrak{K}'\to\mathfrak{L}$, satisfying the condition $G = FG'$. As a corollary we get that every pre-additive category $\mathfrak{K}$ can be embedded as a complete subcategory into an additive category; we remark that this same result ensues from the papers by Burmistrovich [3] and Tsalenko [30].

Jurchescu and Lascu [145] have shown that every small category $\mathfrak{K}$ can be embedded as a complete subcategory into some so-called Cantor category. Every Cantor category is a bicategory relative to the categories of all epimorphisms and of all monomorphisms, and in it any finite diagram possesses direct and inverse limits.

An additive category $\mathfrak{K}$ is called pre-Abelian if for every morphism α there exists a kernel and a cokernel and the canonical morphism $\mathrm{Coim}(\alpha)\to\mathrm{Im}(\alpha)$ is a bimorphism. A pre-Abelian category $\mathfrak{K}$ is called special if in every couniversal (universal) square (1), constructed for any two morphisms α and β, the morphism α' is an epimorphism (monomorphism) as soon as α is an epimorphism (monomorphism). Bănică and Popescu [45] proved that for every special pre-Abelian category $\mathfrak{K}$ there exists an exact embedding functor of category $\mathfrak{K}$ into some Abelian category $\mathfrak{A}$, such that any functor $F:\mathfrak{K}\to\mathfrak{A}_1$ of category $\mathfrak{K}$ into the Abelian category $\mathfrak{A}_1$, taking bimorphisms into isomorphisms, is uniquely extended up to some functor $F_1:\mathfrak{A}\to\mathfrak{A}_1$.

§ 5. Representation of Categories

A great interest has been evoked in connection with the possibility of an isomorphic representation of some category or other as a subcategory of one of the "well-known" categories such as, for example, the category of sets, the category of topological spaces, the category of groups, the category of Abelian groups, the category of modules, etc. In individual cases of this kind the representations allow us to assert the validity of a certain result for a given represented category only on the basis that this result is true for one of the well-known categories mentioned. The first representation theorem was the well-known Eilenberg — MacLane theorem [14] on the fact that every small category is isomorphically representable as some subcategory of the category of sets. The papers of Trnkovâ [24] and Drbohlav [61] were devoted to a certain generalization and refinement of the Eilenberg — MacLane theorem. Trnkovâ considered an arbitrary small category $\mathfrak{K}$ with two fixed subsets I and P of monomorphisms and epimorphisms, respectively, satisfying the following conditions: 1) if $\sigma\nu \in I$ and $\sigma \in P$, then $\nu \in I$; 2) if $\alpha\mu = \sigma\nu$, $\mu \in I$, $\sigma \in P$, then $\alpha = \sigma\chi$. The existence was proved of the isomorphic mapping F of category $\mathfrak{K}$ onto some subcategory of the category of sets such that: a) $F(\alpha)$ is a one-to-one "into" mapping for every monomorphism $\alpha \in I$; b) $F(\alpha)$ is an "onto" mapping for every epimorphism $\alpha \in P$; c) $F(\alpha)$ is a one-to-one "onto" mapping if and only if α is an isomorphism in category $\mathfrak{K}$. Drbohlav [61] denoted by $\mathfrak{S}_m$ the complete subcategory of the category of sets, generated by all sets whose cardinality does not exceed a given cardinal number m, and analyzed the following question: under what conditions imposed on the cardinality of the set of all objects of a small category $\mathfrak{K}$ and on the cardinality of the upper bound of the cardinalities of all sets of the form H(A, B), $A, B \in \mathfrak{K}$, can the category $\mathfrak{K}$ be represented as some subcategory of category $\mathfrak{S}_m$. Drbohlav also proved the following theorem. For any cardinal number m there exists a small category $\mathfrak{K}$ with the following properties: 1) the cardinality of the set of objects of category $\mathfrak{K}$ equals m; 2) the cardinality of the sets H(A, B) equals $\aleph_\omega$ for any pair of objects $A, B \in \mathfrak{K}$; 3) category $\mathfrak{K}$ is not representable as a subcategory of category $\mathfrak{S}_m$.

Sufficient conditions for the isomorphic representation of a category as a subcategory of the category of topological spaces have been derived by Bukovsky, Hedrlin and Pultr [55] and by Boero [51]. In the paper by the three authors it is proved that every small category $\mathfrak{K}$, the cardinality of whose morphisms is less than the first inaccessible cardinal number, is isomorphic with some subcategory $\mathfrak{K}^*$ of the category of topological spaces. Here the subcategory $\mathfrak{K}^*$ can be chosen as being contained either in a complete subcategory of metric spaces or in a complete subcategory of locally compact spaces. Boero proved that every quasi-exact category (see **9**, §3 for the definition of a quasi-exact category) with distributive structure of subobjects of an arbitrary object is isomorphically embedded, with the exact sequences preserved (in what follows we shall call such an embedding exact), into the category of topological spaces and of the partial homeomorphisms between them.

The possibility of embedding a given category as a complete subcategory of some well-described category was considered by Pultr and Hedrlin [22]. In particular, they prove that every small category $\mathfrak{K}$ can be isomorphically embedded as a complete subcategory in a certain category of the form $\mathfrak{K}(A)$, where A is a fixed set: the objects of this category are the families $\{(X, R_a), a \in A\}$, where R_a is a binary relation on the set X, while as the morphisms serve the mappings $f: X \to Y$ such that $f(x)S_a f(x')$ ensues from $xR_a x'$.

Pultr [193] and Trnkovâ [225] gave an affirmative answer to the question of the existence of a category $\mathfrak{U}$, into which any small category can be embedded as a complete subcategory; such a category $\mathfrak{U}$ is termed universal for small categories.

A category $\mathfrak{K}$ with null morphisms and with direct and free products for every finite family of objects was named MacLanian by Isbell [140] if the functors of the taking of a free product of a pair of objects and of a direct product of a pair of objects are naturally equivalent. In the paper cited, Isbell proved that every small MacLanian category can be isomorphically represented, with direct products preserved, as a complete subcategory of the category of commutative semigroups with operators.

The theorem proved by Lyubkin ([17], p. 101] on the possibility of an exact embedding of a small Abelian category as a subcategory of the category of Abelian groups, was reproved by several authors (see Gabriel [118], Freyd [116], Mitchell [185]). Roux

[207] proved that every Abelian category with exact direct limits and a generating object is isomorphically embedded as a complete subcategory into the category of unitary left modules over some associative ring with unity. Independently of Roux, Mitchell [185] proved this same theorem and, moreover, Mitchell proved that if the generating object is simultaneously a projective object, then the stated embedding of the Abelian category into the category of modules is exact. Relying on this result Mitchell proved that every small Abelian category can be embedded exactly as a complete subcategory into the category of unitary left modules over some associative ring R. This theorem of Mitchell is the crowning point of Freyd's monograph [116]; it is also contained in Mitchell's book [184]. Gabriel and Oberst [119] called an Abelian category $\mathfrak{A}$ with exact direct limits and a generating object a spectral category if in it every monomorphism is right-invertible and every epimorphism is left-invertible. It turns out that an Abelian category $\mathfrak{A}$ is spectral if and only if it is isomorphic with the complete subcategory of the category of modules over some regular self-injective ring R, generated by all R-modules of the form $\prod\limits_{i\in I}^{\times} R_i$, where $R_i \simeq R$. Certain results connected with the representation of Abelian categories as complete subcategories or as factor categories of the category of modules are contained also in the paper by Roos [203].

§6. Axiomatic Characteristics of Algebraic Categories

The well-known Birkhoff theorem on the structure of the varieties of universal algebras has determined the particular role of three fundamental operations over algebras: the operations of taking subalgebras, of taking factor algebras, and of forming the direct (Cartesian) product of algebras. In connection with this theorem there arose the problem of the axiomatic description of the categories of universal algebras closed relative to some of the indicated operations; here it is assumed that with every two algebras in a given category there occur all the homomorphisms of these algebras one to the other. Isbell [139] showed that the category of universal algebras closed relative to subalgebras and direct products is characterized, to within equivalence, as a complete category with small projective generators. An object P of a category $\mathfrak{K}$ is called a small projective generator if it possesses the following three properties: a) for every monomorphism $\mu: A \to B$ which is not an isomorphism, there exists a morphism $\gamma: P \to B$ not "passing through" μ; b) P possesses the projectivity property relative to the morphism $\alpha: X \to Y$ not "passing through" the proper monomorphism $\mu: A \to Y$; c) for every morphism $P \xrightarrow{\alpha} \prod_{i \in I}^{*} P_i$, where $P_i = P$, we can find a number n such that α can be decomposed into the product

$$P \xrightarrow{\alpha_n} P_{1*} \ldots {}_*P_n \xrightarrow{\sigma_n} \prod_{i \in I}^{*} P_i,$$

where σ_n is the natural embedding.

Isbell analyzed the independence of the theorem's hypotheses. Another formulation of this theorem was given by Isbell at the International Congress of Mathematicians in Stockholm. We should note, however, that the assertion on the boundedness of the arities* of the operations turned out to be incorrect.

Lawvere [161] described varieties of universal algebras by adding on to the hypotheses of Isbell's theorem one more condition whose formulation requires certain additional definitions. A pair of morphisms $\langle \alpha_1, \alpha_2 : R \to A \rangle$ is called a relation if the morphism $R \xrightarrow{\alpha_1 \times \alpha_2} A \times A$ is a monomorphism. A relation is reflexive if there exists a morphism $\delta: A \to R$ for which $\delta\alpha_1 = \delta\alpha_2 = 1_A$; a relation is symmetric if there exists a morphism $\lambda: R \to R$ for which $\lambda\alpha_1 = \alpha_2$ and $\lambda\alpha_2 = \alpha_1$; a relation is transitive if for every pair of morphisms $\beta_1, \beta_2: X \to R$ for which $\beta_1\alpha_2 = \beta_2\alpha_2$ there exists a morphism $\gamma: X \to R$ such that $\gamma\alpha_1 = \beta_1\alpha_1$ and $\gamma\alpha_2 = \beta_2\alpha_2$. A symmetric, reflexive and transitive relation is called a precongruence; a precongruence $\langle \alpha_1, \alpha_2 \rangle$ is called a congruence if $\alpha_1 \times \alpha_2$ is the kernel of the pair of morphisms $\pi_1\rho$ and $\pi_2\rho$, where $\pi_i: A \times A \to A$, $i = 1, 2$, are projections and ρ is the cokernel of the pair α_1, α_2. Lawvere

---

*Translator's note: The Russian word used here, obviously coined, is "arnost'." I am indebted to Professor John R. Isbell for supplying me with the corresponding English word "arity" used by him in the paper cited.

proved that a complete category with a small projective generator is equivalent to a variety of universal algebras if and only if every precongruence is a congruence.

Lawvere's theorem simplifies for the case of Abelian categories and goes over into the theorem of Mitchell [185] and Gabriel [118]; a right-complete Abelian category is equivalent to the category of all unitary modules over an associative ring with unity if and only if it possesses a small projective generator. A characterization of the category of right-Noetherian modules over right-Noetherian rings was given by Tsit-Yuen Lam [156]. Lawvere [162] has presented supplementary axioms leading to the description, to within equivalence, of the category of sets.

§7. Reflective Subcategories; Varieties

1. The concept of a reflective subcategory was first introduced by Mal'tsev [20] under the term an R-complete subcategory. A complete subcategory $\mathfrak{L}$ of category $\mathfrak{K}$ is called reflective if for each object A of category $\mathfrak{K}$ there exist an object L(A) of category $\mathfrak{L}$ and a morphism $\theta_A: A \to L(A)$ such that for any morphism $\alpha: A \to B$ of the object A into an arbitrary object B of subcategory $\mathfrak{L}$ there exists an identity morphism $\alpha': L(A) \to B$ such that $\alpha = \theta_A \alpha'$; the pair $(\theta_A, L(A))$ is called the $\mathfrak{L}$-reflector of object A. In other words, using MacLane's definition (see **3** of §3) of the free object relative to a functor, the object L(A) is the free object generated by object A relative to the functor of natural embedding $I_{\mathfrak{L},\mathfrak{K}}: \mathfrak{L} \to \mathfrak{K}$ of the complete subcategory $\mathfrak{L}$ into the category $\mathfrak{K}$. From the general results formulated at the end of **3** of §3 it ensues at once that a complete subcategory $\mathfrak{L}$ is a reflective subcategory of category $\mathfrak{K}$ if and only if the functor of natural embedding $I_{\mathfrak{L},\mathfrak{K}}: \mathfrak{L} \to \mathfrak{K}$ possesses a left-adjoint functor $L_{\mathfrak{L}}: \mathfrak{K} \to \mathfrak{L}$; the functor $L_{\mathfrak{L}}$ and the functor $L_{\mathfrak{K}} = L_{\mathfrak{L}} I_{\mathfrak{L},\mathfrak{K}}: \mathfrak{K} \to \mathfrak{K}$ will be called reflective functors corresponding to the reflective subcategory $\mathfrak{L}$. As is well known, for a pair of adjoint functors $L_{\mathfrak{L}}$ and $I_{\mathfrak{L},\mathfrak{K}}$ there exists a natural transformation of functors $\theta: I_{\mathfrak{K}} \to L_{\mathfrak{K}}$, where $I_{\mathfrak{K}}$ is the identity functor of category $\mathfrak{K}$, and the reflector of an arbitrary object A of category $\mathfrak{K}$ is none other than the value of the natural transformation θ and of the functor $L_{\mathfrak{K}}$ on the object A. Hušek [136] proved that a complete subcategory $\mathfrak{L}$, in which for any family of objects their direct product exists, is a reflective subcategory of category $\mathfrak{K}$ if and only if: 1) the embedding functor $I_{\mathfrak{L},\mathfrak{K}}: \mathfrak{L} \to \mathfrak{K}$

preserves direct products; 2) for every object X of category $\mathfrak{K}$ there exists at least one object Y of subcategory $\mathfrak{L}$, such that the set H(X, Y) is not empty; 3) for every object X of category $\mathfrak{K}$ in the class $\mathfrak{M}_X$ of all epimorphisms $\delta: X \to C$, where $C \in \mathfrak{L}$, we can choose a subset $\mathfrak{N}_X$, such that for any morphism $\gamma: X \to D$, where $D \in \mathfrak{L}$, there exists at least one epimorphism $\delta: X \to C$ belonging to the subset $\mathfrak{N}_X$, such that $\gamma = \delta\gamma'$ for some morphism $\gamma': C \to D$. Livshits, Tsalenko and Shul'geifer [18] essentially proved the following assertion. Let $\mathfrak{K}$ be a locally right-small and right-complete bicategory with a null object and let $\mathfrak{L}$ be a reflective subcategory of category $\mathfrak{K}$ with reflector $L_{\mathfrak{K}}$. Then the following properties are equivalent: 1) if $\mu: A \to B$ is an admissible monomorphism and $B \in \mathfrak{L}$, then $\mathfrak{L}$ contains an object A' isomorphic to object A; 2) in the $\mathfrak{L}$-reflector $(\theta_A, L(A))$ of an arbitrary object $A \in \mathfrak{K}$ the morphism θ_A is an admissible epimorphism. When condition 2) is satisfied we shall say that the pair $(\theta, L_{\mathfrak{K}})$ is a factor functor of the identity functor $I_{\mathfrak{K}}$ of bicategory $\mathfrak{K}$.

Let $\mathfrak{K}$ be a category in which every morphism possesses a kernel and let $\eta: I_{\mathfrak{K}} \to P$ be an arbitrary natural transformation of the identity functor $I_{\mathfrak{K}}$ of category $\mathfrak{K}$ into some functor $P: \mathfrak{K} \to \mathfrak{K}$ of category $\mathfrak{K}$ into itself. Then, by associating with each object $A \in \mathfrak{K}$ the kernel $(K(A), \mu_A)$ of the morphism $\eta_A: A \to P(A)$, under a specific stipulation on the choice of the representative of the kernel $(K(A), \mu_A)$ we obtain a functor $K: \mathfrak{K} \to \mathfrak{K}$ and a specific natural transformation $\mu: K \to I_{\mathfrak{K}}$, where for every object $A \in \mathfrak{K}$ $(K(A), \mu_A)$ will be an ideal of object A, being the kernel of the morphism η_A. In this case we shall say that (K, μ) is a normal subfunctor of the identity functor $I_{\mathfrak{K}}$, being the kernel of the natural transformation $\eta: I_{\mathfrak{K}} \to P$ and that the sequence of natural transformation of functors

$$0 \to K \xrightarrow{\mu} I_{\mathfrak{K}} \xrightarrow{\eta} P \tag{2}$$

is exact. In the special case when $L_{\mathfrak{K}}: \mathfrak{K} \to \mathfrak{K}$ is a reflector corresponding to the reflective subcategory $\mathfrak{L}$ of category $\mathfrak{K}$, there holds the exact sequence of natural transformations of functors

$$0 \to K \xrightarrow{\mu} I_{\mathfrak{K}} \xrightarrow{\theta} L_{\mathfrak{K}}, \tag{3}$$

where (K, μ) is the kernel of the natural transformation θ. Livshits, Tsalenko, and Shul'geifer [18] have proved that for every object A of category $\mathfrak{K}$ the ideal $(K(A), \mu_A)$ of object A is a completely characteristic ideal, i.e., for any morphism $\alpha: A \to A$ there exists a natural morphism $\alpha': K(A) \to K(A)$ such that $\mu_A \alpha = \alpha' \mu_A$.

Reflective subcategories have been treated also in the following papers: Gray [123] in which, under definite restrictions on the category $\mathfrak{K}$ there is proved the reflectivity of the subcategory of sheaves in the category of presheaves $F(T, \mathfrak{K})$ on a topological space X with values in category $\mathfrak{K}$; Vazquez [226], which in many ways coincides with Gray's paper; Kennison [153], in which he mainly considers the reflective subcategories of the category of topological spaces; Tsalenko [32], which is devoted chiefly to the reflective subcategories of the category of groups; and also Jurchescu and Lascu [145], Goblot [121], and Hocquemiller [129]. In his book [116], Freyd proved the reflectivity of a complete subcategory of monomorphism-preserving functors of the category of additive covariant functors from a small category $\mathfrak{A}$ into the category $\mathfrak{G}$ of Abelian groups.

2. A more general situation is considered by Sonner [218, 219]. Let there be given two categories $\mathfrak{K}$ and $\mathfrak{K}'$ and a class of elements Φ called quasimorphisms. It is assumed that $\mathfrak{K}$ is the category of left operators on Φ (in the sense of Ehresmann [95]) while $\mathfrak{K}'$ is the category of right operators on Φ and, moreover, $(\alpha \circ f) \circ \alpha' = \alpha \circ (f \circ \alpha')$, $\alpha \in \mathfrak{K}$, $f \in \Phi$, $\alpha' \in \mathfrak{K}'$, every time the products indicated have meaning. For every quasimorphism f there exists a natural object $A \in \mathfrak{K}$, such that the product $1_A \circ f$ is defined (in this case, $1_A \circ f = f$) and there exists a natural object $A' \in \mathfrak{K}'$, such that $f \circ 1_{A'} = f$. It is assumed that for any two objects $A \in \mathfrak{K}$ and $A' \in \mathfrak{K}'$ the collection $\Phi(A, A')$ of all quasimorphisms f from Φ for which $1_A \circ f = f$ and $f \circ 1_{A'} = f$, is a set. It is said that the object $A \in \mathfrak{K}$ possesses a left universal solution if there exists a quasimorphism $f \in \Phi$ such that $1_A \circ f = f$ and if for any quasimorphism f' for which $1_A \circ f' = f'$ there exists a natural morphism $\alpha' \in \mathfrak{K}'$, such that $f' = f \circ \alpha'$. The right universal solution of object $A' \in \mathfrak{K}'$ is defined analogously. A number of properties of such concepts are derived, as well as conditions under which every object of category $\mathfrak{K}$ possesses a left universal solution. In the case when Φ is the class of morphisms whose ends belong to $\mathfrak{L}$ and when $\mathfrak{K}' = \mathfrak{L}$, where $\mathfrak{L}$ is a complete subcategory of category $\mathfrak{K}$, the existence of a left universal solution for each object of category $\mathfrak{K}$ is equivalent to $\mathfrak{L}$ being a reflective subcategory of category $\mathfrak{K}$.

3. Varieties are a special case of reflective subcategories. Livshits, Tsalenko, and Shul'geifer [18] analyzed a category $\mathfrak{K}$, satisfying the following conditions: 1) null morphisms exist; 2) category $\mathfrak{K}$ is locally small; 3) category $\mathfrak{K}$ is complete; 4) every morphism possesses a normal image; 5) for any normal monomorphism $\mu: U \to A$ and for any normal epimorphism $\nu: A \to B$ the monomorphism μ_1 in the normal image (ν_1, L, μ_1) of the morphism μ_ν is normal. From condition 4) it ensues that $\mathfrak{K}$ is a bicategory with isolated subcategories of all normal epimorphisms and of all monomorphisms. A complete subcategory $\mathfrak{B}$ of category $\mathfrak{K}$ is called a variety if $\mathfrak{B}$ is closed relative to subobjects, to normal factor objects, and to the direct product of any family of objects belonging to $\mathfrak{B}$. Every variety $\mathfrak{B}$ is a reflective subcategory with a reflector $M_{\mathfrak{K}}: \mathfrak{K} \to \mathfrak{K}$, and since $\mathfrak{B}$ is closed relative to subobjects, it ensues from the assertions made in **1** that for the variety $\mathfrak{B}$ the pair $(\theta, M_{\mathfrak{K}})$ is a normal factor functor of the identity functor $I_{\mathfrak{K}}$ and, thus, the exact sequence (3) in the case being considered has the form

$$0 \to V \xrightarrow{\sigma} I_{\mathfrak{K}} \xrightarrow{\theta} M_{\mathfrak{K}} \to 0;$$

a normal subfunctor (V, σ) is called a subfunctor of variety $\mathfrak{B}$. It turns out that a normal subfunctor (F, λ) of the identity functor $I_{\mathfrak{K}}$ of category $\mathfrak{K}$ is naturally equivalent to the subfunctor (V, σ) of some variety $\mathfrak{M}$ if and only if the functor $F: \mathfrak{K} \to \mathfrak{K}$ takes normal epimorphisms into normal epimorphisms. A normal factor functor (η, S) of the identity $I_{\mathfrak{K}}$ is naturally equivalent to the reflector $(\theta, M_{\mathfrak{K}})$ of some variety $\mathfrak{B}$ of category $\mathfrak{K}$ if and only if the functor $S: \mathfrak{K} \to \mathfrak{K}$ is right-exact. Analogous results, but under the stricter Hoffmann restrictions [132] on category $\mathfrak{K}$, were obtained by Lecouturier [163].

In [18] an object P is said to be normally projective if it is projective relative to normal epimorphisms, i.e., for any normal epimorphism $\nu: A \to B$ and for any morphism $\alpha: P \to B$ there exists a morphism $\alpha': P \to A$ such that $\alpha = \alpha'\nu$. Under the above-mentioned restrictions 1)-5) on category $\mathfrak{K}$ the following theorem is valid. Let (V, λ) be any completely characteristic ideal of a normally projective object P of category $\mathfrak{K}$. Then the complete subcategory $\mathfrak{B}$ of category $\mathfrak{K}$, generated by all objects $A \in \mathfrak{K}$, such that the kernel of any morphism $\alpha: P \to A$ contains the ideal (V, λ) is a variety and, moreover, the $\mathfrak{B}$-reflector of object P is a normal factor object of object P with respect to the ideal (V, λ). From the well-known Birkhoff theorem it ensues that any variety can be constructed in the category of universal algebras of given type by the method indicated above or by a natural generalization of it in the case when the category of universal algebras does not satisfy condition 1) or 4). Drbohlav's paper [62] is devoted to the generalization of Birkhoff's theorem to the case of categories. Drbohlav considered a locally right-small and right-complete bicategory $\mathfrak{K} = (\mathfrak{K}, \mathfrak{E}, \mathfrak{M})$. A complete subcategory $\mathfrak{B}$ of category $\mathfrak{K}$ is called a variety if it is closed relative to direct products, to admissible subobjects, and to admissible factor objects. If A is some object and $\eta: A \to B$ is an arbitrary admissible epimorphism, then $\mathfrak{K}(A, \eta)$ denotes the complete subcategory of category $\mathfrak{K}$, generated by all objects C such that every morphism $\alpha: A \to C$ is representable in the form $\alpha = \eta\alpha'$ for some morphism α'. Every subcategory of the form $\mathfrak{K}(P, \eta)$, where P is an admissible projective object, is a variety. Certain conditions were derived, including, for example, the condition for the existence of a sufficient number of admissible projective objects for which every variety of category $\mathfrak{K}$ has the form $\mathfrak{K}(P, \eta)$.

Livshits, Tsalenko, and Shul'geifer [18] also considered the operations of multiplication and intersection over varieties.

§ 8. Radicals in Categories

1. The construction of a general theory of radicals in categories, analogous to the general theory of radicals in associative rings, the fundamentals of which were established by Kurosh [11] and by Amitsur [40], was initiated by Amitsur [40] and by Shul'geifer [34]. Shul'geifer considered a category $\mathfrak{K}$, satisfying the conditions 1) and 5) in **3** of § 7 and the conditions: 2) every morphism possesses a kernel and a normal image; 3) for every ascending sequence of ideals of an arbitrary object A there exists a smallest subobject of object A, containing all the ideals of the given sequence, and this subobject is an ideal of object A; 4) the collection of all ideals of any object is a set. It is said that some property of the objects of category $\mathfrak{K}$ is a radical, while all objects possessing property $\mathfrak{R}$, are called $\mathfrak{R}$-radical objects or, more concisely, $\mathfrak{R}$-objects, if the following conditions are fulfilled: a) if $\theta: A \to B$ is a normal epimorphism and A is an $\mathfrak{R}$-object, then B is also an

$\mathfrak{R}$-object (hence, in particular, it follows that an object isomorphic with an $\mathfrak{R}$-object is an $\mathfrak{R}$-object and, thus, it makes sense to speak of the $\mathfrak{R}$-objects and $\mathfrak{R}$-ideals of an arbitrary object A of category $\mathfrak{K}$); b) every object A possesses a largest $\mathfrak{R}$-ideal $(R(A), \mu_A)$, called the $\mathfrak{R}$-radical of object A; c) in the short exact sequence

$$0 \to R(A) \overset{\mu_A}{\to} A \overset{\theta_A}{\to} S(A) \to 0$$

the object S(A) possesses only the null $\mathfrak{R}$-ideal [such an object S(A) is said to be $\mathfrak{R}$-semisimple]. Shul'geifer showed that conditions a)-c) are equivalent to condition a) and the following condition: if for any nonnull normal epimorphism $\eta: A \to C$ the object C possesses a nonnull $\mathfrak{R}$-ideal, then the object A is an $\mathfrak{R}$-radical object. The establishment of the equivalence of the stated $\mathfrak{R}$-radicality conditions allows us to construct completely, just as was done in the above-mentioned paper by Kurosh, the upper and lower radicals defined by a certain class of objects of category $\mathfrak{K}$.

The papers of Ryabukhin [23] and Sulinski [221] are directly contiguous to that of Shul'geifer [34]. The above-mentioned conditions 1)-5) being fulfilled in category $\mathfrak{K}$, Ryabukhin continued the study of the general properties of radicals in categories. A radical $\mathfrak{R}$ is called hereditary if every ideal of an $\mathfrak{R}$-radical object is an $\mathfrak{R}$-ideal. A lower hereditary radical was constructed, defined by an arbitrary class of objects of category $\mathfrak{K}$. It was proved that for any class M of objects of category $\mathfrak{K}$ there exists a smallest one among all hereditary radicals $\mathfrak{R}_M$ in category $\mathfrak{K}$, for which every $\mathfrak{R}_M$-semisimple object is a subdirect sum of some family of objects from class M. Further, it was assumed that in addition to the conditions 1)-5) the following conditions were also fulfilled in category $\mathfrak{K}$: 6) for any set (K_i, μ_i), $i \in I$, of the ideals of any object A there exists the intersection (K, μ) and, moreover, the ideal (K, μ) is the smallest one among all the subobjects of object A, contained in each ideal (K_i, μ_i), $i \in I$ (hence it follows that the collection of all ideals of an arbitrary object constitutes a complete structure); 7) for any ascending chain of ideals (K_i, μ_i), $i \in I$, and any ideal (U, σ) of object A there holds the equality $(U,\sigma) \cap (\bigcup_{i\in I} (K_i,\mu_i)) = \bigcup_{i\in I}((U,\sigma) \cap (K_i,\mu_i))$.

A radical $\mathfrak{R}$ in category $\mathfrak{K}$ is called ideally hereditary if for any object A and any ideal (U, σ) of it there holds the equality $(R(A),\mu_A)$

$(U,\sigma)\cap(R(A),\mu_A)=(R(U),\mu_U\sigma)$, where $(R(A), \mu_A)$ and $(R(U), \mu_U)$ are $\mathfrak{R}$-radicals of the objects A and U, respectively. It was proved that every ideally hereditary radical in category $\mathfrak{K}$ is an upper radical $\mathfrak{R}_M$, defined by some generalized special class M of objects of category $\mathfrak{K}$, and, here, every $\mathfrak{R}_M$-semisimple object is a subdirect sum of some family of objects from M. The class M of objects of category $\mathfrak{K}$ is called generalized special if it satisfies the following conditions: a) if $\mu: U \to A$ is a normal monomorphism and $A \in M$, then $U \in M$; b) if (U, σ) is an ideal of object A such that $U \in M$ and if the ideal (U, σ) has a nonnull intersection with any nonnull ideal of object A, then $A \in M$; c) if (U, λ) is an ideal of object A, then for any M-ideal (V, σ) of object U there exists an ideal $(W, \tau\sigma\lambda)$ of object A such that $(W, \tau\sigma)$ is an M-ideal of object U [an ideal (V, λ) of object A is called an M-ideal if in the normal factor object (η, B) of object A with respect to the ideal (V, λ), the object B belongs to M]. The results of Ryabukhin are new also in the theory of radicals of associative rings.

Sulinski [221] considered a locally small category $\mathfrak{K}$ with conditions 1)-3), 5) and with direct and free products for every family of objects. The question was posed on how to pick out in such a category classes of simple objects, i.e., objects not having proper ideals, such that the upper radical $\mathfrak{R}_M$ in category $\mathfrak{K}$, defined by the class M, would possess many of the properties that the Brown—McCoy radical, i.e., the upper radical $\mathfrak{R}_M$, defined by the class of simple rings with unity, does in the category of associative rings. It turned out that such so-called modular classes M of simple objects of category $\mathfrak{K}$ are generalized special and possess the property that every ideal of the form (P, σ), where $P \in M$, of object A is a retract of object A and there exists a single maximal M-ideal of object A, intersecting the ideal (P, σ) along a null ideal.

Kurosh, in his paper [13] on radicals in groups, noted the definite duality between the properties of the class of $\mathfrak{R}$-radical groups and of the class of $\mathfrak{R}$-semisimple groups for an arbitrary $\mathfrak{R}$-radical in the groups. Livshits' paper [16] was devoted to giving a rigid foundation to the duality between the classes of $\mathfrak{R}$-radical objects and of $\mathfrak{R}$-semisimple objects for an arbitrary radical property $\mathfrak{R}$ in a category $\mathfrak{K}$, satisfying specific self-duality conditions.

The concept of a strict radical was introduced in categories first in the cited Kurosh paper on radicals in groups and then in the just-mentioned Livshits paper. A radical $\mathfrak{R}$ in category $\mathfrak{K}$ is called strict if all $\mathfrak{R}$-radical subobjects of any object $A \in \mathfrak{K}$ are contained in the $\mathfrak{R}$-radical $(R(A), \mu_A)$ of object A. Note, for example, that in an Abelian category every $\mathfrak{R}$-radical is a strict $\mathfrak{R}$-radical, and that in any arbitrary category the concepts of heredity and of ideal heredity in the sense of Ryabukhin [23] coincide for a strict $\mathfrak{R}$-radical. Shul'geifer [38] considered a category $\mathfrak{K}$, satisfying conditions 1) and 2) (see the first paragraph). It was noted that the property that an $\mathfrak{R}$-radical be strict is equivalent to the fact that the association with every object A of category $\mathfrak{K}$ of the $\mathfrak{R}$-radical $(R(A), \mu_A)$ and of the $\mathfrak{K}$-semisimple normal factor object $(\theta_A, S(A))$ of object A by the ideal $(R(A), \mu_A)$ defines, respectively, the normal subfunctor (R, μ) and the normal factor functor (θ, S) of the identity functor $I_{\mathfrak{K}}$ or category $\mathfrak{R}$ or, in other words, the complete subcategory $\mathfrak{P}$ of all $\mathfrak{R}$-semisimple objects and the complete subcategory $\mathfrak{R}$ of all $\mathfrak{K}$,-radical objects are, respectively, the reflective subcategory with reflector $S_{\mathfrak{K}} : \mathfrak{K} \to \mathfrak{K}$ and the coreflective subcategory (a concept dual to reflectivity) with coreflector $R_{\mathfrak{K}} : \mathfrak{K} \to \mathfrak{K}$ of category $\mathfrak{K}$, and there holds the exact sequences of natural transformations

$$0 \to R_{\mathfrak{K}} \overset{\mu}{\to} I_{\mathfrak{K}} \overset{\theta}{\to} S_{\mathfrak{K}} \to 0. \tag{4}$$

Necessary and sufficient conditions are derived in the paper for an exact sequence of form (4) to correspond, in the sense indicated above, to some strict $\mathfrak{R}$-radical in category $\mathfrak{K}$. These conditions are that the functor $R : \mathfrak{K} \to \mathfrak{K}$, considered as a functor from category $\mathfrak{K}$ into a complete subcategory $R(\mathfrak{K})$ of category $\mathfrak{K}$, generated by all objects of form R(A), $A \in \mathfrak{K}$, should preserve kernels, while the functor $S : \mathfrak{K} \to \mathfrak{K}$, considered as a functor from category $\mathfrak{K}$ into a complete subcategory $S(\mathfrak{K})$, generated by all objeots of form S(A), $A \in \mathfrak{K}$, should preserve cokernels. The exact sequence of form (4) corresponding to a hereditary strict $\mathfrak{R}$-radical is characterized in similar fashion.

Tsalenko [32] proved that a class (but not a set) of strict radicals occurs in the category of groups. Wheat's dissertation [230] also was devoted to the theory of radicals in categories.

2. Several papers have been devoted to the theory of radicals (or torsions) in Abelian categories. Dickson [58] considered a locally small complete Abelian category $\mathfrak{K}$. For an arbitrary class $\mathfrak{A}$ of objects the upper radical class $\mathfrak{R}(\mathfrak{A})$, defined by $\mathfrak{A}$, consists of all objects B from $\mathfrak{K}$, for which $H(B, A) = 0$ for any $A \in \mathfrak{A}$. If $\mathfrak{P}$ is the class of all pairwise nonisomorphic simple objects of category $\mathfrak{K}$, then $\mathfrak{R}(\mathfrak{P})$ is called the class of periodic objects of category $\mathfrak{K}$; if, however, $\mathfrak{P}$ is the class consisting of one simple object P, then $\mathfrak{R}(P)$ is called the class of P-primary objects of category $\mathfrak{K}$. As noted above, both $\mathfrak{R}(\mathfrak{P})$, and $\mathfrak{R}(P)$ are radical classes of objects, so that every object $A \in \mathfrak{K}$ contains a maximal periodic ideal $A_{\mathfrak{P}}$ and a maximal P-primary ideal A_P. It was proved that $A_{\mathfrak{P}}$ contains the direct sum ΣA_P (the summation is over all simple objects $P \in \mathfrak{P}$) and that $A_{\mathfrak{P}}$ is the essential extension of ΣA_P. Necessary and sufficient conditions were derived under which $A_{\mathfrak{P}} = \Sigma A_P$. Upper radicals in an Abelian Grothendieck category were considered also in the last chapter of Freyd's book [116].

A complete subcategory $\mathfrak{R}$ of an Abelian category $\mathfrak{K}$, generated by all $\mathfrak{R}$-radical objects relative to some hereditary radical $\mathfrak{R}$, constitutes a dense subcategory in the sense of Grothendieck [6] (we shall call such dense subcategories radical subcategories), and, consequently, with respect to it we can take the factor category $\mathfrak{K}/\mathfrak{R}$ and construct the canonical functor $T : \mathfrak{K} \to \mathfrak{K}/\mathfrak{R}$, where, moreover, the category $\mathfrak{K}/\mathfrak{R}$ is Abelian and the functor T is exact. Gabriel [118] treated the special case when for the canonical functor $\mathfrak{K} \to \mathfrak{K}/\mathfrak{R}$ there exists a right-adjoint functor $S : \mathfrak{K}/\mathfrak{R} \to \mathfrak{K}$. Such radical subcategories are called subcategories of localization. Necessary and sufficient conditions were derived for a radical subcategory to be a subcategory of localizations. If in an Abelian category $\mathfrak{K}$ every object possesses an injective envelope, then every radical subcategory of category $\mathfrak{K}$ is a subcategory of localizations and every hereditary $\mathfrak{R}$-radical is an upper radical defined by some class of injective objects.

Starting from this Gabriel derived, in the language of definite subsets of the set of right ideals of an associative ring R with unity, a description of all hereditary radicals in the category of right unitary R-modules. We note that the same description was obtained by Maranda [177] who arrived at radicals in the category of right R-modules by starting from the so-called injective structures. Ross [203] also has considered the subcategory of localizations of an Abelian category.

§ 9. Categories with Involution

The systematic investigation of the algebra of binary relations from the category point of view was initiated by MacLane [173] and carried on by Puppe [195] and Tsalenko [31]; also see Levy-Bruhl [168, 169, 170]. In this section we present the fundamental results of Puppe and Tsalenko, which also cover the results of MacLane. Categories were defined in these papers, which are the categories of binary relations over near-Abelian categories.

A category $\mathfrak{K}$ is called a category with involution or an I-category if the following conditions are fulfilled: a) in every set H(A, B), $A, B \in \mathfrak{K}$, there is given an order relation $\subset$, relative to which the multiplication of morphisms is right-monotonic, i.e., for any $\alpha_1, \alpha_2 \in H(A, B)$ and $\beta \in H(B, C)$, $\alpha_1 \subset \alpha_2 \Rightarrow \alpha_1\beta \subset \alpha_2\beta$; b) for every set H(A, B) there is given the mapping $\alpha \in H(A, B) \to \alpha^* \in H(B, A)$, called an involution and possessing the following three properties: b1) $\alpha_1 \subset \alpha_2 \Rightarrow \alpha_1^* \subset \alpha_2^*$; b2) $\alpha^{**} = \alpha$; b3) $(\alpha\beta)^* = \beta^*\alpha^*$.

Further axioms for I-categories stemmed from the solution of the problem formulated above.

Axiom K1. In an I-category $\mathfrak{K}$ there exists an object 0 such that the set H(0, 0) consists only of 1_0, while every set H(0, A), $A \in \mathfrak{K}$, contains the smallest and the largest morphisms, ω_A and Ω_A, respectively.

The object 0 is called an I-null object. For every $\alpha \in H(A, B)$ we set $I\alpha = \omega_A\alpha$, $B\alpha = \Omega_A\alpha$, $K\alpha = \omega_B\alpha^*$, $D\alpha = \Omega_B\alpha^*$. The morphism α is called I- (B-, K-, D-) regular if $I\alpha = \omega_B$ ($B\alpha = \Omega_B$, $K\alpha = \omega_A$, $D\alpha = \Omega_A$); α is called a proper morphism if it is ID-regular. A proper morphism is called a projection if it is B-regular and is called an injection if it is K-regular.

Axiom K2. a) $I\alpha \subset I\beta \Leftrightarrow \beta\alpha^*\alpha \subset \beta$; b) $B\alpha \supset B\beta \Leftrightarrow \beta\alpha^*\alpha \supset \beta$.

Axiom K3. For every $\alpha \in H(0, A)$, $A \in \mathfrak{K}$, there exist: a) an injection $\mu: U \to A$ such that $B\mu = \alpha$; b) a projection $\nu: A \to Q$ such that $K\nu = \alpha$.

Puppe showed that when axioms K1-K3 are satisfied in an I-category all the proper morphisms form a quasiexact category (see 9 of § 3). For every quasiexact category Tsalenko constructed an I-category satisfying axioms K1-K3. Since every exact functor between subcategories of proper morphisms is uniquely prolongable up to a functor between I-categories, consistent with the I-structure, then by the same token a one-to-one correspondence is established between quasiexact categories, on the one hand, and I-categories with axioms K1-K3, on the other.

Axiom K4. The sets H(A, B) are structures.

Axiom K5. a) $I\alpha\subset K\beta_1 \Rightarrow \alpha(\beta_1\cap\beta_2)=\alpha\beta_1\cap\alpha\beta_2$; b) $B\alpha\supset D\beta_1 \Rightarrow$ $\Rightarrow \alpha\cdot(\beta_1\cup\beta_2)=\alpha\beta_1\cup\alpha\beta_2$.

Axiom K6. For any two objects $A_1, A_2 \in \mathfrak{K}$ there exist: a) an object P and proper morphisms $\pi_i: P \to A_i$, i = 1, 2, for which $\pi_1^*\pi_2$ is the largest element of the set $H(A_1, A_2)$; b) an object S and proper morphisms $\sigma_i: A_i \to S$, i = 1, 2, for which $\sigma_1\sigma_2^*$ is the smallest element in $H(A_1, A_2)$.

Puppe showed that when axioms K1-K6 are satisfied, the category of proper morphisms of the I-category turns out to be Abelian and, conversely, for every Abelian category the category with involution, constructed for it, satisfies axioms K1-K6. When axioms K1-K4 are satisfied, the two groups of axioms K5a, K6a and K5b, K6b are equivalent. The system of axioms K1-K4, K5a, K6a is independent. The possibility of the wide use of the category of binary relations in homological algebra, in particular, for describing spectral sequences, was noted by Puppe and implemented by MacLane [19].

§ 10. Universal Algebras in Categories

1. The previous survey [17] had presented the fundamental results of Eckmann and Hilton [66] who studied objects with composition in arbitrary categories, i.e., objects A equipped with the morphisms $\mu: A \times A \to A$; in particular, they showed that the group-theoretic axioms can be formulated as the properties of morphism μ. A logical development of this idea is the passage made by Lawvere [161] and Freyd [117] to the study of universal "algebras" in arbitrary categories.

Lawvere called any category $\mathfrak{U}$ whose objects are the integers 0, 1, 2, . . . , n, . . . and in which every object $n \geq 2$ is a direct product of n copies of the object 1, an algebraic theory $\mathfrak{U}$. If $\mathfrak{K}$ is an arbitrary category with finite direct products, then a unary covariant functor $F:\mathfrak{U}\to\mathfrak{K}$, commuting with direct products, is called a $\mathfrak{K}^{\mathfrak{U}}$-algebra. In other words, a $\mathfrak{K}^{\mathfrak{U}}$-algebra is the object $A = F(1)$ together with the morphisms $\omega_n: A \times \ldots \times A$ (n times) $= A^n \to A$; the set of morphisms ω_n is not empty for a fixed n; in every case it contains all the projections of the direct product. These morphisms induce definite morphisms between A^n and A^m and all of them together are subject to equalities ensuing from the fact that the mapping $F:\mathfrak{U}\to\mathfrak{K}$ is a functor. A homomorphism of two $\mathfrak{K}^{\mathfrak{U}}$-algebras F_1 and F_2 is the natural transformation $\varphi: F_1 \to F_2$. The results stated below are due to Freyd. Let $\mathfrak{U}_1$ and $\mathfrak{U}_2$ be two algebraic theories and let $\mathfrak{S}$ be the category of sets. The categories of $(\mathfrak{S}^{\mathfrak{U}_1})^{\mathfrak{U}_2}$-algebras and of $(\mathfrak{S}^{\mathfrak{U}_2})^{\mathfrak{U}_1}$-algebras are isomorphic. It turns out that we can construct an algebraic theory $\mathfrak{U}$, for which $\mathfrak{S}^{\mathfrak{U}} \simeq (\mathfrak{S}^{\mathfrak{U}_1})^{\mathfrak{U}_2}$. Then, $\mathfrak{U}$ is called the tensor product of theories $\mathfrak{U}_1$ and $\mathfrak{U}_2$; here it turns out that $\mathfrak{K}$ $\mathfrak{K}^{\mathfrak{U}} \simeq (\mathfrak{K}^{\mathfrak{U}_1})^{\mathfrak{U}_2}$. Thus, the tensor product of algebraic theories is a commutative and, as can be shown, also an associative operation. If theories $\mathfrak{U}_1$ and $\mathfrak{U}_2$ have 0-ary operations, then theory $\mathfrak{U}$ has a single 0-ary operation. If there are binary operations with a two-sided unity in theories $\mathfrak{U}_1$ and $\mathfrak{U}_2$ then theory $\mathfrak{U}$ is the theory of "modules" over a semiring (a semiring is a ring without subtraction).

Let $\mathfrak{U}$ be a theory and A a $\mathfrak{K}^{\mathfrak{U}}$-algebra. Then every set H(X, A) can be converted into an $\mathfrak{S}^{\mathfrak{U}}$-algebra $\overline{X}$: if $\alpha_1, \ldots, \alpha_n: X \to A$, $\omega_n: A^n\ \omega\ A$, then, by definition, $(\alpha_1\alpha_2 \ldots \alpha_n)\omega_n = (\alpha_1 \times \alpha_2 \times \ldots \times \alpha_n)\omega_n$. Now if $\beta: X \to Y$, then $H(\beta, 1_A) = \overline{\beta}$ is a homomorphism of $\mathfrak{S}^{\mathfrak{U}}$-algebras: $\overline{\beta}: \overline{Y} \to \overline{X}$. Consequently, every $\mathfrak{K}^{\mathfrak{U}}$-algebra induces a contravariant functor $H^A: \mathfrak{K} \to \mathfrak{S}^{\mathfrak{U}}$ functors naturally equivalent to the functor H^A are called representable. A $(\mathfrak{K}^*)^{\mathfrak{U}}$-algebra is called a $\mathfrak{K}^{\mathfrak{U}}$-coalgebra. In any set H(A, X) every $\mathfrak{K}^{\mathfrak{U}}$-coalgebra A once again induces a structure of the $\mathfrak{S}^{\mathfrak{U}}$-algebra and defines a covariant functor $H_A: \mathfrak{K} \to \mathfrak{S}^{\mathfrak{U}}$; functors naturally equivalent to the functor H_A also are called representable. We present examples of coalgebras. If $\mathfrak{U}$ is a theory and F is a free $\mathfrak{S}^{\mathfrak{U}}$-algebra with one generator x, then for each $\alpha_n \in \mathfrak{U}$ the homomorphism $\omega_n: F \to F_1^* \ldots {}^* F_n$, where F_i is a free algebra with a generator x_i, takes x into the word $\alpha_n(x_1, \ldots, x_n)$. If $\mathfrak{U}$ is a group theory and $\mathfrak{K}$ is the category of topological spaces with base point and with classes of homotopic mappings as morphisms, then the sphere S^n are cogroups, while the group structure induced in the spaces $H(S^n, X)$ is nothing more than the usual multiplication in homotopy groups.

For every category $\mathfrak{S}^{\mathfrak{U}}$ there exists the forgetful functor $F: \mathfrak{S}^{\mathfrak{U}} \to \mathfrak{S}$. If the category $\mathfrak{S}^{\mathfrak{U}}$ is an $\mathfrak{S}^{\mathfrak{U}}$-category (see § 11), then we can say that every $\mathfrak{S}^{\mathfrak{U}}$-algebra is an $\mathfrak{S}^{\mathfrak{U}}$-coalgebra and that if there is a binary operation with unity in the theory $\mathfrak{U}$ then $\mathfrak{U}$ is the theory of modules over commutative semirings. This result describes within known limits those algebraic categories in which the duality of the functors can be constructed. Finally, we state the following theorem. Let $\mathfrak{K}$ be a complete category. The covariant functor $F: \mathfrak{K} \to \mathfrak{S}^{\mathfrak{U}}$ is representable if and only if it possesses the left-adjoint functor $G: \mathfrak{S}^{\mathfrak{U}} \to \mathfrak{K}$.

Let A be an $(\mathfrak{S}^{\mathfrak{U}_1})^{\mathfrak{U}_2}$-coalgebra and B and $(\mathfrak{S}^{\mathfrak{U}_2})^{\mathfrak{U}_3}$-algebra. Then the functor composition

$$\mathfrak{S}^{\mathfrak{U}_1} \overset{H_A}{\to} \mathfrak{S}^{\mathfrak{U}_2} \overset{H_B}{\to} \mathfrak{S}^{\mathfrak{U}_3}$$

is a representable functor by virtue of the preceding theorem, i.e., in $\mathfrak{S}^{\mathfrak{U}_1}$ there exists an $(\mathfrak{S}^{\mathfrak{U}_1})^{\mathfrak{U}_3}$-coalgebra $B \otimes A$, such that the indicated functor composition is naturally equivalent to the functor $H_{B \otimes A}$. Generating and defining relations have been described for this algebra.

2. In the paper [68] by Eckmann and Hilton, over a certain category $\mathfrak{K}$ there were specially investigated categories of algebras with one binary operation possessing some of the four properties mentioned in the previous survey ([17], p. 103). The following theorem was established with respect to all such categories.

Theorem. If a category $\mathfrak{K}$ with null morphisms is finitely left-complete, then any category of algebras of the above-mentioned type possesses those same properties. Furthermore, the inverse limits in category $\mathfrak{K}$ and in any category of algebras over $\mathfrak{K}$ coincide.

If $\mathfrak{U}$ is a group theory, then category $\mathfrak{S}$ turns out to be equivalent to the cogroup category over the category $\mathfrak{S}^{\mathfrak{U}}$. A large part of the results of the mentioned paper is contained in some form or other in the results of Freyd cited above.

3. In [67], Eckmann and Hilton generalize the concept of an operation in another direction (also see [127]). Let $\mathfrak{K}$ be a finitely complete category with null morphisms. For every family of objects $X_1, X_2, \ldots, X$, numbered in some specified manner, an "intermediate-direct product" $X^{r,s}$, $0 \leq s \leq r \leq n$, is defined in the following way: if $I = \{i_1, i_2, \ldots, i_n\}$ is an ordered subset of the set $\{1, 2, \ldots, n\}$, then $X_I = X_{i_1} * X_{i_2} * \ldots * X_{i_k}$; if $I \subseteq J$, then there exists the natural projection $\pi^I_J : X_J \to X_I$, $\pi^I_I = 1$; $X^{r,s} = \varprojlim \{X_I, \pi^J_I\}$, where I and J range, independently of each other, over all subsets of r elements and over all subsets with not more than s elements; in particular, $X^{n,n} = X_1^* \ldots {}^*X_n$ and $X^{1,1} = X_1 \times \ldots \times X_n$. We denote the product $X^{r,r}$ simply by X^r. For every r, $1 \leq r \leq n$, there exists the natural morphism $\varkappa^{r-1} : X^r \to X^{r-1}$. It turns out that for the canonical morphism $\varkappa : X_1^* \ldots {}^*X_n \to X_1 \times \ldots \times X_n$ there holds the factorization $\varkappa = \varkappa^{n-1}\varkappa^{n-2} \ldots \varkappa^1$. The product ${}^{r,s}X$ and the dual factorization for $\varkappa$ are defined dually. Now let $X_1 = X_2 = \ldots = X_n = X$ and let $\nabla : X_1^* \ldots {}^*X_n \to X$ be a codiagonal. By definition the object X has a length (the weak length) $n-1$ if n is the smallest integer for which there exists a morphism $\mu : X^{n-1} \to X$ such that $\nabla = \varkappa^{n-1}\mu$ [for which $(\ker \varkappa^{n-1})\nabla = 0$]. The colength (the weak colength) is defined dually. In the category of groups the length and the colength equal zero for a single group; for nontrivial groups the length coincides with the nipotency class; the colength equals unity only for free groups and equals two for all the remaining groups. In the category of sets with base point the length and colength equal zero for one-element sets; for the remaining sets the length equals unity and the colength equals infinity. In the category of topological spaces with classes of homotopy mappings as morphisms the colength is a homotopy invariant known by the name of the Schnirelmann — Lyusternik category. It turned out that the homotopy ratio between morphisms can be determined in a wide class of categories; in the factor category generated by this ratio there arise the concepts of homotopy length and colength; see [67] for details.

It was noted above that the category of groups over a finitely complete category $\mathfrak{K}$ is finitely left-complete. If in addition finite free products exist in the category of groups, then in the factorization in this category of the morphism $\varkappa = \varkappa^{n-1}\varkappa^{n-2} \cdot \ldots \cdot \varkappa^1$: $X_1 * \ldots * X_n \to X_1 \times \ldots \times X_n$ all the $\varkappa_i$, $i = 1, \ldots, n-1$, are epimorphisms left-invertible in $\mathfrak{K}$.

4. Closely contiguous to the papers of Eckmann and Hilton are those of Gemignani [120] and Freyd [114]. Freyd generalized the results of Eckmann and Hilton on the connection between the category of sets with base point and the category of groups. Let $\mathfrak{K}$ be a category with null morphisms and with direct and free products of finite families of objects. If $\mathfrak{K}^{\mathfrak{M}}$ is the category of monoids over $\mathfrak{K}$, then the forgetful functor $F:\mathfrak{K}^{\mathfrak{M}}\to\mathfrak{K}$ is defined naturally. Every object $A\in\mathfrak{K}$ generates some F-free object in the sense of Mersch (see §3, **3**) if and only if the functor F possesses a left-adjoint functor $G:\mathfrak{K}\to\mathfrak{K}^{\mathfrak{M}}$. The comultiplication $\mu^*: X\to X*X$ is induced naturally on every F-free object $(X, \mu)\in\mathfrak{K}^{\mathfrak{M}}$. If (X, μ) is a group, then (X, μ^*) is a cogroup. It turns out that free products of finite families of objects exist in category $\mathfrak{K}^{\mathfrak{M}}$ if and only if the following conditions are fulfilled: a) in $\mathfrak{K}^{\mathfrak{M}}$ every morphism possesses a kernel and a cokernel; b) the functor F possesses a left-adjoint functor G; c) for every $A\in\mathfrak{K}^{\mathfrak{M}}$ the canonical morphism $G(F(A))\to A$ is a normal epimorphism. When conditions a)-c) are satisfied every object G(A) is F-free and the association $A\to G(A)$ is a functor $\bar{G}$ from $\mathfrak{K}$ into the category $\mathfrak{K}_{\mathfrak{M}}$ of comonoids over $\mathfrak{K}$, which possesses a right-adjoint functor $\bar{F}:\mathfrak{K}_{\mathfrak{M}}\to\mathfrak{K}$. If $A_1 = \bar{F}(B_1)$ and $A_2 = \bar{F}(B_2)$, then a one-to-one correspondence exists between the sets $H_{\mathfrak{K}}(A_1, A_2)$ and $H_{\mathfrak{K}_{\mathfrak{M}}}(\bar{G}(A_1), \bar{G}(A_2))$ This result also is a generalization of an Eckmann—Hilton result in **2**.

§11. Categories with Multiplication

1. The pair $(\mathfrak{K}, \Pi)$, consisting of a category $\mathfrak{K}$ and a covariant bifunctor $\Pi: \mathfrak{K}\times\mathfrak{K}\to\mathfrak{K}$, is called a category with multiplication [notation: $\Pi(A, B) = A(\Pi)B$ or simply AB; $\Pi(\alpha, \beta) = \alpha(\Pi)\beta$]. Let $\mathfrak{K}^n$ denote the category $\mathfrak{K}\times\mathfrak{K}\times\ldots\times\mathfrak{K}$ (n times). If $F:\mathfrak{K}^n\to\mathfrak{K}$, and $G:\mathfrak{K}^m\to\mathfrak{K}$, then their product is defined in a natural way as the functor $F(\Pi)G: \mathfrak{K}^{n+m}\to\mathfrak{K}$. In particular, $\Pi=I_{\mathfrak{K}}(\Pi)I_{\mathfrak{K}}$. The set of iterations of the functor Π is the smallest set of functors $F:\mathfrak{K}^n\to\mathfrak{K}$, which contains $I_{\mathfrak{K}}$ and is closed relative to the product of functors.

The functor Π is associative, while the category $\mathfrak{K}$ itself is a category with associative multiplication, if there exists the natural equivalence

$$\alpha:\Pi(\Pi)I_{\mathfrak{K}}\simeq I_{\mathfrak{K}}(\Pi)\Pi.$$

Let $T:\mathfrak{K}^2\to\mathfrak{K}^2$ be a transposition. The functor Π is commutative, while the category $\mathfrak{K}$ is a category with commutative multiplication, if there exists the natural equivalence

$$\tau:\Pi\simeq\Pi T.$$

Every object $A\in\mathfrak{K}$ defines the functors ${}_A\Pi$, $\Pi_A:\mathfrak{K}\to\mathfrak{K}$: ${}_A\Pi(B)=A(\Pi)B$; $\Pi_A(B)=B(\Pi)A$. The object Λ is a left (right) unit relative to Π if there exists the natural equivalence $\varepsilon_l:{}_\Lambda\Pi\simeq I_{\mathfrak{K}}$, (the natural equivalence $\varepsilon_r:\Pi_\Lambda\simeq I_{\mathfrak{K}}$).

Every set consisting of certain natural equivalences α, τ, ε_l, ε_r and their inverses, generates certain equivalences between the iterations of functor Π (in the iterations we admit any permutation of the arguments and the identification of certain arguments with the object Λ). Let the set of equivalences be fixed. This set is called coherent if between any two iterations of functor Π it generates not more than one natural equivalence. MacLane's paper [174] was devoted to finding the coherency conditions for various groups of natural transformations. MacLane's results were improved by Kelly [152] by delineating independent groups of necessary and sufficient coherency conditions.

A natural equivalence α is coherent if and only if for any objects $A, B, C, D\in\mathfrak{K}$ the diagram

$$\begin{array}{ccccc} A[B(CD)] & \xrightarrow{\alpha} & (AB)(CD) & \xrightarrow{\alpha} & [(AB)C]D \\ \downarrow{\scriptstyle 1(\Pi)\alpha} & & & & \uparrow{\scriptstyle \alpha(\Pi)1} \\ A[(BC)D] & & \xrightarrow{\qquad\alpha\qquad} & & [A(BC)]D \end{array} \tag{5}$$

is commutative.

The natural equivalences α and τ are coherent if and only if for any objects $A, B, C, D\in\mathfrak{K}$ diagram (5) and the diagrams

$$\begin{array}{ccc} AB & \xrightarrow{\tau} & BA \\ & {\scriptstyle 1}\searrow & \downarrow{\scriptstyle\tau} \\ & & AB, \end{array} \tag{6}$$

$$\begin{array}{ccccc} A(BC) & \xrightarrow{\alpha} & (AB)C & \xrightarrow{\tau} & C(AB) \\ \uparrow{\scriptstyle 1(\Pi)\tau} & & & & \uparrow{\scriptstyle\alpha} \\ A(CB) & \xrightarrow{\alpha} & (AC)B & \xrightarrow{\tau(\Pi)1} & (CA)B \end{array} \tag{7}$$

are commutative.

The natural equivalences α, ε_l, ε_r are coherent if and only if diagram (5) and the diagram

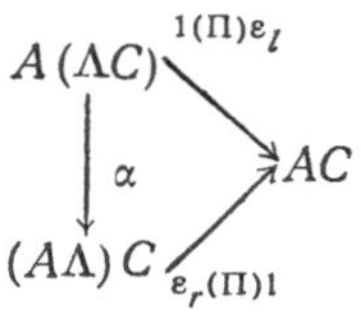

(8)

are commutative.

The natural equivalences α, τ, ε_l, ε_r are coherent if and only if diagrams (5)-(8) are commutative.

Now let $\mathfrak{K}$ and $\mathfrak{L}$ be two categories with coherent associative and commutative multiplications Π and $\hat{\Pi}$, respectively. Every functor $F:\mathfrak{K}\to\mathfrak{L}$ defines a pair of functors ΠF, $(F\times F)\hat{\Pi}:\mathfrak{K}\times\mathfrak{K}\to\mathfrak{L}$. Epstein [109] introduced the concept of the coherency of the natural transformation π: $(F \times F)\hat{\Pi} \to \Pi F$ relative to Π and $\hat{\Pi}$ and found the necessary and sufficient conditions for such coherence. A new type of coherence turns out to be necessary in the study of cup-multiplication and of the Steenrod operations in the derived functors of functor F. Epstein also found the necessary and sufficient coherency conditions for the natural equivalence

$$\beta:R((A_1(\Pi)A_2),A_3)\simeq R(A_1,R(A_2,A_3)),$$

where $R:\mathfrak{K}^*\times\mathfrak{K}\to\mathfrak{K}$ is a covariant bifunctor.

2. Benabou [48, 49] began the systematic study of categories with coherent associative multiplication Π with two-sided unity. In particular, he showed that for every category $\mathfrak{K}$ the category $H(\mathfrak{K}, \mathfrak{K})$ of functors $F:\mathfrak{K}\to\mathfrak{K}$ (if it has meaning) is a category with multiplication relative to the operation of successive realization of the functors. Benabou defined the concept of a morphism between categories with multiplication and for each category with multiplication $\mathfrak{K}$ constructed a left (and a right) representation in the category $H(\mathfrak{K}, \mathfrak{K})$: a left (right) representation is the association of the functor ${}_A\Pi$ (the functor Π_A) with every object $A\in\mathfrak{K}$ Further, Benabou [49], following Eckmann and Hilton [66], studied an elementary algebra in categories with multiplication by providing the object $A\in\mathfrak{K}$ the structure $\mu: A(\Pi)A \to A$. Every object A with an associative comultiplication and with counity defines a semisimplicial object of category $\mathfrak{K}$. As an example of the application of the results obtained Benabou constructed over each set S the category of graphs over S with associative multiplication. One result on adjoint functors between categories with multiplication can be found in Maranda [178].

3. Kelly [151] answered the following questions: what multiplication in categories can be called a "tensor" multiplication, and are associativity, commutativity, and the presence of one-sided (two-sided) unities general and independent properties of tensor multiplications?

Let a category $\mathfrak{P}$ be furnished with a covariant functor E in the category of sets $\mathfrak{S}$. The pair $(\mathfrak{K}, \mathrm{Hom}_{\mathfrak{K}})$, consisting of the category $\mathfrak{K}$ and the covariant functor $\mathrm{Hom}_{\mathfrak{K}}:\mathfrak{K}^*\times\mathfrak{K}\to\mathfrak{P}$, is called a $\mathfrak{P}$-category if $\mathrm{Hom}_{\mathfrak{K}} E = H:\mathfrak{K}^*\times\mathfrak{K}\to\mathfrak{S}$. Let $(\mathfrak{K}, \mathrm{Hom}_{\mathfrak{K}})$ and $(\mathfrak{L}, \mathrm{Hom}_{\mathfrak{L}})$ be two $\mathfrak{P}$-categories, $\mathfrak{M}$ an arbitrary category, and let there be given the functors $F: \mathfrak{M}^*\times\mathfrak{K}\to\mathfrak{L}$, $G:\mathfrak{L}\times\mathfrak{M}\to\mathfrak{K}$. These functors are said to be adjoint ($\mathfrak{P}$-adjoint) if there exists the natural equivalence

$$\alpha: H(G(Y,Z),X)\simeq H(Y,F(Z,X))$$
$$(\alpha: \mathrm{Hom}_{\mathfrak{K}}(G(Y,Z),X)\simeq \mathrm{Hom}_{\mathfrak{L}}(Y,F(Z,X))),$$
$$X\in\mathfrak{K},\ Y\in\mathfrak{L},\ Z\in\mathfrak{M}.$$

Now let the category $\mathfrak{P}$ itself be a $\mathfrak{P}$-category, i.e., there exists a functor $\mathrm{Hom}_{\mathfrak{P}}:\mathfrak{P}^*\times\mathfrak{P}\to\mathfrak{P}$, for which $\mathrm{Hom}_{\mathfrak{P}}E=H$. If the functor $\mathrm{Hom}_{\mathfrak{P}}$ possesses a left-adjoint functor, then this adjoint functor is called a tensor multiplication and is denoted by $\otimes$. A tensor multiplication need not possess even one of the above-mentioned properties. A tensor multiplication is associative if and only if the functors $\otimes$ and $\mathrm{Hom}_{\mathfrak{P}}$ are $\mathfrak{P}$-adjoint. Kelly derived the necessary and sufficient conditions for a tensor multiplication to commutate and for the presence of a two-sided unity, as well as the coherency conditions for the corresponding natural equivalence.

§12. Duality of Functors

Fuks' papers [25, 26] on the duality of functors in the category of topological spaces laid the ground for analogous investigations for other categories and led Shvarts [33] to the delineation of a class of categories in which the duality of functors can be constructed.

Let ($\mathfrak{K}$, $\mathrm{Hom}_{\mathfrak{K}}$) and ($\mathfrak{L}$, $\mathrm{Hom}_{\mathfrak{L}}$) be two $\mathfrak{P}$-categories. The functor $F:\mathfrak{K}\to\mathfrak{L}$ is called a $\mathfrak{P}$-functor if there exists a natural transformation $\tau:\mathrm{Hom}_{\mathfrak{K}}(A, B)\to\mathrm{Hom}_{\mathfrak{L}}(F(A), F(B))$ such that the mapping $E(\tau): H(A, B) \to H(F(A), F(B))$ is induced by functor F. We shall call the category $\mathfrak{P}$ almost concrete, if the functor $E:\mathfrak{P}\to\mathfrak{S}$ takes different morphisms into different ones. The class of objects A_λ, $\lambda \in \Lambda$, and of sets $\Pi_\lambda^{\lambda'}\subseteq H(A_\lambda, A_{\lambda'})$ is called a spectrum if $1_{A_\lambda}\in\Pi_\lambda^\lambda$ and if $\Pi_\lambda^{\lambda'}\Pi_{\lambda'}^{\lambda''}\subseteq\Pi_\lambda^{\lambda''}$. The inverse (direct) limit $\varprojlim(A_\lambda, \Pi_\lambda^{\lambda'})$ ($\varinjlim(A_\lambda, \Pi_\lambda^{\lambda'})$) of the spectrum is determined by the usual couniversal (universal) property. An almost concrete category $\mathfrak{P}$ is called a category with inverse limits if $\varprojlim(A_\lambda, \Pi_\lambda^{\lambda'})$ exists if and only if $\varprojlim(E(A_\lambda), E(\Pi_\lambda^{\lambda'}))$ exists and the functor E commutes with the inverse limits. Shvarts showed that the duality of functors can be constructed in a category $\mathfrak{P}$, satisfying the following conditions: 1) $\mathfrak{P}$ is an almost concrete category with inverse limits; 2) $\mathfrak{P}$ is a $\mathfrak{P}$-category; 3) an associative tensor multiplication $\otimes$ with unity exists in $\mathfrak{P}$; 4) the functors $\mathrm{Hom}_{\mathfrak{P}}(A, X)=\Omega_A$, $A\otimes X=\Sigma_A$, A is a fixed object, are $\mathfrak{P}$-functors.

Let $F, G:\mathfrak{P}\to\mathfrak{P}$ be two functors; Λ is the class of morphisms of category $\mathfrak{P}$. For $\lambda: X \to Y$ we set $A_\lambda=\mathrm{Hom}_{\mathfrak{P}}(F(X), G(Y))$; $\varphi\in\Pi_\lambda^{\lambda'}$, if $\lambda' = \alpha\lambda\beta$, where $\alpha: X' \to X$, $\beta: Y \to Y'$, and $\varphi = H(F(\alpha), G(\beta))$: $H(F(X), G(Y)) \to H(F(X'), G(Y'))$. If the class of natural transformations of F into G is a set, then the inverse limit of the indicated spectrum exists and is denoted Hom(F, G); here E(Hom(F, G)) = H(F, G). In categories with a cogenerating object the object Hom(F, G) exists for any pair of functors.

The functor G is called the dual functor of F (G = DF) if for any object $A\in\mathfrak{P}$ $G(A) = \mathrm{Hom}(F, \Sigma_A)$ and for any morphism $\alpha\in H$ $\alpha\in H(A, B)$, $G(\alpha) = H(1, \tilde{\alpha})$, where $\tilde{\alpha}: \Sigma_A \to \Sigma_B$ is induced by α. For any functors R and F there holds the isomorphism Hom(R, DF) $\simeq$ Hom(F, DR). By substituting the functor DF in the place of R we obtain the isomorphism Hom(DF, DF) $\simeq$ Hom(F, DDF). By virtue of this isomorphism the natural transformation $\varkappa: F \to DDF$ corresponds to the unity 1_{DF}. If $\varkappa$ is a natural equivalence, then the functor F is called reflexive. The functors Ω_A and Σ_A are dual to each other and reflexive. For every functor F there exists a spectrum of functors $\{F_\lambda\}$ of the form $\Omega_A \Sigma_B$ such that $F(X) = \varinjlim F_\lambda(X)$ for any X. The natural equivalence $D(RF) \simeq DRDF$ for a fixed functor R and for any functor F holds if and only if this equivalence holds for functors F of the form Ω_A. Under certain restrictions on the functors F and R the equivalence mentioned was established by Fuks [26] for the category of topological spaces.

Linton [171] extended the duality described above to functors acting from a category $\mathfrak{L}$ into category $\mathfrak{K}$, assuming that categories $\mathfrak{L}$ and $\mathfrak{K}$ satisfy the following conditions: 1) duality can be defined for the category $\mathfrak{K}$; 2) for any objects $A, B \in \mathfrak{L}$ the set H(A, B) can be treated as an object in category $\mathfrak{K}$; 3) the "tensor" multiplication $\Pi: \mathfrak{L} \times \mathfrak{L} \to \mathfrak{K}$ is given.

Kuznetsov [10], Berman [2], Mityagin and Shvarts [21], and Kissina and Shvarts [9] have described a dual functor in various categories of topological spaces. In particular, Kissina and Shvarts showed the relation $D^3 \simeq D$ is not fulfilled for the operator D in the category of sets.

§ 13. Homotopy Theory

Attempts to give a firm foundation to the intuitive duality between homotopy groups and cohomology groups, noted by Eckmann and Hilton led, on the one hand, to the construction of a duality theory for functors and, on the other, to the construction of homotopy theory in categories (Huber [133, 134], Kleisli [154], Meier [180], Adams [39], and Eckmann and Kleisli [69]).

Huber's construction was based on the concept of a standard construction. Let $\mathfrak{K}$ be some category. Then the triple $(C, \varkappa, \pi)$, consisting of a covariant functor $C: \mathfrak{K} \to \mathfrak{K}$ and of the natural transformations $\varkappa: C \to I_{\mathfrak{K}}$, and $\pi: C \to C^2$, is called a standard construction if the following relations are fulfilled for each object

$$1_{C(X)} = \pi(X) C(\varkappa(X)) = \pi(X) \varkappa(C(X)),$$
$$\pi(X) \pi(C(X)) = \pi(X) C(\pi(X)).$$

Every standard construction generates the simplicial functor $F = \{(F_n, d_n^i, s_n^i), n \geq -1\}$, i.e., the sequence of functors $F_n: \mathfrak{K} \to \mathfrak{K}$ and of the natural transformations $d_n^i: F_n \to F_{n-1}$ (the face operators) and $s_n^i : F_n \to F_{n+1}$ (the degeneracy operators), where $F_n = C^{n+1}$ $(C^0 = I_{\mathfrak{K}})$,

$$d_n^i = C^i(\varkappa(C^{n-i}(X))),\ s_n^i(X) = C^i(\pi(C^{n-i}(X))).$$

Every functor T from $\mathfrak{K}$ into the category of sets associates a simplicial set with any object $X \in \mathfrak{K}$ and, by the same token, permits the determination of the homotopy groups of object X. The construction $(C, \varkappa, \pi)$ is called T-trivial if there exists a natural transformation $\chi: T \to CT$ such that $1_{T(X)} = \chi(X)T(\varkappa(X))$. Every pair of adjoint functors $F \dashv G: \mathfrak{K} \to \mathfrak{L}$ and every standard construction $(C, \varkappa, \pi)$ in $\mathfrak{K}$ induces a standard construction $(\bar{C}, \bar{\varkappa}, \bar{\pi})$ in $\mathfrak{L}$ in a defined manner. If the original construction is T-trivial, then the induced construction is GT-trivial. As the functor T we can take, for example, the functor $H_A(X) = H(A, X)$, A is a fixed object.

The standard construction $(C, \varkappa, \pi)$ in category $\mathfrak{K}$ generates a construction $(C', \varkappa', \pi')$ in the category of morphisms over $\mathfrak{K}$: if $\alpha_1 : A_1 \to B_1$ and $\alpha_2 : A_2 \to B_2$,

$$\beta_1 : A_1 \to A_2,\ \beta_2 : B_1 \to B_2, \text{ and } \alpha_1\beta_2 = \beta_1\alpha_2,$$

then

$$C'(\alpha) = C(\alpha),\ C'(\beta_1, \beta_2) = (C(\beta_1), C(\beta_2)),$$
$$\varkappa'(\alpha) = (\varkappa(A_1), \varkappa(A_2)),\ \pi'(\alpha) = (\pi(A_1), \pi(A_2)).$$

Under certain additional assumptions it was proved that the homotopy groups for objects and for morphisms are embedded in one exact homotopy sequence. The concepts of fibration and cofibration were defined and the corresponding exact sequences constructed.

In every category $\mathfrak{K}$ there exists a trivial construction $(I_{\mathfrak{K}}, 1_{I_{\mathfrak{K}}}, 1_{I_{\mathfrak{K}}})$. Huber [134] constructed the following nontrivial standard construction in an Abelian category $\mathfrak{A}$ with a projective generator U and with infinite direct sums. Let $\mathfrak{S}$ be the category of sets; we set

$$G(X) = H(U, X) = H_U : \mathfrak{A} \to \mathfrak{S},\ F(M) = \bigoplus_{\alpha \in M} U_\alpha,$$

where $U_\alpha = U$ for all $\alpha \in M$, $M \in \mathfrak{S}$. The functors F and G are adjoint and the trivial construction in $\mathfrak{S}$ induces a G-trivial construction in $\mathfrak{A}$. The homotopy groups $\pi_n(X, Y)$ which arise coincide with the projective homotopy groups $\Pi_{n+1}(X, Y)$ defined by Kleisli [154] for Abelian categories with a sufficient number of projective objects.

Eilenberg and Moore [108] and Kleisli [155] have shown that every standard construction is in fact induced by some pair of adjoint functors. The structure which arises here is of independent interest.

§ 14. Homological Algebra in Categories

1. MacLane's lectures [175] and his book [19] contain the latest accounts concerning homological algebra within the framework of Abelian categories. Let $\mathfrak{A}$ be an Abelian category and let a class $\mathfrak{P}$ be picked out in it, of short sequences $0\to\cdot\xrightarrow{\varkappa}\cdot\xrightarrow{\sigma}\cdot\to 0$, or, in brief, $(\varkappa, \sigma)$. The class $\mathfrak{P}$ is called proper if it satisfies the following conditions:

(P1) if $(\varkappa, \sigma)\in\mathfrak{P}$, then any isomorphic sequence also belongs to $\mathfrak{P}$; (P2) for any objects $A, B\in\mathfrak{A}$ the sequence $0\to A\to A\oplus B\to B\to 0\in\mathfrak{P}$; (P3) if $(\varkappa, \sigma)\in\mathfrak{P}$, and $(\varkappa', \sigma')\in\mathfrak{P}$ and if the product $\varkappa\varkappa'$ is defined, then there exists the sequence $(\varkappa\varkappa', \tau)$ belonging to $\mathfrak{P}$, and dually; (P4) if $\varkappa$ and $\varkappa'$ are morphisms and if $(\varkappa\varkappa', \tau)\in\mathfrak{P}$, then there exists the sequence $(\varkappa, \sigma)\in\mathfrak{P}$ and dually.

A monomorphism $\varkappa$ is said to be admissible if some sequence $(\varkappa, \sigma)\in\mathfrak{P}$; an admissible epimorphism is defined dually. A morphism α is admissible if it factors into a product of an admissible epimorphism and an admissible monomorphism. An object P is called an admissible projective object if it is projective relative to admissible epimorphisms. Every proper class $\mathfrak{P}$ defines a sequence of functors $\operatorname{Ext}^n_{\mathfrak{P}}(C, A)$, $n=1, 2, \ldots, n, \ldots$ Following Yoneda, $\operatorname{Ext}^n_{\mathfrak{P}}(C, A)$ is defined as the group of congruent classes of lengths of the exact sequences

$$0\to A_0 = A\xrightarrow{\alpha_0} A_1\xrightarrow{\alpha_1}\ldots\to A_n\xrightarrow{\alpha_n} A_{n+1} = C\to 0,$$

in which every morphism α_i is admissible. Without using projective objects it was proved that for every sequence $(\varkappa, \sigma): 0\to A\to B\to C\to 0\in\mathfrak{P}$ and for every object D there exists the exact sequence

$$\ldots \to \mathrm{Ext}^{n-1}_{\mathfrak{P}}(A, D) \to \mathrm{Ext}^{n}_{\mathfrak{P}}(C, D) \to \mathrm{Ext}^{n}_{\mathfrak{P}}(B, D) \to$$
$$\to \mathrm{Ext}^{n}_{\mathfrak{P}}(A, D) \to \mathrm{Ext}^{n+1}_{\mathfrak{P}}(C, D) \to \ldots$$

If enough admissible projective objects occur in a category $\mathfrak{A}$ i.e., for every object $A \in \mathfrak{A}$ there exists an admissible epimorphism σ: P $\to$ A of the projective object P, then every object C possesses an admissible resolvent ε: X $\to$ C. The usual isomorphism

$$\mathrm{Ext}^{n}_{\mathfrak{P}}(C, A) \simeq H^{n}(H_{\mathfrak{A}}(X, A)).$$

holds in this case. The derived functors for functor $T:\mathfrak{A} \to \mathfrak{B}$ are determined by the usual method, and the whole sequence of derived functors is characterized by a certain universal property.

In any Abelian category as the class $\mathfrak{P}$ we can take the class of all short exact sequences. A second typical example arises when we consider resolvent pairs of categories. The pair of categories $\mathfrak{A}$ and $\mathfrak{M}$ is called resolvent if, first, there are a pair of adjoint functors $F:\mathfrak{M} \to \mathfrak{A}$ and $\square:\mathfrak{A} \to \mathfrak{M}$ with the conjugation $\alpha:F \dashv \square$ and, second, the functor $\square$ is exact and takes only null morphisms into null morphisms. As the class $\mathfrak{P}$ in category $\mathfrak{A}$ we can take those short exact sequences whose images are decomposed in $\mathfrak{M}$. In this case every object F(M), $M \in \mathfrak{M}$, is an admissible projective object, and there turn out to be enough admissible projective objects in category $\mathfrak{A}$. Every admissible resolvent possesses a contracting homotopy in $\mathfrak{M}$ whose square equals zero. In the category $\mathfrak{A}$ it turns out to be possible to realize a functorial structure of two types of resolvents: unnormalized and normalized B-resolvent, and, moreover, these resolvents are chain-equivalent and an $\mathfrak{M}$-contracting homotopy can be chosen functorially for each of them. The homology and cohomology of different algebraic systems arise from appropriate resolvent pairs of categories. In his lectures MacLane also described various types of categories which prove to be useful in the study of higher homological operations and higher homotopies.

Freyd [115] showed that a relative homological algebra, generated by a class $\mathfrak{P}$ of proper short exact sequences of category $\mathfrak{A}$, is equivalent to the absolute homological algebra for the category $\mathfrak{L}_{\mathfrak{P}}(\mathfrak{A})$ of contravariant functors, left-exact for sequences of $\mathfrak{P}$ from $\mathfrak{A}$ into the category of Abelian groups.

Other possible methods for defining an Abelian cohomology for objects of both Abelian and non-Abelian categories, as well as for constructing an extension theory in various non-Abelian categories, can be found in Ehresmann [94, 101], Inasaridze [7, 8], Badescu [43], Lavendhomme [160], Evrard [110, 111], Hacque [124], Pareigis [189], and others.

2. The homology HA of a complex A of free Abelian groups is a complete system of invariants for A in the case when the chain equivalence of complexes A and B ensues from the isomorphism HA $\simeq$ HB. Kelly's papers [149, 150] are devoted to giving a precise meaning to the expression "a complete system of invariants" and to finding such systems for various important examples. The functor $T:\mathfrak{A}\to\mathfrak{B}$ is called complete if it possesses the three properties: 1) for any objects $A, A'\in\mathfrak{A}$ the mapping H(A, A') $\to$ H(T(A), T(A')) induced by the functor T is an "onto" mapping; 2) for any object $B\in\mathfrak{B}$ there exists an object $A\in\mathfrak{A}$, such that T(A) $\simeq$ B; 3) if the morphism T(α) is invertible, then the morphism α is invertible. Of course, many complete functors exist for a category $\mathfrak{A}$; it is apparent that a complete functor from $\mathfrak{A}$ into a concrete category $\mathfrak{B}$ must be regarded as a "faithful" complete functor. If categories $\mathfrak{A}$ and $\mathfrak{B}$ are additive, and if the functor T is additive and possesses properties 1) and 2), then property 3) is fulfilled if and only if for every object $A\in\mathfrak{A}$ the set of morphisms α : A $\to$ A with T(α) = 0 is contained in the Jacobson radical of the ring H(A, A).

For categories of complexes of free Abelian groups with classes of homotopy mappings as morphisms, the functor of taking the homology is complete. From every complete functor $T:\mathfrak{A}\to\mathfrak{B}$ we can construct a complete functor between the categories of morphisms over $\mathfrak{A}$ and $\mathfrak{B}$. In particular, by this means Kelly obtained complete systems of invariants for a chain transformation. In [150] Kelly established the completeness of the functor of taking the homology for the category of pairs (A, B) of complexes of Abelian groups, where B is a subcomplex in A and A/B is the complex of free Abelian groups. Here the close connection between the category indicated and the category of morphisms over the category of complexes of free Abelian groups was established.

§ 15. Concrete Categories

1. The category of groups has been considered by Lavendhomme [160] and Tsalenko [32]. Lavendhomme proved that the category of groups $\mathfrak{G}$ can be embedded, with direct products preserved, as a complete subcategory $\bar{\mathfrak{G}}$ in a category $\mathfrak{H}$, such that in the category $\mathfrak{H}$ every morphism possesses a kernel and a cokernel, every monomorphism and every epimorphism of the subcategory $\bar{\mathfrak{G}}$ is, respectively, a normal monomorphism and a normal epimorphism, and the image Im α and the coimage Coim α are isomorphic for every morphism α of subcategory $\bar{\mathfrak{G}}$. The author uses this embedding to construct a homology theory in the category of groups $\mathfrak{G}$ Tsalenko proved that every natural transformation $\sigma: U \to V$ between two subfunctors U and V of the identity functor $I_{\mathfrak{G}}$ of the category of groups either is zero or $\sigma(A)$ is the natural embedding of the subgroup U(A) in the group V(A) for every group A. Tsalenko called a reflective subcategory $\mathfrak{L}$ a quasivariety if it is closed relative to subgroups. The product $\mathfrak{L}_1\mathfrak{L}_2$ of two quasivarieties $\mathfrak{L}_1$ and $\mathfrak{L}_2$ is a quasivariety consisting of all groups which are the extensions of a certain group from $\mathfrak{L}_2$ with the aid of a certain group from $\mathfrak{L}_1$. The product of quasivarieties is associative. Certain properties of the semigroup P_r (I) of quasivarieties were described.

Javier [141, 142] studied categories as objects which serve certain subsemigroups of a given semigroup, while as morphisms certain homomorphisms between the subsemigroups selected.

Definite relations between categories whose objects are Boolean lattices, Boolean spaces, Boolean rings, and Boolean semirings, were derived by Doctor [59]. Jurie [146] has considered the category of Boolean rings with unity and their homomorphisms taking unity into unity. The direct limit of a family of homomorphisms $\rho_i : A \to B_i$, $i \in I$, is called the universal amalgam of rings B_i relative to the homomorphisms ρ_i, $i \in I$. Several necessary and sufficient conditions were derived so that in the category being considered the ring S and the family of homomorphisms $\sigma_i : B_i \to S$, $i \in I$, satisfying the condition $\rho_i\sigma_i = \rho_j\,\sigma_j$ for any $i, j \in I$, would determine a universal amalgam of rings B_i relative to the homomorphisms ρ_i, $i \in I$.

Gel'fand and Shilov [4] and Khelemskii [28] have considered additive subcategories of the category of finite-dimensional vector spaces over the complex number field. The following question was posed and, in individual cases, completely answered: if for every object E_α of the category $\mathfrak{K}$ being considered, of finite-dimensional linear spaces, the ring $H(E_\alpha, E_\alpha)$ is known, then are each of the Abelian groups $H(E_\alpha, E_\beta)$ determined by them ? Complete answers to this question were obtained, for example, for the case when the category $\mathfrak{K}$ consists of two linear spaces E_1 and E_2 and both rings $H(E_1, E_1)$ and $H(E_2, E_2)$ are either semisimple or maximal nilpotent subalgebras of complete matrix algebras.

Concrete categories have been studied also by Kennison [153], Burgess [56], and Laplaza [158].

2. Several papers have been devoted to categories of modules over associative rings and to the functors between them. Solian [214, 215, 216] treated additive functors between the categories ${}_\Lambda\mathfrak{M}$ and ${}_\Gamma\mathfrak{M}$ of left unitary modules over the rings Λ and Γ, respectively. The family of Λ-homomorphisms $\gamma_i : A \to C$, $i \in I$, of module A into module C is called a local epimorphism (local monomorphism) by Solian if $\bigcup_{i \in I} \mathrm{Im}(\gamma_i) = C$ (if $\bigcap_{i \in I} \mathrm{Ker}\, \gamma_i = 0$). In [215, 216], Solian studied the additive covariant (contravariant) functors $T: {}_\Lambda\mathfrak{M} \to {}_\Gamma\mathfrak{M}$, taking local epimorphisms into local epimorphisms (into local monomorphisms).

The category ${}_R\mathfrak{M}$ of left R-modules was considered by Faith [112]. If $E \in {}_R\mathfrak{M}$, then we set $R' = \mathrm{End}_R(E)$ and $R'' = \mathrm{End}_{R'}(E)$. The mapping $\lambda: R \to R''$, defined by the equality $\lambda(\xi)v = \xi v$, where $\xi \in R$ and $v \in E$, is a ring homomorphism. A module E is called balanced if the homomorphism λ is an isomorphism. A module E is a generating object of category ${}_R\mathfrak{M}$ if and only if E is a balanced R-module and a finitely-generated projective R'-module. Gabriel's paper [118] on Abelian categories, which is long and very rich in content, contains many applications to categories of modules. Maranda's papers [177, 179] are mainly devoted to categories of modules.

§ 16. Generalizations

1. An extremely broad generalization of the concept of category is due to Benabou [50]. Let $\mathfrak{K}$ be a category with associative multiplication Π and with unity Λ (see §11). A relative, or formal, category over $\mathfrak{K}$ is the quadruple $(\mathfrak{D}, M, \mu, \varkappa)$, consisting of the class of objects $A, B, \ldots \in \mathfrak{D}$, the function M with domain $\mathfrak{D} \times \mathfrak{D}$ and with values on the objects of category $\mathfrak{K}$ and the system of morphisms $\mu = \{\mu_{ABC} : M(A, B)(\Pi) M(B, C) \to M(A, C)\}$ and $\varkappa = \{\varkappa_A : \Lambda \to M(A, A)\}$, where A, B, C are arbitrary objects. The following conditions should be fulfilled here: 1) $(\mu_{ABC}(\Pi) M(C, D)) \mu_{ACD} = (M(A, B)(\Pi)\mu_{BCD})\mu_{ABD}$ for any objects $A, B, C, D \in \mathfrak{D}$; 2) $(M(A, B)(\Pi)\varkappa_B)\mu_{ABB} = 1_{M(A,B)} = (\varkappa_A (\Pi) M(A, B))\mu_{AAB}$ for any objects $A, B \in \mathfrak{D}$. Every category (with null morphisms) is relative over the category of sets (with base point) with cartesian product as multiplication; every additive category is relative over the category of Abelian groups with tensor product as multiplication; the category of topological spaces with cartesian product is relative over itself. If $(\mathfrak{D}, M, \mu, \varkappa)$ is a relative category over $\mathfrak{K}$, then $\mathfrak{K}$, every mapping $f : \mathfrak{D}' \to \mathfrak{D}$ allows us to convert $\mathfrak{D}'$ into a relative category by setting $M'(A', B') = M(f(A'), f(B'))$, $\mu'_{A'B'C'} = \mu_{f(A')f(B')f(C')}$, $\varkappa'_{A'} = \varkappa_{f(A')}$ for any objects $A', B', C' \in \mathfrak{D}'$. Every functor between the categories $\mathfrak{K}$ and $\mathfrak{L}$ with multiplication allows us to convert every relative category over $\mathfrak{K}$ into a relative category over $\mathfrak{L}$. Benabou determined the functors between relative categories and their natural transformations and pointed out a method for constructing such functors and transformations.

Relative categories over the "category" of Maranda categories [178] are called categories of the second kind. An example of such a category is the class of categories $\mathfrak{K}_\alpha$, to each pair of which there corresponds a "category" of functors and their natural transformations. Of course, a very broad axiomatic of set theory is needed for the consideration of similar formations to be legitimate. In [178] Maranda investigated the question of the existence of limits in relative categories. It should be noted that Maranda's category of the second kind is a special case of a double category in the sense of Ehresmann [77]. Ehresmann calls the class $\mathfrak{K}$ of elements a double category if in $\mathfrak{K}$ there are defined the two operations $\circ$, $\perp$, relative to which $\mathfrak{K}$ forms the categories $\mathfrak{K}^0$ and $\mathfrak{K}^{\perp}$ respectively. The following relations should be satisfied here: 1) if the products $\alpha\circ\beta$, $\gamma\circ\delta$, $\alpha\perp\gamma$, $\beta\perp\delta$ are defined, then $(\alpha\circ\beta)\perp(\gamma\circ\delta) = (\alpha\perp\gamma)\circ(\beta\perp\delta)$; 2) the functions f and g, which associate with each morphism α of category $\mathfrak{K}^0$ its left unity $f(\alpha)$ and its right unity $g(\alpha)$, are functors of the category $\mathfrak{K}^{\perp}$ into itself; analogously, the corresponding functions $f^{\perp}$ and $g^{\perp}$ are functors of the category $\mathfrak{K}^0$ into itself and, here, the following relations should be satisfied for any $\alpha \in \mathfrak{K}$

$$f(f^{\perp}(\alpha)) = f^{\perp}(f(\alpha)),\quad g(g^{\perp}(\alpha)) = g^{\perp}(g(\alpha)),$$
$$f(g^{\perp}(\alpha)) = g^{\perp}(f(\alpha)),\quad f^{\perp}(g(\alpha)) = g(f^{\perp}(\alpha)).$$

In [107], Ehresmann treated n-ary categories as a natural generalization of double categories.

2. In [57], Dedecker and Mersch introduced the concept of a precategory. The definition of a precategory differs from the definition of a category in that in the former we do not require that the product of morphisms α and β, belonging respectively to H(A, B) and H(A, C), be always defined. An example of a precategory is the image $T(\mathfrak{K})$ of an arbitrary functor $T:\mathfrak{K}\to\mathfrak{L}$ from the category $\mathfrak{K}$ into category $\mathfrak{L}$. It was natural to define the concept of a functor between precategories. It was proved that for any precategory $\mathfrak{P}$ there exists a category $\overline{\mathfrak{P}}$, such that: 1) the precategory $\mathfrak{P}$ is embedded in the category $\overline{\mathfrak{P}}$ by the embedding functor $i:\mathfrak{P}\to\overline{\mathfrak{P}}$ and, 2) any functor $T:\mathfrak{P}\to\mathfrak{K}$ from the precategory $\mathfrak{P}$ into an arbitrary category $\mathfrak{K}$ can be uniquely represented in the form T = iT' for some functor $T':\overline{\mathfrak{P}}\to\mathfrak{K}$.

Other generalizations of the concept of a category were considered by Rosinger [204, 205] and Ehresmann [100].

LITERATURE CITED

1. G. D. Berishvili, "Direct and inverse limits," Soobshch. Akad. Nauk GruzSSR, 43(3): 535-542 (1966).
2. G. Kh. Berman, "Functors in the category of locally convex spaces," Dokl. Akad. Nauk SSSR, 154(3): 497-499 (1964).
3. I. E. Burmistrovich, "Embedding of an additive category into a category with direct products," Dokl. Akad. Nauk SSSR, 132(6): 1235-1237 (1960).
4. I. M. Gel'fand and G. E. Shilov, "Categories of finite-dimensional spaces," Vestn. Mosk. Univ., Mat., Mekh., No. 4, 27-48 (1963).
5. N. I. Glebov, "The structure of the class of R-tests for equivalence," in: Problems of Cybernetics, No. 9, Fizmatgiz, Moscow (1963), pp. 103-122.
6. A. Grothendieck, "Sur quelques points d'algebre homologique," Tohoku Math. J., 9: 119-221 (1957).
7. Kh. N. Inasaridze, "On the theory of extensions in categories," Soobshch. Akad. Nauk GruzSSR, 30(5): 537-544 (1963).
8. Kh. N. Inasaridze, "Universal functors," Soobshch. Akad. Nauk GruzSSR, 38(3): 513-520 (1965).

9. L. M. Kissina and A. S. Shvarts, "On the description of duality of functors," Dokl. Akad. Nauk SSSR, 167(2): 282-285 (1966).
10. V. V. Kuznetsov, "Duality of functors in the category of sets with base points," Dokl. Akad. Nauk SSSR, 159(4): 738-741 (1964).
11. A. G. Kurosh, "Radicals of rings and algebras," Mat. Sb., 33(1): 13-26 (1953).
12. A. G. Kurosh, "Direct decompositions in algebraic categories," Tr. Mosk. Mat. Obshch., Vol. 8, pp. 391-412 (1959); Vol. 9, p. 562 (1960).
13. A. G. Kurosh, "Radicals in group theory," Sibirsk. Mat. Zh., 3(6): 912-931 (1962).
14. A. G. Kurosh, A. Kh. Livshits, and E. G. Shul'geifer, "Foundations of the theory of categories," Usp. Mat. Nauk, 15(6): 3-52 (1960).
15. A. Kh. Livshits, "Summation of mappings and the concept of a center in categories," Mat. Sb., 60(2): 159-184 (1963).
16. A. Kh. Livshits, "Category-theoretical foundations of duality, radicality, and semisimplicity," Sibirsk. Mat. Zh., 5(2): 319-336 (1964).
17. A. Kh. Livshits, M. S. Tsalenko, and E. G. Shul'geifer, "Category theory," in: Algebra. Topology. 1962, Progress in Science, VINITI Akad. Nauk SSSR, Moscow (1963), pp. 90-106.
18. A. Kh. Livshits, M. S. Tsalenko, and E. G. Shul'geifer, "Varieties in categories," Mat. Sb., 63(4): 554-581 (1964).
19. S. MacLane, Homology, Springer-Verlag, Berlin (1963), 422 pp.
20. A. I. Mal'tsev, "Determining relations in categories," Dokl. Akad. Nauk SSSR, 119(6): 1095-1098 (1958).
21. B. S. Mityagin and A. S. Shvarts, "Functors in categories of Banach spaces," Usp. Mat. Nauk, 19(2): 65-130 (1964).
22. A. Pultr and Z. Hedrlin, "The representation of small categories," Dokl. Akad. Nauk SSSR, 160(2): 284-286 (1965).
23. Yu. M. Ryabukhin, "Radicals in categories," Izv. Akad. Nauk MoldSSR, Ser. Fiz.-Mat. i Tekhn. Nauk, No. 6, 58-74 (1964).
24. V. Trnková, "On the theory of categories," Comment. Math. Univ. Carolinae, 3(4): 9-35 (1962).
25. D. B. Fuks, "Homotopy duality," Dokl. Akad. Nauk SSSR, 141(4): 818-821 (1961).
26. D. B. Fuks, "Natural mappings of functors in the category of topological spaces," Mat. Sb., 62(2): 160-179 (1963).
27. D. B. Fuks and A. S. Shvarts, "On the homotopy theory of functors in the category of topological spaces," Dokl. Akad. Nauk SSSR, 143(3): 543-546 (1962).
28. A. Ya. Khelemskii, "Algebras of nilpotent operators and the categories connected with them," Vestn. Mosk. Univ., Mat., Mekh., No. 4, 49-55 (1963).
29. M. S. Tsalenko, "Proper unions and special subdirect sums in categories," Mat. Sb., 57(1): 75-94 (1962).
30. M. S. Tsalenko, "The completion of categories by free and direct unions of objects," Mat. Sb., 60(2): 235-256 (1963).
31. M. S. Tsalenko, "Correspondences over a quasiexact category," Dokl. Akad. Nauk SSSR, 155(2): 292-294 (1964).

32. M. S. Tsalenko, "R-complete subcategories of the category of groups," Usp. Mat. Nauk, 21(1): 174-175 (1966).
33. A. S. Shvarts, "Duality of functors," Dokl. Akad. Nauk SSSR, 148(2): 288-291 (1963).
34. E. G. Shul'geifer, "On the general theory of radicals in categories," Mat. Sb., 51(4): 487-500 (1960).
35. E. G. Shul'geifer, "Proper embeddings of categories," Usp. Mat. Nauk, 17(3): 212-213 (1962).
36. E. G. Shul'geifer, "Proper embeddings of categories," Mat. Sb., 61(4): 467-503 (1963).
37. E. G. Shul'geifer, "The structure of the ideals of an object of a category. II," Mat. Sb., 62(3): 335-344 (1963).
38. E. G. Shul'geifer, "Functorial characterization of strict radicals in categories," Sibirsk. Mat. Zh., 7(6): 1412-1421 (1966).
39. J. M. Adams, "Abstract homotopy theory in categories," Doctoral dissertation, New Mexico State University, 1963, 35 pp. Dissertation Abstr., 24(4): 1625 (1963).
40. S. A. Amitsur, "A general theory of radicals. II," Am. J. Math, 76: 100-125 (1954).
41. P. C. Baayen, "Opmerkingen over Kowalsky's artikel: 'Kategorien topologischer Räume'," Rapp. Math. Centrum., No. 14, p. 6 (1962).
42. L. Badescu, "Functori Adjuncti Contravarianti," Studii Si Cercetări Mat. Acad. RSR, 17(7): 1145-1152 (1965).
43. L. Badescu, "Extension de la definit on des foncteurs derives pour les categories abeliennes," Rev. Roumaine Math. Pures et Appl., 11(1): 103-107 (1966).
44. C. Bănică and N. Popescu, "Quelques considerations sur l'exactitude des foncteurs," Bull. Math. Soc. Sci. Math. et Phys. (RPR), 7(3-4): 143-147 (1963).
45. C. Bănică and N. Popescu, "Sur les catégories préabeliennes," Rev. Roumaine Math. Pures. et Appl. (RPR), 10(5): 621-633 (1965).
46. C. Bănică and N. Popescu, "Asupra categoriilor preabeliene," Studii Si Cercetări Mat. Acad. RSR, 17(4): 563-575 (1965).
47. L. Bégueri and G. Poitou, "'Diagram-chasing' dans les catégories abéliennes," Bull. Soc. Math. France, 93(4): 323-332 (1965).
48. J. Benabou, "Catégories avec multiplication," Compt. Rend. Acad. Sci., 256(9): 1887-1890 (1963).
49. J. Benabou, "Algèbre élémentaire dans les catégories avec multiplication," Compt. Rend. Acad. Sci., 258(3): 771-774 (1964).
50. J. Benabou, "Catégories relatives," Compt. Rend. Acad. Sci., 260(14): 3824-3827 (1965).
51. P. Boero, "Categorie omologiche," Rend. Semin. Mat. Univ. Padova, 35(2): 267-298 (1965).
52. R. Bonyun, "On the square of a homological monoid," Can. Math. Bull., 9(1): 49-55 (1966).
53. J. Braconnier, "Les catégories," Cahiers Rhodaniens, No. 10, pp. 1/1-1/12 (1961).

54. J. Bucur, "Fonctions définies sur le spectre d'une catégorie et théories de décompositions," Rev. Roumaine Math. Pures et Appl. (RPR), 9(6): 583-588 (1964).
55. L. Bukovský, Z. Hedrlin, and A. Pultr, "On topological representation of semigroups and small categories," Mat.-Fyz. Časop., 15(3): 197-199 (1965).
56. W. Burgess, "The meaning of mono and epi in some familiar categories," Can. Math. Bull., 8(6): 759-769 (1965).
57. P. Dedecker and J. Mersch., "Précatégories et relations d'équivalence dans les catégories," Compt. Rend. Acad. Sci., 256(23): 4811-4814 (1963).
58. S. E. Dickson, "A torsion theory for Abelian categories," Trans. Am. Math. Soc., 121(1): 223-235 (1966).
59. H. P. Doctor, "The categories of Boolean lattices, Boolean rings, and Boolean spaces," Can. Math. Bull, 7(2): 245-252 (1964).
60. K. Drbohlav, "A note on epimorphisms in algebraic categories," Comment. Math. Univ. Carolinae, 4(2): 81-85 (1963).
61. K. Drbohlav, "Concerning representations of small categories," Comment. Math. Univ. Carolinae, 4(4): 147-151 (1963).
62. K. Drbohlav, "A categorical generalization of a theorem of G. Birkhoff on primitive classes of universal algebras," Comment. Math. Univ. Carolinae, 6(1): 21-41 (1965).
63. L. Dubikajtis and J. Iwański,. "An independent system of axioms defining the Ehresmann groupoid," Bull. Acad. Polon. Sci., Ser. Sci. Math. Astron. et Phys., 12(111): 687-693 (1964).
64. A. Duma, "Asupra categoriilor de diagrame," Studii Si Cercetări Mat. Acad. RSR, 17(5): 803-807 (1965).
65. A. Duma, "Sur les catégories des diagrammes," Rev. Roumaine Math. Pures et Appl. (RPR), 10(5): 653-657 (1965).
66. B. Eckmann and P. J. Hilton, "Grouplike structures in general categories. I," Math. Ann., 145(3): 227-255 (1962).
67. B. Eckmann and P. J. Hilton, "Grouplike structures in general categories. II," Math. Ann., 151(2): 150-186 (1963).
68. B. Eckmann and P. J. Hilton, "Grouplike structures in general categories. III," Math. Ann., 150(2): 165-187 (1963).
69. B. Eckmann and H. Kleisli, "Algebraic homotopy groups and Frobenius algebras. III," J. Math., 6(4): 533-552 (1962).
70. C. Ehresmann, "Catégories inductives et pseudogroupes," Ann. Inst. Fourier, Vol. 10, pp. 307-332 (1960).
71. C. Ehresmann, "Especes des structures locales," Semin. Topol. et Geom. Different., Fac. Sci. Paris, Vol. 3, pp. 1-24 (1960-1962).
72. C. Ehresmann, "Elargissements des categories," Semin. Topol. et Geom. Différent., Fac. Sci. Paris, Vol. 3, pp. 25-73 (1960-1962).
73. C. Ehresmann, "Catégories des foncteurs types," Rev. Union. Math. Argent. y Asoc. Fis. Argent, Vol. 20, pp. 194-209 (1962).
74. C. Ehresmann, "Catégories structurées. III. Quintettes et applications covariantes," Cahiers Semin. Topol. et Géom. Différent., Fac. Sci. Paris, Vol. 5 (1963).

75. C. Ehresmann, "Catégories structurées quotients," Cahiers Semin. Topol. et Geom. Différent., Fac. Sci. Paris, Vol. 5 (1963).
76. C. Ehresmann, "Prolongements des catégories différentiables," Cahiers Semin. Topol. et Geom. Différent., Fac. Sci. Paris, Vol. 6 (1964).
77. C. Ehresmann, "Catégories doubles et catégories structurées." Compt. Rend. Acad. Sci., 256(6): 1198-1201 (1963).
78. C. Ehresmann, "Catégorie double des quintettes; applications covariantes," Compt. Rend. Acad. Sci., 256(9): 1891-1894 (1963).
79. C. Ehresmann, "Catégories structurées d'operateurs," Compt. Rend. Acad. Sci., 256(10): 2080-2083 (1963).
80. C. Ehresmann, "Sous-structures et applications K-covariantes," Compt. Rend. Acad. Sci., 256(11): 2280-2283 (1963).
81. C. Ehresmann, "Structures quotients et catégories quotients," Compt. Rend. Acad. Sci., 256(24): 5031-5034 (1963).
82. C. Ehresmann, "Catégories structurées," Ann. Sci. Ecole Norm. Supér., 80(4): 349-426 (1963).
83. C. Ehresmann, "Groupoides sous-inductifs," Ann. Inst. Fourier, 13(2): 1-60 (1963).
84. C. Ehresmann, "Complétion des catégories ordonnées," Compt. Rend. Acad. Sci., 257(26): 4110-4113 (1963).
85. C. Ehresmann, "Catégories et structures: Extraits," Cahiers Semin. Topol. et Géom. Différent., Fac. Sci. Paris, Vol. 6 (1964).
86. C. Ehresmann, "Structures quotients," Comment. Math. Helv., 38(3): 219-242 (1964).
87. C. Ehresmann, "Structures quotients," Comment. Math. Helv., 38(4): 243-283 (1964).
88. C. Ehresmann, "Catégories ordonnées, holonomie et cohomologie," Ann. Inst. Fourier, 14(1): 205-268 (1964).
89. C. Ehresmann, "Complétion des catégories ordonnées," Ann. Inst. Fourier, 14(2): 89-144 (1964).
90. C. Ehresmann, "Sous-structures et catégories ordonnées," Fundam. Math., 54(2): 211-228 (1964).
91. C. Ehresmann, "Produit croisé des catégories," Compt. Rend. Acad. Sci., 258(9): 2461-2464 (1964).
92. C. Ehresmann, "Completion des catégories sous-prélocales," Compt. Rend. Acad. Sci., 259(4): 701-704 (1964).
93. C. Ehresmann, "Expansion d'homomorphismes et foncteurs," Compt. Rend. Acad. Sci., 259(7): 1372-1375 (1964).
94. C. Ehresmann, "Sur une notion générale de cohomologie," Compt. Rend. Acad. Sci., 259(13): 2050-2053 (1964).
95. C. Ehresmann, "Catégories et structures," Dunod, Paris (1965), 358 pp.
96. C. Ehresmann, "Structures quasiquotients," Cahiers Semin. Topol. et Géom. Différent., Fac. Sci. Paris, Vol. 7, Oct.(1965).
97. C. Ehresmann, "Expansion générale des foncteurs," Compt. Rend. Acad. Sci., 260(1): 30-33 (1965).
98. C. Ehresmann, "Catégorie quotient d'une catégorie par une sous-catégorie," Compt. Rend. Acad. Sci., 260(8): 2116-2119 (1965).

99. C. Ehresmann, "Quasi-surjections et structures quasi-quotient," Compt. Rend. Acad. Sci., 261(7): 1577-1580 (1965).
100. C. Ehresmann, "Quasi-catégories structurées," Compt. Rend. Acad. Sci., 261(9): 1932-1935 (1965).
101. C. Ehresmann, "Groupoides structurés quasi-quotient et quasicohomologie," Compt. Rend. Acad. Sci., 261(22): 4583-4586 (1965).
102. C. Ehresmann, "Expansion des systèmes des structures dominés," Compt. Rend. Acad. Sci., AB262(1): A8-A11 (1966).
103. C. Ehresmann, "Adjonction des limites aux catégories structurées," Compt. Rend. Acad. Sci., 263(19): A655-A658 (1966).
104. C. Ehresmann, "Quasi-élargissement d'un système des structures structurés," Compt. Rend. Acad. Sci., 263(21): A762-A765 (1966).
105. C. Ehresmann, "Premier théorème d'expansion structurée," Compt. Rend. Acad. Sci., 263(23): A863-A866 (1966).
106. C. Ehresmann, "Théorème de quasi-expansion régulière," Compt. Rend. Acad. Sci., 264(2): A56-A59 (1967).
107. C. Ehresmann, "Problèmes universels relatifs aux catégories n-aires," Compt. Rend. Acad. Sci., 264(6): A273-A276 (1967).
108. S. Eilenberg and J. C. Moore, "Adjoint functors and triples. III," J. Math., 9(3): 381-398 (1965).
109. D. B. A. Epstein, "Functors between tensored categories," Invent. Math., 1(3): 221-228 (1966).
110. M. Evrard, "Homologie dans les catégories nonabéliennes; exactitude dans C," Compt. Rend. Acad. Sci., 260(3): 749-751 (1965).
111. M. Evrard, "Homologie dans les catégories nonabéliennes; définition de l'homologie," Compt. Rend. Acad. Sci., 260(4): 1049-1051 (1965).
112. C. Faith, "A general Wedderburn theorem," Bull. Am. Math. Soc., 73(1): 65-67 (1967).
113. A. U. Fregoso, "Local categories," Doctoral Dissertation, Indiana University, 1963, 59 pp. Dissertation Abstr., 25(1): 495 (1964).
114. A. Frei, "Objets libres dans une catégorie primitive et comultiplications," Compt. Rend. Acad. Sci., AB262(8): A421-A424 (1966).
115. P. Freyd, "Relative homological algebra made absolute," Proc. Nat. Acad. Sci. USA, 49(1): 19-20 (1963).
116. P. Freyd, "Abelian categories: An introduction to the theory of functors," Harper and Bros., New York (1964), 164 pp.
117. P. Freyd, "Algebra-valued functors in general and tensor products in particular," Colloq. Math., Vol. 14, pp. 89-106 (1964).
118. P. Gabriel, "Des catégories abéliennes," Bull. Soc. Math. France, 90(3): 323-448 (1962).
119. P. Gabriel and U. Oberst, "Spektralkategorien und reguläre Ringe im Von-Neumannschen Sinn," Math. Z., 92(5): 389-395 (1966).
120. G. Gemignani, "Gruppi e cogruppi in una categoria," Ann. Scuola Norm. Super. Pisa. Sci. Fis. e Mat., 20(1): 139-171 (1966).
121. R. Goblot, "Foncteur de séparation," Compt. Rend. Acad. Sci., AB262(19): A1034-A1036 (1966).

122. E. M. Goldberg, "Relations in categories," Doctoral Dissertation, Columbia University, 1965, 77 pp. Dissertation Abstr., 26(5): 2775-2776 (1965).

123. J. W. Gray, "Sheaves with values in a category," Topology, 3(1): 1-18 (1965).

124. M. Hacque, "Satellites des foncteurs limites projectives et cohomologie abélienne des petites catégories," Compt. Rend. Acad. Sci., 260(6): 1532-1535 (1965).

125. M. Hasse, "Über Erzeugendensysteme schlichter Kategorien," Wiss. Z. Techn. Univ. Dresden, 13(4): 1035-1036 (1964).

126. M. Hasse and L. Michler, "Über die Einbettbarkeit von Kategorien in Gruppoide," Math. Nachr., 25(3): 169-177 (1963).

127. P. J. Hilton, "Note on free and direct products in general categories," Bull. Soc. Math. Belg., 13(1-2): 38-49 (1961).

128. P. J. Hilton and W. Ledermann, "Remark on the l.c.m. in a homological ringoid," Quart. J. Math., 11(44): 287-294 (1960).

129. M. Hocquemiller, "Probléme universel de catégorie," Bol. Soc. Mat. São Paulo, 18(1-2): 49-55 [1963 (1966)].

130. H.-J. Hoehnke, "Einige Bemerkungen zur Einbettbarkeit von Kategorien in Gruppoide," Math. Nachr., 25(3): 179-190 (1963).

131. H.-J. Hoehnke, "Zur Theorie der Gruppoide. IX," Monatsber. Dtsch. Akad. Wiss. Berlin, 5(7): 405-411 (1963).

132. F. Hoffmann, "Über eine die Kategorie der Gruppen umfassende Kategorie," Sitzsber. Bayer. Acad. Wiss. Math.-Naturwiss. Kl., 1960, Munich (1961), pp. 163-204.

133. P. J. Huber, "Homotopy theory in general categories," Math. Ann., 144(5): 361-385 (1961).

134. P. J. Huber, "Standard constructions in Abelian categories," Math. Ann., 146(4): 321-325 (1962).

135. M. Hušek, "S-categories," Comment. Math. Univ. Carolinae, 5(1): 37-46 (1964).

136. M. Hušek, "Remarks on reflections," Comment. Math. Univ. Carolinae, 7(2): 249-253 (1966).

137. J. R. Isbell, "Adequate subcategories," Ill. J. Math., 4(4): 541-552 (1960).

138. J. R. Isbell, "Two set-theoretical theorems in categories," Fundam. Math., 53(1): 43-49 (1963).

139. J. R. Isbell, "Subobjects, adequacy, completeness, and categories of algebras," Rozpr. Mat., No. 36, 33 pp (1964).

140. J. R. Isbell, "Natural sums and abelianizing," Pacif. J. Math., 14(4): 1265-1281 (1964).

141. E. J. Javier, "Categorias de semigrupos principales," Rev. Mat. Hisp.-Amer., 24(6): 207-217 (1964).

142. E. J. Javier, "Una categoria de semigrupos," Acta Salmanticensia, Ser. Cienc., 6(4): 79-88 (1965).

143. G. Joubert, "Extensions de foncteurs ordonnés," Compt. Rend. Acad. Sci., 260(12): 3251-3254 (1965).

144. G. Joubert, "Extensions des foncteurs ordonnés et applications," Compt. Rend. Acad. Sci., 261(3): 623-626 (1965).

145. M. Jurchescu and A. Lascu, "Morfisme stricte, categorii cantoriene, functori de completare," Studii Si Cercetări Mat. Acad. RSR, 18(2): 219-234 (1966).

146. P.-F. Jurie, "Sur quelques caractérisations des sommes amalgamées d'homomorphismes bpp-éiens," Compt. Rend. Acad. Sci., 264(5): A217-A220 (1967).
147. L. Kauffman, "Categories," Tech. Engng. News, 47(8): 41-43 (1966).
148. G. M. Kelly, "On the radical of a category," J. Austral. Math. Soc., 4(3): 299-307 (1964).
149. G. M. Kelly, "Complete functors in homology. I. Chain maps and endomorphisms," Proc. Cambridge Phil. Soc., 60(4): 721-735 (1964).
150. G. M. Kelly, "Complete functors in homology. II. The exact homology sequence," Proc. Cambridge Phil. Soc., 60(4): 737-749 (1964).
151. G. M. Kelly, "Tensor products in categories," J. Algebra, 2(1): 15-37 (1965).
152. G. M. Kelly, "On MacLane's conditions for coherence of natural associativities, commutativities, etc.," J. Algebra, 1(4): 397-402 (1964).
153. J. F. Kennison, "Reflective functors in general topology and elsewhere," Trans. Am. Math. Soc., 118(6): 303-315 (1965).
154. H. Kleisli, "Homotopy theory in Abelian categories," Can. J. Math., 14(1): 139-169 (1962).
155. H. Kleisli, "Every standard construction is induced by a pair of adjoint functors," Proc. Am. Math. Soc., 16(3): 544-546 (1965).
156. Lam Tsit-Yuen, "The category of Noetherian modules," Proc. Nat. Acad. Sci. USA, 55(5): 1038-1040 (1966).
157. G. B. Lampton, Jr., "Abstract Abelian categories," Doctoral Dissertation, University of South Carolina, 1964, 67 pp. Dissertation Abstr., 25(11): 6658 (1965).
158. M. L. Laplaza, "Nota sobre los epimortismos en la categoria de los anillos," Rev. Mat. Hisp.-Amer., 26(3): 67-72 (1966).
159. R. Lavendhomme, "La notion d'idéal dans la théorie des catégories," Ann. Soc. Sci. Bruxelles, Ser. 1, 79(1): 5-25 (1965).
160. R. Lavendhomme, "Un plongement pleinement fidèle de la catégorie des groups. Application à l'algébre homologique nonabelienne," Bull. Soc. Math. Belg., 17(2): 153-185 (1965).
161. F. W. Lawvere, "Functorial semantics of algebraic theories," Proc. Nat. Acad. Sci., 50(5): 869-872 (1963).
162. F. W. Lawvere, "An elementary theory of the category of sets," Proc. Nat. Acad. Sci. USA, 52(6): 1506-1511 (1964).
163. P. Lecouturier, "Quelques propriétés des catégories hofmanniennes," Ann. Soc. Sci. Bruxelles, Ser. 1, 80(1): 19-72 (1966).
164. S. Legrand, "Construction des catégories ordonnées, projections de graphes multiplicatifs ordonnés," Compt. Rend. Acad. Sci., 260(12): 3255-3258 (1965).
165. S. Legrand, "Catégorie double projection d'un graphe multiplicatif double," Compt. Rend. Acad. Sci., AB262(7): A381-A383 (1966).
166. J. B. Leicht, "On commutative squares," Can. J. Math., 15(1): 59-79 (1963).
167. J. B. Leicht, "Über die elementaren Lemmata der homologischen Algebra in quasiexacten Kategorien," Monatsh. Math., 68(3): 240-254 (1964).
168. J. Levy-Bruhl, "Sur les catégories ordonnées," Compt. Rend. Acad. Sci., 258(6): 1669-1671 (1964).
169. J. Levy-Bruhl, "Demi-groupe et catégorie modulaires," Compt. Rend. Acad. Sci., 260(15): 4134-4136 (1965).

170. J. Levy-Bruhl, "Sur les notions d'image, noyau, coimage, conoyau," Compt. Rend. Acad. Sci., 262(25): A1381-A1384 (1966).
171. F. E. J. Linton, "Autonomous categories and duality of functors," J. Algebra, 2(3): 315-349 (1965).
172. S. MacLane, "Locally small categories and the foundations of set theory," in: Infinitistic Methods, Pergamon Press, Oxford; Warszawa. P. W. N. (1961), pp. 25-43.
173. S. MacLane, "An algebra of additive relations," Proc. Nat. Acad. Sci. USA, 47(7): 1043-1051 (1961).
174. S. MacLane, "Natural associativity and commutativity," Rice Univ. Studies, 49(4): 28-46 (1963).
175. S. MacLane, "Categorical algebra," Bull. Am. Math. Soc., 71(1): 40-106 (1965).
176. J. M. Maranda, "Some remarks on limits in categories," Can. Math. Bull., 5(2): 133-146 (1962).
177. J. M. Maranda, "Injective structures," Trans. Am. Math. Soc., 110(1): 98-135 (1964).
178. J. M. Maranda, "Formal categories," Can. J. Math., 17(5): 758-801 (1965).
179. J. M. Maranda, "Completions of modules and rings," Trans. Roy. Soc. Canada, Sec. 1-3(3): 271-291 (1965).
180. W. Meier, "Beiträge zur algebraischen Homotopietheorie der Moduln," Diss., Dokt. Math., Eidgenöss. Techn. Hochschule Zürich (1962), 54 pp.
181. J. Mersch., "Chasses sur les diagrammes des catégories abéliennes," Bull. Soc. Roy. Sci. Liége, 33(7-8): 411-450 (1964).
182. J. Mersch, "Structures quotients," Bull. Soc. Roy. Sci. Liège, 33(1-2): 45-58 (1964).
183. J. Mersch, "Le problème du quotient dans les catégories," Mem. Soc. Roy. Sci. Liège, Vol. 11, No. 1 (1965).
184. B. Mitchell, Theory of Categories, Academic Press, New York (1965), 273 pp.
185. B. Mitchell, "The full imbedding theorem," Am. J. Math., 86(3): 619-637 (1964).
186. C. Năstăsescu and N. Popescu, "Sur la structure des objets de certaines catégories abéliennes," Compt. Rend. Acad. Sci., AB262(24): A1295-A1297 (1966).
187. Y. Nouazé, "Catégories localement de type fini et catégories localement noethériennes," Compt. Rend. Acad. Sci., 257(4): 823-824 (1963).
188. L. Oubina and J. Bosch, "Sous-jacence, union et intersection en catégories," Compt. Rend. Acad. Sci., 260(9): 2389-2391 (1965).
189. B. Pareigis, "Cohomology of groups in arbitrary categories," Proc. Am. Math. Soc., 15(5): 803-819 (1964).
190. N. Popescu, "Elemente de teoria fasciculelor. I," Studii Si Cercetari Mat. Acad. RSR, 18(2): 267-296 (1966).
191. N. Popescu, "Elemente de teoria fasciculelor. II," Studii Si Cercetari Mat. Acad. RSR, 18(3): 407-456 (1966).
192. N. Popescu and P. Gabriel, "Caractérisation des catégories abéliennes avec générateurs et limites inductives exactes," Compt. Rend. Acad. Sci., 258(17): 4188-4190 (1964).
193. A. Pultr, "Concerning universal categories," Comment. Math. Univ. Carolinae, 5(4): 227-239 (1964).

194. R. Pupier, "Sur les décompositions de morphismes dans les catégories à sommes ou à produits fibres," Compt. Rend. Acad. Sci., 258(26): 6317-6319 (1964).
195. D. Puppe, "Korrespondenzen in abelschen Kategorien," Math. Ann., 148(1): 1-30 (1962).
196. J. Riquet, "Categorisation de la notion des structures et des structures locales chez N. Bourbaki et C. Ehresmann. I," Semin. P. Dubreil, M.-L. Dubreil-Jacotin et C. Pisot. Fac. Sci. Paris, 1960-1961, 14 Annee, Fasc. I, Paris (1963), pp. 5/01-5/32.
197. P. Robert, "Méthodes fonctionnelles sur l'axiomatique des systèmes de générateurs, des rangs, etc.," Compt. Rend. Acad. Sci., 260(16): 4291-4294 (1965).
198. P. Robert, "Méthodes fonctionnelles sur l'axiomatique des systèmes de générateurs des rangs," Compt. Rend. Acad. Sci., 260(18): 4661-4664 (1965).
199. P. Robert, "Methodes fonctionnelles sur l'axiomatique des systèmes de générateurs, des rangs, etc.," Compt. Rend. Acad. Sci., 260(23): 5983-5986 (1965).
200. J.-E. Roos, "Introduction à l'étude de la distributivité des foncteurs $\varprojlim$ par rapport aux $\varinjlim$ dans des catégories des faisceaux (topos)," Compt. Rend. Acad. Sci. 259(5): 969-972 (1964).
201. J.-E. Roos, "Sur la distributivité des foncteurs $\varprojlim$ par rapport aux $\varinjlim$ dans les catégories des faisceaux (topos)," Compt. Rend., 259(9): 1605-1608 (1964).
202. J.-E. Roos, "Complément à l'étude de la distributivité des foncteurs $\varprojlim$ par rapport aux $\varinjlim$ dans les catégories des faisceaux (topos)," Compt. Rend. Acad. Sci., 259(11): 1801-1804 (1964).
203. J.-E. Roos, "Caractérisation des catégories qui sont quotients des catégories des modules par des souscatégories bilocalisantes," Compt. Rend. Acad. Sci., 261(23): 4954-4957 (1965).
204. T. Rosinger, "Une extension de la notion de catégorie Eilenberg — MacLane," Rev. Roumaine Math. Pures et Appl. (RPR), 9(9): 881-885 (1964).
205. T. Rosinger, "O extindere a notiunii de categorie in sens Eilenberg — MacLane," Studii Si Cercetări Mat. Acad. RPR, 16(8): 997-999 (1964).
206. A. Roux, "Objets injectifs et projectifs," Cahiers Rhodaniens, No. 10, pp. 3/1-3/7 (1961).
207. A. Roux, "Sur une équivalence des catégories abéliennes," Compt. Rend. Acad. Sci., 258(23): 5566-5568 (1964).
208. A. Roux, "Un théorème de plongement des catégories," Compt. Rend. Acad. Sci., 258(19): 4646-4647 (1964).
209. R. Rubio, "Sobre los fundamentos de la geometria diferencial," Casop. Pestov. Mat., 89(3): 278-295 (1964).
210. S. Salvioli and J. Bosch, "Sous-objets quotients et images en catégories," Compt. Rend. Acad. Sci., 260(8): 2111-2113 (1965).
211. Z. Semadeni, "Free and direct objects," Bull. Am. Math. Soc., 69(1): 63-66 (1963).
212. Z. Semadeni, "Projectivity, injectivity, and duality," Rozpr. Mat., No. 35, (1963).
213. Z. Semadeni, "Free objects in the theory of categories," Colloq. Math., Vol. 14, pp. 107-110 (1964).

214. A. Solian, "A canonical decomposition of additive functors of modules," Atti Accad. Naz. Lincei. Rend. Cl. Sci. Fis., Mat. e Natur., 36(2): 129-135 (1964).
215. A. Solian, "Foncteurs qui conservent les épimorphismes locaux," Compt. Rend. Acad. Sci., 259(7): 1376-1379 (1964).
216. A. Solian, "Foncteurs qui transforment les épimorphismes locaux en monomorphismes locaux," Rev. Roumaine Math. Pures et Appl., (RPR), 11(4): 401-410 (1966).
217. J. Sonner, "On the formal definition of categories," Math. Z., 80(2): 163-176 (1962).
218. J. Sonner, "Universal and special problems," Math. Z., 82(3): 200-211 (1963).
219. J. Sonner, "Universal solutions and adjoint homomorphisms," Math. Z., 86(1): 14-20 (1964).
220. R. Stärk, "Nullsysteme in allgemeinen Kategorien," Diss. Eidgenöss. Techn. Hochschule Zürich. Dokt. Math. Zürich (1963).
221. A. Suliński, "The Brown—McCoy radical in categories," Fundam. Math., 59(1): 23-41 (1966).
222. J. C. Taylor, "Weak families of maps," Can. Math. Bull., 8(6): 771-781 (1965).
223. P. Tondeur, "Categories y functores," Cursos y Semin. Mat. Univ. Buenos Aires, No. 21 (1965).
224. V. Trnková, "Sum of categories with amalgamated subcategory," Comment. Math. Univ. Carolinae, 6(4): 449-474 (1965).
225. V. Trnková, "Universal categories," Comment. Math. Univ. Carolinae, 7(2): 143-206 (1966).
226. R. Vazquez, "Seminario de algebra homologica. Gavillas con valores en categorias," An. Inst. Mat. Univ. Nac. Autónoma México, Vol. 3, pp. 37-77 (1963).
227. I. Vidav, "O kategorijah in funktorjih," Obz. Mat. in Fiz., 11(1): 1-12 (1964).
228. W. Waliszewski, "Categories, groupoids, pseudogroups and analytical structures," Rozpr. Mat., No. 45 (1965).
229. B. Wegener, "Factorgruppoide und homomorphe Abbildungen bei Gruppoiden," Wiss. Z. Techn. Hochschule Dresden, 10(3): 423-431 (1961).
230. M. L. Wheat, "Radical subcategories," Doctoral Dissertation, University of Kansas, 1964, 68 pp. Dissertation Abstr., 26(2): 1076-1077 (1965).
231. O. Wyler, "Weakly exact categories," Arch. Math., 17(1): 9-19 (1966).
232. O. Wyler, "Direct sums in weakly exact categories," Arch. Math., 17(3): 216-225 (1966).

Class Field Theory. Field Extensions

S. P. Demushkin

In this brief survey on class field theory and related questions we mainly present the papers reviewed in the "Mathematics" section of Referativnyi Zhurnal during 1958-1967. Among the books published during this time we note those by Chevalley [20] (a systematic exposition and application of cohomology groups), Artin and Tate [12] (the most modern exposition of class field theory with the application of cohomology groups and of class systems; the ground field is either the field of algebraic numbers of finite degree or the field of functions of one variable over a finite field of constants), and Serre [99] (the connection of class field theory with algebraic curves). In 1963 there appeared a survey by Ribenboim [93] who set forth the results in the Hilbert — Takagi theory and the reciprocity law of Artin and who examined the problem of finding functions whose values could be generated by any Abelian extension of the field of algebraic numbers. On local class field theory we have the survey by Hochschild [43] (also see Samuel's report [94] on Hochschild's results: G. Hochschild, "Local class field theory," Ann. Math., 51(2):331-347 (1950), where local class field theory is presented with the help of cohomologies). The address by Heilbronn [42] to the London Mathematical Society is also of the nature of a survey. The elementary information needed for a further study of class field theory can be found in Holzer [46], Weiss [122] (introducing the fundamental concepts with the exception of cohomology groups), and Lang [7] (idèles and adèles).

1. Class field theory continues to attract the attention of many mathematicians. During the last ten years it has been developed in the following directions: the extension of the domain of objects over which it is constructed; the generalization of the theory itself (the introduction of class formations); the development of the theory of complex multiplication; the sharpening of the principal ideal theorem, of the reciprocity laws [11, 33, 40, 76, 108, 110, 124], as well as of the results on the norm residue symbol [24, 35, 75, 85, 87, 109]. Later on we touch on the directions listed (we shall not dwell in detail on the last two aspects).

While studying the extensions of the field of algebraic numbers with specified ramification points, Golod and Shafarevich [2] succeeded in solving negatively the famous class field tower problem. Therefore, later on we shall mention certain results on extensions with specified ramification points. At the end of the survey we present the papers on the growth of the class number of ideals of cyclotomic fields, and also on Galois groups of local fields.

2. In this paragraph we dwell on papers in which the ground field is assumed to be more general. A definite achievement in this area was Lang's construction [82] of a class field theory over the field K of algebraic functions of several variables with a finite field of constants k. If L/K is a finite Abelian extension of such a field K, $\mathfrak{o}$ is a local ring of prime points on some model of the field K/k, $\mathfrak{p}$ is its maximal ideal, and $\mathfrak{o}$ is unramified in L, then to the ring $\mathfrak{o}$ (or to the ideal $\mathfrak{p}$) we can add the element $(\mathfrak{p}, L/K)$ (the Frobenius symbol) of the group G(L/K) = G. This mapping is continued up to a homomorphism φ (the reciprocity mapping) into the group G of the free group $\mathfrak{Z}$, generated by the set of local rings of field K, unramified in L. The fundamental problem of class field theory is the study of the reciprocity mapping. The kernel of the mapping φ was determined for certain fields L in [82].

Mori [88] has presented a class field theory over a field which is the extension of the infinite power of a local field. The theory of Abelian extensions of the field of algebraic functions of one variable can be reduced to the theory of certain special homomorphisms, called isogenies, of commutative algebraic groups. Serre [97] has constructed an analogous theory for local fields where the role of the algebraic groups is played by the proalgebraic groups introduced by the author (also see [100]). In [100], Serre applied proalgebraic groups to describe the group $\mathfrak{A}_K$ of the maximal Abelian extension of a complete relatively-discrete normed field K with an algebraically closed residue field. The structure of a proalgebraic group can be introduced on the unit group U_K of such a field K. Let $\pi_1(U_K)$ be a fundamental group.

The isomorphism $\theta\colon \pi_1(U_K)\to\mathfrak{A}_K$ was established in [101]. The Abelian extensions of field K correspond to the isogenies of the group U_K.

3. We pass on to the following generalizations, namely, class formations and abstract class formations. Class formations were introduced in a series of papers by Kawada [60-65]. They are defined as follows. Let $\mathfrak{K}$ be the collection of all finite extensions of a field k_0, contained in some infinite algebraic normal separable extension Ω/k_0. Let an Abelian group E (k) be associated with every field $k\in\mathfrak{K}$ where: 1) if $k\subset K$, then $E(k)\subset E(K)$; 2) if K/k is normal, then G(K/k) is the operator group for E (K) and the collection $E(K)^G = E(k)$; 3) if K_1/k is a normal subfield of the field K/k, then for any $x\in E(K_1)$ we have $\sigma x = \overline{\sigma x}$, where $\overline{\sigma}$ is an automorphism of field K_1/k, induced by σ. The system $\{E(k)\}$ is a class formation for the extension Ω/k_0 if for every normal extension of K/k (k, $K\in\mathfrak{K}$) with the Galois group G, the group $H^1(G, E(K))$ is trivial but $H^2(G, E(K))$ is a finite cyclic group whose order equals the order of group G. Class formations are known for the fields of $\mathfrak{p}$-adic numbers and of formal power series over a finite field [E (k)) is the multiplicative group of field k], and also for the fields of algebraic numbers and of algebraic functions of one variable over the finite field of constants [E(k) is the class group of ideal Chevalley elements]. Kawada also constructed class formations for certain other fields. Next he showed that there exist the so-called abstract class formations, i.e., those constructed without the use of the structure of the fields $k\subset K$ (see [36-39, 65]).

4. A rather large number of papers have been devoted to complex multiplication. The problem was posed long ago and consists in finding functions whose values would generate Abelian extensions of the fields of algebraic numbers. Three reports [104, 112, 120] were devoted to complex multiplication at the 1955 Symposium on Algebraic Number Theory in Tokyo. A good exposition of this theory can be found in the lectures given at the Seminar on Complex Multiplication [96]. Meyer [86] has given formulas of an analytical nature for the class number of the fields with complex multiplication and their subfields. The articles by Deuring [22] and by Ramachandra [90] are devoted to this same group of questions. By refining the results of [105], Shimura and Taniyama [106] explicitly indicated class fields which may be obtained by the complex multiplication of Abelian varieties. Lubin and Tate [84] gave an analytical derivation and interpretation of the reciprocity law for a local field and showed how the Abelian extensions of such a field are generated by means of the zeros of certain analytic functions. The results of [84] are stated also in [115]. Kubota [77] defined the complex multiplication system purely algebraically. Two papers [13, 23] have examined the fields of moduli of Abelian varieties. Doi [23] found the field of moduli of the direct product $B_1 \times \ldots \times B_h$ of the prime Abelian varieties B_i with complex multiplication of one and the same type. He showed that the field of moduli of such a product is contained in the class field obtained from the field of moduli of one factor B_i. The complex multiplication of Abelian varieties with application to number theory has been presented in the book by Shimura and Taniyama [107].

Novikov [9] showed that if Σ/Ω (Ω is an imaginary quadratic field) is a class field with respect to the class group H of ideals with a conductor modulus f and, moreover, if $(f(2)) = 1$ and $(2) \in H$, then Meyer's formula on the class number of a field with complex multiplication can be transformed into a formula having an algebraic meaning. Chevalley [21] proved that for an imaginary quadratic field Ω the class field for it is generated by the values of the modular function at points which are determined by the ideal classes of field Ω.

5. As is well known, every ideal in a class field becomes principal. In [48], this theorem was strengthened in the following way: the numbers which make the ideals principal are the values of elliptic modular forms for the imaginary quadratic field Ω. For just such a field Reichardt [91] explicitly found the number γ from the class field for which $(\gamma)=\mathfrak{P}$, where $\mathfrak{P}$ is the prime ideal of degree one of the field Ω, not dividing d (d is the discriminant of field Ω; also see [92]). Thus, in the given case the principal ideal theorem was proved without the use of the reciprocity law. Terada [116] gave an explicit formula for the number γ with $(\gamma)=\mathfrak{m}$, where the ideal $\mathfrak{m}$ is relatively prime with the discriminant of field Ω and $N(\mathfrak{m}) \equiv 5 \pmod{12}$.

Several papers have also been devoted to the generalization of the principal ideal theorem [113, 114, 117, 118]. For example, Terada [117] proved that for the case $K \supset \Omega \supset k$, where K/k is a class field and Ω/k is a cyclotomic field, every ideal in Ω, invariant relative to the action of the Galois group $G(\Omega/k)$, becomes principal in K. The aim of Takahashi's article [111] was to give an explicit representation of the Tannaka—Terada result in the case $Q(\sqrt{m}) \subset Q(\zeta_n)$, $2 \nmid n$ or $4|n$. The principal ideal theorem was treated in cohomological terms in [119].

The general principal ideal theorem was proved by Kempfert [59] with the aid of complex multiplication.

6. We now pass to a group of papers on extensions with specified ramification points. These questions attracted a lot of attention until Golod and Shafarevich [2] proved that fields exist for which the class field tower does not terminate. It was proved in [2] that the number of generators d(G) and the number of relations r(G) in a finite p-group G are connected by the relation $r(G) > \left(\frac{d(G)-1}{2}\right)^2$. Earlier, Shafarevich [10] had proven that the relation $r \leq d + \rho$, where ρ is the rank of the unit group of field k, is true for a minimal number of generators d and a number of relations r of the Galois group of a maximal unramified p-extension of the field of algebraic numbers k. From the last result it ensues that if the number of generators γ of the class group of ideals of the field k and ρ are related by $\gamma \geq 3 + 2\sqrt{\rho + 2}$, then over k there exist infinite unramified extensions. For example, such a field will be $k = Q(\sqrt{-3 \cdot 5 \cdot 7 \cdot 11 \cdot 13 \cdot 17 \cdot 19})$ (also see Serre [103] and Hoechsmann [44, 45] on this). Using the results of [2], Lamprecht [81] has proven that for a given field k of algebraic numbers and for a prime number p there exists a positive integer $l = l(k, p)$ with the property: there are infinitely many algebraic extensions K of the field k of degree p^l such that K has an infinite class field tower. We can set $l(k, p) = f + 1$, where $f \geq 0$ is the exponent with which p occurs in the class number of field k.

An interesting connection between the number of prime numbers ramified in a normal extension and the class field tower was established by Brumer [18]. If K is a normal extension of degree n of the rational number field Q, $\omega(n)$ is the number of prime numbers $p|n$, s is the number of prime numbers ramified in K, and r is the number of generators of the ideal class group of field K, then $r \geq s/\omega(n) - 2n$. From this the author gets that the class field tower over field K does not terminate when $s \geq \omega(n)(3 + 2n + 2\sqrt{\rho + 2})$, where ρ is the number of generators in the unit group of field K.

In a series of papers [69–73], Koch has considered groups of maximal p-extensions of the algebraic number field with specified ramification points. The group $G(K_{\bar{S}}/k)$, where $\bar{S}$ is the set of all divisors of field k and $K_{\bar{S}}/k$ is the maximal p-extension of field k, was described in [69]. This description then made it possible now to find precisely certain partial groups $G(K_S/k)$, where S is some set of prime divisors of field k. In [70] there was considered the class field 2-tower over the quadratic field of algebraic numbers k: $k = k^{(0)}$, $k^{(n+1)}$ $(n \geq 0)$ is the composite of all unramified quadratic extensions of the field $k^{(n)}$, central over k, and $k^{(\infty)} = \bigcup_{n=0}^{\infty} k^{(n)}$. Necessary and sufficient conditions were found for $k^{(\infty)} \neq k^{(1)}$ when the discriminant of field k has three prime divisors, and it was proved that the 2-tower terminates after the second stage only in the case when $G(k^{(2)}/k)$ is a quarternion group.

The class field tower up to the second stage for the field $K = Q(\sqrt[l]{m})$ was investigated by Fröhlich [30]. Conditions were derived that $K_2 \neq K_1$.

An example of the break in a 3-tower with two ramification points can be found in Browkin [1, 17]. The group of a maximal 2-extension of the rational number field with a single critical number 2 has been described by Markshaitis [8]. It is isomorphic with the topological 2-group with two generators σ and τ and a single relation $\sigma^2 = 1$.

7. Let us now dwell on papers studying the Galois groups of local fields. Here, almost all the extreme cases have been investigated. However, although Koch [68] has also obtained some general information on the group of the maximal algebraic extension of a local field, it has not yet been described completely. In 1955, Iwasawa [49] determined the Galois group of the composite of all completely ramified extensions of a local field k: $G(V/k) = \langle \alpha, \beta \mid \alpha\beta\alpha^{-1} = \rho^q \rangle$, where q is the number of elements of the residue field. Extensions of degree relatively prime to p were obtained by Feit [25]. The group of the maximal p-extension of a local field containing the p-th root of unity was described by Demushkin [3–5]. It was shown that in this case the group of the maximal p-extension $(p \neq 2)$ is isomorphic with a topological p-group with an even number of generators $\sigma_1, \ldots, \sigma_\nu$ and with a single relation of the form $\sigma_1^q [\sigma_1, \sigma_2] \ldots [\sigma_{v-1}, \sigma_v] = 1$, where $q = p^m$ $(m \geq 1)$. These results were strengthened and generalized to the case of groups with a countable number of generators by Labute [78-80].

8. Finally, we pass on to papers which study the Γ-extensions of algebraic number fields, which estimate the growth of the class number of the ideals of such fields, and which determine the ideal group itself. Here, the papers by Iwasawa [50-55] are very interesting (also see [98]). Iwasawa showed that the p-component of the class number of the ideals of a cyclotomic field $Q(\zeta_n)$, where ζ_n is the p -th root of unity, can be expressed, beginning with some number n, in the form p^{l_n} with $l_n = \lambda_n + \mu p^n + \nu$. Iwasawa and Sims [56] verified on a computer that $\mu = 0$ for all $p \leq 4001$. The condition for $\mu > 0$ can be expressed in terms of an infinite series of congruences on the Bernoulli number [50]. Jehne [58] noted that λ and μ may not vanish simultaneously.

An interesting dependence of the class number of ideals on the ramification points was detected by Brumer and Rosen [19]. They showed that if the algebraic number field K is fixed and if L ranges over the normal extensions of K of relative degree n, then the class number of field L tends to ∞ as the number of prime ideals, ramified in L, of the field K increases.

The problem of describing the ideal class groups of algebraic number fields remains of interest and unsolved. Iwasawa [55] proved that the first factor of the class number of the ideals of a cyclotomic field can be expressed as the index of some additive group in an integral group ring Z[G], where G is the Galois group of the cyclotomic field.

9. Remarks. a) Schilling [95] proved the fundamental theorems in the local class field theory without the computation of the norm group index in the multiplicative group of the field.

b) Yamamoto [123] proved a theorem on isomorphism in local class field theory without using Herbrand's lemma, the theory of prime algebras, and $\mathfrak{P}$-adic logarithms. The basic investigation tool here was a certain endomorphism of the additive group of the residue field of the fundamental field, generalizing $x \to x^p - x$.

c) Iwasawa [54] described the structure of the Galois group $G(M/Q^p)$, where M is the maximal Abelian extension of the field $F = \bigcup_n F_n$, $F_n = Q_p(\zeta_{p^n})$.

d) Serre [102] has considered the cohomology groups of profinite groups.

e) In [31], Fröhlich proved the following theorem: there exist infinitely many normal extensions K/Q with the properties: 1) G(K/Q) is a symmetric group of n symbols; 2) $\Lambda \cap K = Q$, where Λ is a fixed normal extension of field Q; 3) $\Lambda K/K$ is unramified.

f) Hasse [41] effectively constructed the absolute class field of the imaginary quadratic field $Q(\sqrt{d})$ with d = −23, −31, −47.

g) Certain generalizations of local class field theory in a non-Abelian aspect occur in [74].

h) Latt [83] made a detailed study of the structure of the Galois group of the maximal Abelian extension of the rational number field. All its closed subgroups and continuous characters were found.

i) An example is given in [32] of a one-class field K/k with the property: every prime ideal of field k is unramified in K/k.

LITERATURE CITED

1. J. Browkin, "Examples of maximal 3-extensions with two ramification points," Izv. Akad. Nauk SSSR, Ser. Mat., 27(3): 613-620 (1963).
2. E. S. Golod and I. R. Shafarevich, "The class field tower," Izv. Akad. Nauk SSSR, Ser. Mat., 28(2): 261-272 (1964).
3. S. P. Demushkin, "The group of maximal p-extension of a local field," Dokl. Akad. Nauk SSSR, 128(4): 657-660 (1959).
4. S. P. Demushkin, "The group of maximal p-extension of a local field," Izv. Akad. Nauk SSSR, 25(3): 329-346 (1961).
5. S. P. Demushkin, "The 2-extensions of a local field," Sibirsk. Mat. Zh., 4(4): 951-955 (1963).
6. Y. Kawada, "The connection between the various theories of Abelian extension and class field theory," Sugaku, 6(3): 129-150 (1954).
7. S. Lang, Algebraic Numbers, Addison-Wesley, Reading, Mass. (1964).
8. G. I. Markshaitis, "p-Extensions with one critical number," Izv. Akad. Nauk SSSR, Ser. Mat., 27(2): 463-466 (1963).
9. A. P. Novikov, "The class number of a field with complex multiplication," Izv. Akad. Nauk SSSR, Ser. Mat., 26(5): 677-686 (1962).
10. I. R. Shafarevich, "Extensions with specified ramification points," Publ. Math. Inst. Hautes Études Scient., No. 18, pp. 71-95 (1963).
11. N. C. Ankeny, "Reciprocity theorem in cyclotomic fields," Studies Math. Analysis and Related Topics, University Press, Stanford, Calif. (1962), pp. 8-15.
12. E. Artin and J. Tate, Class Field Theory, W. A. Benjamin, Inc., New York (1967).

13. W. L. Baily, "On the moduli of Abelian varieties with multiplications," J. Math. Soc. Japan, 15(4): 367-386 (1963).

14. A. Borel, "Some properties of adele groups attached to algebraic groups," Bull. Am. Math. Soc., 67(6): 583-585 (1961).

15. A. Borel, "Some finiteness properties of adele groups over number fields," Publ. Math. Inst. Hautes Études Scient., No. 16, pp. 101-126 (1963).

16. R. Brauer, "Relations between class numbers," in: Algebr., Geom. Conf. Notes, University of Chicago.

17. J. Browkin, "On the generalized class field tower," Bull. Acad. Polon. Sci., Sér. Sci. Math., Astron. et Phys., 11(4): 143-145 (1963).

18. A. Brumer, "Ramification and class towers of number fields," Mich. Math. J., 12(2): 129-131 (1965).

19. A. Brumer and M. Rosen, "Class number and ramification in number fields," Nagoya Math. J., Vol. 23, pp. 91-101 (1963).

20. C. Chevalley, Class Field Theory, Nagoya University (1954), 104 pp.

21. C. Chevalley, "Le théorème fondamental de la multiplication complexe," Semin. Bourbaki Secrét. Math., 1956-1957. 9th year, 2nd ed., Paris (1959), pp. 138/1-138/13.

22. M. Deuring, "Die Klassenkörper der komplexen Multiplikation," Erg. Math. Wiss., Vol. 1, No. 10, Part II (1958).

23. K. Doi, "On the field of moduli of an Abelian variety with complex multiplication," J. Math. Soc. Japan, 15(3): 237-243 (1963).

24. B. Dwork, "Norm residue symbol in local number fields," Abhandl. Math. Seminar Univ. Hamburg, 22(3-4): 180-190 (1958).

25. W. Feit, "On p-regular extensions of local fields," Proc. Am. Math. Soc., 10(4): 592-595 (1959).

26. A. Fröhlich, "The genus field and genus group in finite number field," Mathematika, 6(1): 40-46 (1959).

27. A. Fröhlich, "The genus field and genus group in finite number field. II," Mathematika, 6(12): 142-146 (1959).

28. A. Fröhlich, "The rational characterization of certain sets of relatively Abelian extensions," Phil. Trans. Roy. Soc. London, A251(998): 385-425 (1959).

29. A. Fröhlich, "A prime decomposition symbol for certain non-Abelian fields," Acta Sci. Math., 21(3-4): 229-246 (1960).

30. A. Fröhlich, "A remark on the classfield tower of the field $p\sqrt[e]{m}$," J. London Math. Soc., 37(2): 193-194 (1962).

31. A. Fröhlich, "On nonramified extensions with prescribed Galois group," Mathematika, 9(18): 133-134 (1962).

32. G. Fujisaki, "On an example of an unramified Galois extension," Sugaku, Vol. 9, pp. 97-99 (1957/58).

33. Y. Furuta, "A reciprocity law of the power residue symbol," J. Math. Soc. Japan, 10(1): 46-54 (1958).

34. L. D. Geissinger, "A generalization of local class field theory," Doctoral Dissertation, Indiana University, 1963, 48 pp. Dissertation Abstr., 24(7): 2921-2922 (1964).

35. K. W. H. Glass, "On extending a norm residue symbol," Doctoral Dissertation, Indiana University, 1963, 41 pp. Dissertation Abstr., 24(7): 2922-2923 (1964).
36. K. L. Grant, "Abstract class formations," Doctoral Dissertation, Indiana University, 1963, 38 pp. Dissertation Abstr., 25(1): 496-497 (1964).
37. K. L. Grant and G. Whaples, "Abstract class formations," Bull. Am. Math. Soc., 67(4): 393-395 (1961).
38. K. L. Grant and G. Whaples, "Correction to 'Abstract class formations'," Bull. Am. Math. Soc., 69(1): 123 (1963).
39. K. L. Grant and G. Whaples, "Abstract class formations," J. Fac. Sci. Univ. Tokyo, Sec. 1, 11(2): 187-194 (1965).
40. H. Hasse, "Zum expliziten Reziprozitäsgesetz," Arch. Math., 13(6): 479-485 (1962).
41. H. Hasse, "Über den Klassenkörper zum quadratischen Zahlkörper mit der Diskriminante — 47," Acta Arithm., 9(4): 419-434 (1964).
42. H. Heilbronn, "Old theorems and new methods in class-field theory," J. London Math. Soc., 37(1): 6-9 (1962).
43. G. Hochschild, "Local class field theory," in: Algebr. Geom. Conf. Notes, University of Chicago.
44. K. Hoechsmann, "Tagungsbericht Arbeitsgemeinschaft über das Klassenkörperturmproblem, 18. bis 23. Oktober, 1964," Math. Forschungsinst. Oberwolfach (1964).
45. K. Hoechsmann, "Zahlentheorie (insbesondere algebraische Zahlentheorie)," Math. Forschungsinst. Oberwolfach (1964).
46. L. Holzer, Zahlentheorie, Vol. II, No. 6, Teubner Verlagsgesellschaft, Leipzig (1959).
47. T. Honda, "On absolute class fields of certain algebraic number fields," J. Reine und Angew. Math., 203(1-2): 80-89 (1960).
48. Hsü Tao-ning, "Über den Hauptidealsatz für imaginäre quadratische Zahlkörper," J. Reine und Angew. Math., 212(1-2): 49-62 (1963).
49. K. Iwasawa, "On Galois groups of local fields," Trans. Am. Math. Soc., 80(2): 448-469 (1955).
50. K. Iwasawa, "On some invariants of cyclotomic fields," Am. J. Math., 80(3): 773-783 (1958).
51. K. Iwasawa, "On the theory of cyclotomic fields," Ann. Math., 70(3): 530-561 (1959).
52. K. Iwasawa, "On Γ-extensions of algebraic number fields," Bull. Am. Math. Soc., 65(4): 183-226 (1959).
53. K. Iwasawa, "On some properties of Γ-finite modules," Ann. Math., 70(2): 291-312 (1959).
54. K. Iwasawa, "On local cyclotomic fields," J. Math. Soc. Japan, 12(1): 16-21 (1960).
55. K. Iwasawa, "A class number formula for cyclotomic fields," Ann. Math., 76(1): 171-179 (1962).
56. K. Iwasawa and C. C. Sims, "Computations of invariants in the theory of cyclotomic fields," J. Math. Soc. Japan, 18(1): 86-96 (1966).
57. W. Jehne, "Zur modernen Klassenkörpertheorie," Sitzsber. Deut. Akad. Wiss., Berlin, Kl. Math. Allgem. Naturwiss., No. 3 (1954).

58. W. Jehne, "Bermerkung über die p-Klassengruppe des p-ten Kreiskörpers," Arch. Math., 10(6): 422-427 (1959).
59. H. Kempfert, "Zum allgemeinen Hauptidealsatz," J. Reine und Angew. Math., 21(1-2): 38-64 (1962).
60. Y. Kawada, "Class formations," Duke Math. J., 22(2): 165-177 (1955).
61. Y. Kawada, "Class formations. III," J. Math. Soc. Japan, Vol. 7, Suppl. Dec., pp. 453-490 (1955).
62. Y. Kawada, "Class formations. IV, Infinite extension of the ground field," J. Math. Soc., Japan, 9(4): 395-405 (1957).
63. Y. Kawada, "Class formations. V. Infinite extensions of the p-adic field or the rational field," J. Math. Soc. Japan, 12(1): 34-64 (1960).
64. Y. Kawada, "Class formations. VI. Restriction to a subfamily," J. Fac. Sci. Univ. Tokyo, Sec. I, 8(2): 229-262 (1960).
65. Y. Kawada, "Abstract class formations," Bol. Soc. Mat. São Paulo, 15(1-2): 7-23 [1960 (1964)].
66. Y. Kawada and I. Satake, "Class formations. II," J. Fac. Sci. Univ. Tokyo, Sec. I, 7(3): 353-389 (1956).
67. E. F. Kennel, "Class field theory in dimension greater than one," Doctoral Dissertation, University of Oregon, 1965, 68 pp. Dissertation Abstr., 26(9): 5462 (1966).
68. H. Koch, "Über galoissche Erweiterungen $\mathfrak{P}$ -adischen Zahlkörper," J. Reine und Angew. Math., 209(1-2): 8-11 (1962).
69. H. Koch, "Über maximale 1-Erweiterungen mit vorgeschriebenen Verzweigungsstellen," Monatsber. Deut. Akad. Wiss. Berlin, 5(6): 341-343 (1963).
70. H. Koch, "Über den 2-Klassenkörperum eines quadratischen Zahlkörpers. I," J. Reine und Angew. Math., Vol. 214-215, pp. 201-206 (1964).
71. H. Koch, "1-Erweiterungen mit vorgegebenen Verzweigungsstellen," J. Reine und Angew. Math., 219(1-2): 30-61 (1965).
72. H. Koch, "1-Erweiterungen mit zwei Verzweigungsstellen," Monatsber. Deut. Akad. Wiss. Berlin, 7(9): 616-623 (1965).
73. H. Koch, "Über beschränkte Gruppen," J. Algebra, 3(2): 206-224 (1966).
74. M. Krasner, "Généralisations non-abéliennes de la théorie locale des corps de classes, "Sémin. Bourbaki. Secrét. Math., 1950-1951, 3rd year, 2nd ed., Paris (1959), pp. 47/1-47/24.
75. T. Kubota, "Factor sets in a number field and the norma residue symbol," J. Math. Soc. Japan, 11(2): 129-138 (1959).
76. T. Kubota, "Reciprocities in Gauss' and Eisenstein's number fields," J. Reine und Angew. Math., 208(1-2): 35-50 (1961).
77. T. Kubota, "On the field extensions by complex multiplication," Trans. Am. Math. Soc., 118(6): 113-122 (1965).
78. J. P. Labute, "Les groupes de Demuškin de rang dénombrable," Compt. Rend. Acad. Sci., AB262(1): A4-A7 (1966).
79. J. P. Labute, "Demuskin groups of rank $\aleph_0$," Bull. Soc. Math. France, 94(3): 211-244 (1966).
80. J. P. Labute, "Classification of Demuškin groups," Can. J. Math., 19(1): 106-132 (1967).

81. E. Lamprecht, "Existenz von Zahlkörpern mit nicht abbrechendem Klassenkörperturm," Arch. Math., 18(2): 140-152 (1967).
82. S. Lang, "Unramified class field theory over function fields in several variables," Ann. Math., 64(2): 285-325 (1956).
83. K. Latt, "Über die Galoisgruppe des in bezug auf den Körper den rationallen Zahlen maximalen abelschen Körpers," Wiss. Z. Humboldt-Univ. Berlin. Math.-Naturwiss. Reihe, 7(5): 583-608 (1957-1958).
84. J. Lubin and J. Tate, "Formal complex multiplication in local fields," Ann. Math., 81(2): 380-387 (1965).
85. K. Masuda, "An application of the generalized norm residue symbol," Proc. Am. Math. Soc., 10(2): 245-252 (1959).
86. K. Meyer, Die Berechnung der Klassenzahl abelscher Körper über quadratischen Zahlkörpern, Berlin (1957).
87. A. Moore, "On extending a norm residue symbol," Doctoral Dissertation, Indiana University, 1962, 35 pp.
88. M. Mori, "Über die Klassenkörpertheorie für unendliche Erweiterungen von einem $\mathfrak{P}$-adischen Zahlkörper," Proc. Japan Acad., 33(7): 376-379 (1957).
89. J. R. Phillips, "Inclusion — transfer theorem of class field theory," Doctoral Dissertation, Indiana University, 1965, 39 pp. Dissertation Abstr., 26(5): 2782 (1965).
90. K. Ramachandra, "Some applications of Kronecker's limit formulas," Ann. Math., 80(1): 104-108 (1964).
91. H. Reichardt, "Ein Beweis des Hauptidealsatzes für imaginärquadratische Zahlkörper," Math. Nachr., 17(3-6): 318-329 (1959).
92. H. Reichardt, "Hauptidealsatz für imaginärquadratische Zahlkörper," in: Atti Sesto Congr. Unione Mat. Ital. Tenuto Napoli, 11-16 Sett. 1959, Ed. Cremonese, Rome (1960).
93. P. Ribenboim, "Résultats et problèmes de la théorie des corps de classes," Rev. Union Mat. Argent. y Asoc. Fis. Argent, Vol. 20, pp. 125-145 (1962).
94. P. Samuel, "Théorie du corps de classes local selon G. P. Hochschild," Sémin. Bourbaki, Secrét. Math., 1950-1951, 3rd year, 2nd ed., Paris (1959), pp. 42/1-42/5.
95. O. F. G. Schilling, "On local class field theory," J. Math. Soc. Japan, 13(3): 234-245 (1961).
96. Seminar on Complex Multiplication. Lecture Notes in Mathematics, Springer-Verlag, Berlin (1966).
97. J.-P. Serre, "Corps locaux et isogénies," Sémin. Bourbaki, Secrét. Math., 1958-1959, 11th year, 3rd Part, 2nd ed., Paris (1959), pp. 185/1-185/9.
98. J.-P. Serre, "Classes des corps cyclotomiques," Sémin. Bourbaki, Secrét. Math. 1958-1959, 11th year, 1st Part, 2nd ed., Paris (1959), pp. 174/1-174/11.
99. J.-P. Serre, "Groupes algébriques et corps de classes. (Actual. Scient. et Industr., No. 1264, publ. Inst. Math. Univ. Nancago, 7). Hermann, Paris (1959).
100. J.-P. Serre, "Groupes proalgébriques," Publ. Math. Inst. Hautes Études Sci., No. 7 [1960 (1961)].
101. J.-P. Serre, "Sur les corps locaux à corps résiduel algébriquement clos," Bull. Soc. Math. France, 89(2): 105-154 (1961).

102. J.-P. Serre, Cohomologie Galoisienne, 2nd ed., Berlin, Springer (1964).
103. J.-P. Serre, "Existence de tours infinies de corps de classes d'après Golod et Safarevič," Colloq. Internat. Centre Nat. Rech. Sci., No. 143, pp. 231-238 (1966).
104. G. Shimura, "On complex multiplication," in: Tokyo Symposium on Algebraic Number Theory (1955), pp. 23-30.
105. G. Shimura, "On the class fields obtained by complex multiplication of Abelian varieties," Osaka Math. J., 14(1): 33-44 (1962).
106. G. Shimura and Y. Taniyama, "Complex multiplication of Abelian varieties and its applications to number theory," Publ. Math. Soc. Japan, No. 6 (1961).
107. G. Shimura and Y. Taniyama, "Complex multiplication of Abelian varieties and its applications to number theory. S. 1," Math. Soc. Japan (1961); Bull. Soc. Math. Belg., 16(1): 79-80 (1964).
108. K. Shiratani, "Note on the Kummer — Hilbert reciprocity law," J. Math. Soc. Japan, 12(4): 412-421 (1960).
109. K. Shiratani, "On the quadratic norm symbol in local fields," J. Math. Soc. Japan, 13(4): 416-430 (1961).
110. Th. Skolem, "Remarks on proofs by cyclotomic formulas of reciprocity laws for power residues," Math. Scand., 9(1b): 229-242 (1961).
111. S. Takahashi, "On Tannaka — Terada's principal ideal theorem for rational ground field," Tohoku Math. J., 17(1): 87-104 (1965).
112. Y. Taniyama, "Jacobian varieties and number fields," in: Tokyo Symposium on Algebraic Number Theory (1955), pp. 31-46.
113. T. Tannaka, "On the generalized principal ideal theorem," in: Proc. Internat. Sympos. Algebr. Number Theory, 1955, Tokyo (1956), pp. 65-70.
114. T. Tannaka, "A generalized principal ideal theorem and a proof of a conjecture of Deuring," Ann. Math., 67(3): 574-589 (1958).
115. J. Tate, "Multiplication complexe formelle dans les corps locaux," Colloq. Internat. Centre Nat. Rech. Sci., No. 143, pp. 257-258 (1966).
116. F. Terada, "Complex multiplication and principal ideal theorem," Tohoku Math. J., Ser. 2, 6(1): 21-25 (1954).
117. F. Terada, "On a generalized principal ideal theorem," Tohoku Math. J., 6(2): 95-100 (1954).
118. F. Terada, "A generalization of the principal ideal theorem," J. Math. Soc. Japan, Vol. 7, Suppl. Dec., pp. 530-536 (1955).
119. F. Terada, "A generalization of the principal ideal theorem," in: Proc. Internat. Sympos. Algebr. Number Theory, 1955, Tokyo (1956), pp. 264-265.
120. A. Weil, "On the theory of complex multiplication," in: Proc. Internat. Sympos. Algebr. Number Theory, 1955, Tokyo (1956), pp. 9-22.
121. A. Weil, "Sur la théorie du corps de classes," Sémin. Bourbaki, Secrét Math., 1952-1953, 5th year, 2nd ed., Paris (1959), pp. 83/1-83/3.
122. E. Weiss, Algebraic Number Theory, McGraw-Hill, New York (1963).
123. K. Yamamoto, "Isomorphism theorem in the local class field theory," Mem. Fac. Sci. Kyushu Univ., A12(2): 67-102 (1958).
124. K. Yamamoto, "On the Kummer — Hilbert reciprocity law," Mem. Fac. Sci. Kyushu Univ., A13(2): 85-95 (1959).

GEOMETRY

Finsler Spaces and Their Generalizations

V. I. Bliznikas

Introduction

Right up to today the ideas of Riemann determine the path of progress in the differential geometry of generalized spaces. These ideas are based on the possibility of the geometrization of the theory of differential invariants relative to some transformation group. This geometrization leans on analogy to the extent that the types of invariants being studied are encountered in the differential geometry of Euclidean space (or of some other Klein space). The generalization consists in the rejection of some relations or others which are characteristic of Klein geometry (in particular, of Euclidean geometry). The first generalization of Riemann geometry (the theory of the invariants of a quadratic differential form) is due to Finsler [203, 204] who considered that the square of the arc length element of a curve is an arbitrary homogeneous function of the differentials of the local coordinates of the point.* Certain aspects of Finsler geometry were considered also by Noether [406, 407].

Cartan [180] proposed the introduction of a metric not for a curve but for the area of a hypersurface, i.e., the construction of a geometry of a manifold starting from the notion of the area on a hypersurface. Subsequently this idea of Cartan was generalized, i.e., many authors started to construct the geometry of a manifold based on the notion of the area of an m-dimensional surface (m <

* Riemann himself had already noted the possibility of introducing the square of the arc length element as an arbitrary homogeneous function of second dimension relative to the differentials of the point coordinates. Special cases of Finsler spaces were considered by Hilbert, Carathéodory, etc. (up to 1918).

n − 1). The geometry of multiple integrals arose as a generalization of Finsler geometry, while the geometry of differential equations, as a generalization of the geometry of the geodesics of Riemann space and of Finsler space. All these types of Finsler spaces and their various generalizations are reduced to the concept of the space of support elements with a given fundamental differential-geometric object [88-91]. A support element space is a fiber space whose layers are the representation spaces of the differential group $GL^p(n, R)$.

The present survey is devoted to papers on Finsler geometry (and on various generalizations of this geometry) published during 1953-1967 (certain papers published prior to 1953 also are surveyed). These works were carried out by various methods, but in presenting their contents we shall adhere to the method of Laptev [92-95].

§ 1. Fundamental Concepts

1. Higher-Order Line Element Space and Extensors. Let V_n be a differentiable manifold whose local coordinates are given by the solution of a completely integrable system $\omega^i = 0$, i.e.,

$$D\omega^i = \omega^k \wedge \omega^i_k \tag{1}$$

and

$$D\omega^i_{j_1\ldots j_a} = \sum_{s=1}^{a} \frac{a!}{s!\,(a-s)!}\, \omega^k_{(j_1\ldots j_s} \wedge \omega^i_{j_{s+1}\ldots j_a)k} + \omega^k \wedge \omega^i_{j_1\ldots j_a k}, \tag{2}$$

$$i, j, k = 1, 2, \ldots, n; \quad a, b, c = 1, 2, \ldots, p.$$

The Pfaffian forms

$$\theta^i_{j_1\ldots j_a} = \omega^i_{j_1\ldots j_a}\big|_{\omega^k=0}$$

are the invariant forms of the p-th order differential group $GL^p(n, R)$ of the differentiable manifold V_n and

$$D\theta^i_{j_1\ldots j_a} = \sum_{s=1}^{a} \frac{a!}{s!(a-s)!}\, \theta^k_{(j_1\ldots j_s} \wedge \theta^i_{j_{s+1}\ldots j_a)k}. \tag{3}$$

The system of forms ω^i and $\vartheta^{(a)i}$ ($a = 1, 2, \ldots, p$), where

$$\vartheta^{(a)i} = dv^{(a)i} + \sum_{s=1}^{a} \frac{a!}{s!} \omega^i_{k_1 \ldots k_s} \sum_{(a_1+\ldots+a_s=a)} \frac{1}{a_1! \ldots a_s!} v^{(a_1)k_1} \ldots v^{(a_s)k_s}, \tag{4}$$

is completely integrable, and the first integrals of the system,

$$\omega^i = 0, \; \vartheta^{(1)i} = 0, \ldots, \vartheta^{(p)i} = 0,$$

are the local coordinates of the p-th order parametrized line element $(x^i, v^{(1)i}, \ldots, v^{(p)i})$. The collection of these p-th order line elements forms a space $L^{(p)}_n$ which is called a higher-order line element space. The structure equations of the space $L^{(p)}_n$ can be represented in the form

$$D\vartheta^{(a)i} = \vartheta^{(c)k} \wedge \vartheta^{(a)i}_{(c)k} \quad (a = 0, 1, 2, \ldots, p), \tag{5}$$

where

$$\vartheta^{(0)i} = \omega^i, \; \vartheta^{(0)i}_{(0)j} = \omega^i_j, \; \vartheta^{(a)i}_{(0)k} = \vartheta^{(a)i}_k,$$

$$\vartheta^{(a)i}_{(c)k} = \frac{\partial \vartheta^{(a)i}}{\partial v^{(c)k}} \quad (a, c = 0, 1, 2, \ldots, p) \tag{6}$$

and ($v^{(0)i} = x^i$) the partial differentiation with respect to x^k is indicated by an additional index when the forms are designated by the letter "ω," i.e., $\partial \omega^i / \partial x^k = \omega^i_k$. It is obvious that

$$\vartheta^{(a)i}_{(c)k} = 0 \quad (c > a).$$

In general, the forms

$$\vartheta^{(a)i}_{(c_1)k_1 (c_2)k_2 \ldots (c_s)k_s} = \frac{\partial^s \vartheta^{(a)i}}{\partial v^{(c_1)k_1} \partial v^{(c_2)k_2} \ldots \partial v^{(c_s)k_s}} \tag{7}$$

have the following structure:

$$D\vartheta^{(a)i}_{(c_1)k_1 \ldots (c_s)k_s} = \sum_{m=1}^{s} \frac{s!}{m!\,(s-m)!} \{\vartheta^{(b)l}_{[(c_1)k_1 \ldots (c_m)k_m} \wedge$$
$$\vartheta^{(a)i}_{(c_{m+1})k_{m+1} \ldots (c_s)k_s\} (b)l} + \vartheta^{(b)l} \wedge \vartheta^{(a)i}_{(c_1)k_1 \ldots (c_s)k_s (b)l}, \tag{8}$$

where the braces $\{\ldots\}$ signify a circulation relative to the indices (c)k. Note that the forms $\vartheta^{(a)i}_{(b)j}$ are connected by the relations

$$\vartheta^{(a)i}_{(b)j} = \frac{a!}{b!\,(a-b)!}\,\vartheta^{(a-b)\,i}_{j}. \tag{9}$$

The Pfaffian forms

$$\theta^{(a)i}_{(b)j} = \vartheta^{(a)i}_{(b)j}\big|_{\omega^i=0,\ \vartheta^{(a)i}=0} \tag{10}$$

are the invariant forms of the Lie group $GL^p(n; 1; R)$ and

$$D\theta^{(a)i}_{(b)j} = \theta^{(c)l}_{(b)j} \wedge \theta^{(a)i}_{(c)l}. \tag{11}$$

The group $GL^p(n; 1; R)$ is the transformation group of the vectors of the tangent space for $L^{(p)}_n$. If $T^{(p)}$ is a tangent space for $L^{(p)}_n$ ($T^{*(p)}$ is the dual tangent space), then the elements of the tensor product

$$\overset{q}{\otimes} T^{(p)} \overset{r}{\otimes} T^{*(p)}$$

are called extensors which are q times excovariant and r time excontravariant. The local coordinates of this extensor can be looked upon as the first integrals of the system of differential equations

$$dT^{(b_1)j_1 \ldots (b_r)j_r}_{(a_1)i_1 \ldots (a_q)i_q} - \sum_{s=1}^{q} T^{(b_1)j_1 \ldots (b_r)j_r}_{(a_1)i_1 \ldots (c)k \ldots (a_q)i_q}\,\theta^{(c)k}_{(a_s)i_s} + \sum_{s=1}^{r} T^{(b_1)j_1 \ldots (c)k \ldots (b_r)j_r}_{(a_1)i_1 \ldots (a_q)i_q}\,\theta^{(b_s)j_s}_{(c)k} = 0.$$

2. Higher-Order Surface Element Space and Multidimensional Extensors. Let V_n and V_m be differentiable manifolds and, moreover, let the forms ω^α and $\omega^\alpha_{\beta_1 \ldots \beta_{\hat{a}}}$ have the very same structure as the forms ω^i and $\omega^i_{j_1 \ldots j_a}$. Then the spray space (in the sense of Ehresmann) or the space of m-dimensional p-th order surface elements $K^{(p)}_{n,m}$ is determined by the completely integrable system

$$\omega^i = 0,\ \omega^\alpha = 0, \ldots,\ \vartheta^i_\alpha = 0, \ldots,\ \vartheta^i_{\alpha_1 \ldots \alpha_p} = 0, \tag{12}$$

where

$$\vartheta^i_{\alpha_1\ldots\alpha_a} = dp^i_{\alpha_1\ldots\alpha_a} + \sum_{s=1}^{a} \frac{a!}{s!} \omega^i_{k_1\ldots k_s} \times \sum_{(j_1+\ldots+j_s=a)} \frac{1}{j_1!\ldots j_s!} p^{k_1}_{(\alpha_1\ldots\alpha_{j_1}} \cdots p^{k_s}_{\alpha_{j_1+\ldots+j_{s-1}+1}\ldots\alpha_a)} - \sum_{s=1}^{a} \frac{a!}{s!\,(a-s)!} \omega^{\gamma}_{(\alpha_1\ldots\alpha_s} p^i_{\alpha_{s+1}\ldots\alpha_a)\gamma}. \tag{13}$$

If the forms $\omega^{\gamma}_{\alpha_1\ldots\alpha_a}$ are the invariants forms of the differential group $GL^p(m, R)$, then the system

$$\omega^i = 0,\ \vartheta^i_{\alpha_1} = 0,\ \ldots,\ \vartheta^i_{\alpha_1\ldots\alpha_p} = 0$$

is completely integrable and the first integrals of this system determine an m-dimensional p-th order line element (together with the center), i.e., a point of the space $\overset{\bullet}{K}{}^{(p)}_{n,m}$ (the space of m-dimensional p-th order line elements). The structure equations of the space $K^{(p)}_{n,m}$ have the form:

$$D\omega^i = \omega^k \wedge \omega^i_k,\quad D\omega^{\alpha} = \omega^{\beta} \wedge \omega^{\alpha}_{\beta},$$

$$D\vartheta^i_{\alpha_1\ldots\alpha_a} = \sum_{c=1}^{a} \vartheta^k_{\beta_1\ldots\beta_c} \wedge \vartheta^{i\beta_1\ldots\beta_c}_{\alpha_1\ldots\alpha_a k} + \omega^k \wedge \vartheta^i_{\alpha_1\ldots\alpha_a k} + \omega^{\gamma} \wedge \vartheta^i_{\alpha_1\ldots\alpha_a,\,\gamma}, \tag{14}$$

where

$$\vartheta^{i\beta_1\ldots\beta_c}_{\alpha_1\ldots\alpha_a k} = \frac{\partial \vartheta^i_{\alpha_1\ldots\alpha_a}}{\partial p^k_{\beta_1\ldots\beta_c}},$$

$$\vartheta^i_{\alpha_1\ldots\alpha_a h} = \sum_{s=1}^{a} \frac{a!}{s!} \omega^i_{k_1\ldots k_s h} \sum_{(j_1+\ldots+j_s=a)} \frac{1}{j_1!\ldots j_s!} p^{k_1}_{(\alpha_1\ldots\alpha_{j_1}} \cdots p^{k_s}_{\alpha_{j_1+\ldots+j_{s-1}+1}\ldots\alpha_a)},$$

$$\vartheta^i_{\alpha_1\ldots\alpha_a,\,\gamma} = -\sum_{s=1}^{a} \frac{a!}{s!(a-s)!} \omega^{\beta}_{\gamma(\alpha_1\ldots\alpha_s} p^i_{\alpha_{s+1}\ldots\alpha_a)\beta}.$$

The forms

$$\theta^{i\beta_1\ldots\beta_c}_{\alpha_1\ldots\alpha_a j} = \vartheta^{i\beta_1\ldots\beta_c}_{\alpha_1\ldots\alpha_a j}\Big|_{\omega^i=0,\,\vartheta^i_{\alpha_1\ldots\alpha_a}=0,\,\omega^{\alpha}=0}$$

determine the Lie group GL(n; m; p). The linear and homogeneous objects relative to this group are called multidimensional extensors (they are determined by differential equations analogous to those for extensors).

The spaces $L_n^{(p)}$ and $K_{n,m}^{(p)}$ are special cases of fiber spaces in the sense that differential groups, i.e., the groups

$$GL^p(n, R),\ GL^p(n, R)\times GL^p(m, R),$$

act in the layers of these spaces.

If on the space $L_n^{(p)}$ (or on the space $K_{n,m}^{(p)}$) we are given the field of some differential-geometric object, called the fundamental differential-geometric object, then by the geometry of this space (rigged by means of the differential-geometric object) we mean the theory of the invariants and the invariant operations constructed with the aid of the fundamental differential-geometric object and its differential continuations. We analyze from this point of view not only Finsler space but also other spaces which are some generalizations or other of Finsler space (we shall explicitly indicate every time the fundamental differential-geometric object and the structural differential equations of this object).

§ 2. Finsler Space

If the scalar function F(x, v) $(\vartheta^{(1)i} \equiv \vartheta^i)$ is given on $L_n^{(1)}$:

$$dF = F_i\omega^i + 'F_i\vartheta^i, \tag{15}$$

whose differential equation is invariant relative to the transformation

$$\bar{v}^i = \rho v^i \ (\rho > 0), \tag{16}$$

i.e.,

$$F(x, \rho v) = \rho F(x, v),\ F_i(x, \rho v) = \rho F_i(x, v),\ 'F_i v^i = F,$$

then the space $L_n^{(1)}$ is called a Finsler space F_n. The sequential continuation of this equation (by this we mean exterior differentiation and the development of the obtained equations by Cartan's lemma), in accordance with the notation

$$\vartheta_j^{(1)i} = \vartheta_j^i,\ \vartheta_{jk}^{(1)i} = \vartheta_{jk}^i,\ \text{etc.},$$

has the form:

$$dF_i - F_k\omega_i^k - {}'F_k\vartheta_i^k = F_{ji}\omega^j + {}'F_{ji}\vartheta^j,$$

$$d'F_i - {}'F_k\omega_i^k = {}'F_{ij}\omega^j + {}''F_{ji}\vartheta^j,$$

$$dF_{ij} - F_{kj}\omega_i^k - F_{ik}\omega_j^k - F_l\omega_{ij}^l - {}'F_l\vartheta_{ij}^l - {}'F_{il}\vartheta_j^l - {}'F_{lj}\vartheta_i^l = F_{kij}\omega^k + {}'F_{kij}\vartheta^k,$$

$$d'F_{ij} - {}'F_{il}\omega_j^l - {}'F_{lj}\omega_i^l - {}'F_l\omega_{ij}^l - {}''F_{lj}\vartheta_i^l = {}'F_{ikj}\omega^k + {}''F_{ikj}\vartheta^k,$$

$$d''F_{ij} - {}''F_{lj}\omega_i^l - {}''F_{il}\omega_j^l = {}''F_{kij}\omega^k + {}'''F_{kij}\vartheta^k,$$

where F_{ijk}, $''F_{ij}$, $'''F_{ijk}$ are symmetric and $'F_{ijk} = {}'F_{ikj}$. It is obvious that $''F_{ij}v^j = 0$.

The fundamentals of the geometry of Finsler spaces were established by Finsler, a student of Caratheodory, in 1918 (see his dissertation [203]), who called them general metric spaces (these spaces were named Finsler spaces at the suggestion of Synge [501]). The construction of a geometry of the space F_n reduces to the study of the fundamental sequence of the differential-geometric objects of this space. The second differential continuation includes the metric tensor

$$g_{ij} = F''F_{ij} + {}'F_i{}'F_j, \tag{17}$$

moreover,

$$\nabla g_{ij} \equiv dg_{ij} - g_{kj}\omega_i^k - g_{ik}\omega_j^k = g_{kij}\omega^k + g'_{kij}\vartheta^k \tag{18}$$

and

$$g'_{ijk} = g'_{ikj} = g'_{jik}.$$

The angle metric of space F_n has the form (the angle between two infinitely close support elements):

$$d\varphi^2 = \frac{1}{F}\,{}''F_{ij}\vartheta^i\vartheta^j \tag{19}$$

and is invariant relative to transformations (16). These angle metrics, connected with a regular variational problem (when the tensor g_{ij} is nonsingular), have been considered by Landsberg [333-335], Bliss [169], Golab [223, 226, 227], and Lippmann [347-349].

The third differential continuation of the metric function includes the object of linear connection and the objects of affine connection and, moreover, the linear connection forms are:

$$\overset{*}{\vartheta}{}^i = \vartheta^i + \Gamma^i_j \omega^j,$$

while the forms of the general affine connection (Γ, C) are

$$\overset{*}{\omega}{}^i_j = \omega^i_j + \Gamma^i_{kj}\omega^k + C^i_{kj}\vartheta^k;$$

the structures of the differential-geometric objects Γ^i_j and (Γ^i_{jk}, C^i_{jk}) are the following:

$$\nabla\Gamma^i_j - \vartheta^i_j \equiv 0, \quad \nabla\Gamma^i_{jk} + C^i_{jl}\vartheta^l_k - \omega^i_{jk} \equiv 0, \quad (\mathrm{mod}\,\omega^k,\ \vartheta^k)$$

$$\nabla C^i_{jk} \equiv 0.$$

The forms $\overset{*}{\omega}{}^i_j$ are invariant relative to transformations (16) if

$$\Gamma^i_{kj}(x,\ \rho v) = \Gamma^i_{kj}(x,\ v),\ C^i_{kj}(x,\ \rho v) = \rho^{-1}C^i_{kj},\ C^i_{kj}v^k = 0.$$

On the components of the tensor C^i_{kj} there are usually imposed additional conditions

$$C^i_{jk}v^k = 0.$$

Then

$$\overset{*}{\Gamma}{}^i_{jk} = \Gamma^i_{jk} - C^i_{lk}\Gamma^l_{pj}v^p \tag{20}$$

form an object of affine connection, i.e.,

$$\nabla\overset{*}{\Gamma}{}^i_{jk} - \omega^i_{jk} \equiv 0.$$

The objects Γ^i_j first appeared in the works of Berwald (for example, see [160]) who obtained a certain specific truncated affine connection (when $C^i_{kj} = 0$) such that its geodesics would coincide with the extremals of the integral

$$\int F dt.$$

The differential equations of the extremals of space F_n can be represented in the form ($v^i = dx^i/ds$):

$$\frac{d^2x^i}{ds^2} + H^i(x, v) = 0,$$

where

$$H^i = g^{ih}\left(F \frac{\partial^2 F}{\partial x^p \partial v^h} v^p + \frac{\partial F}{\partial v^h} \frac{\partial F}{\partial x^p} v^p - F \frac{\partial F}{\partial x^h}\right),$$

while in a nonholonomic coordinate system the differential-geometric object H^i includes the second differential continuation of Eq. (15) by the following formula:

$$H^i = g^{ih}(F'F_{hp}v^p + 'F_h F_p v^p - FF_h),$$

where g^{ij} is the inverse tensor for the tensor g_{ij}. It turns out that

$$\nabla H^i - \varphi^i = H^i_k \omega^k + 'H^i_k \vartheta^k, \tag{21}$$

where $\varphi = \vartheta^i_k v^k$, and another space, i.e., the so-called path space of the Finsler space, is adjoined to the space F_n. It turns out that we can set

$$\Gamma^i_j = \frac{1}{2}\,'H^i_j \tag{22}$$

and the first differential continuation of the differential-geometric object Γ^i_j includes a symmetric object of affine connection (Berwald object), i.e.,

$$\overset{\circ}{\Gamma}{}^i_{jk} = \frac{1}{2}\,''H^i_{kj}. \tag{23}$$

Objects of other connections (Γ, C) differ from the object $\overset{\circ}{\Gamma}{}^i_{jk}$ by two tensors, i.e., the tensor C^i_{jk} and the tensor $T^i_{jk} = -T^i_{jk}$. Thus, any affine connection of space F_n has the form:

$$\left(\frac{1}{2}\,''H^i_{jk} + T^i_{jk},\ C^i_{jk}\right).$$

The tensor C^i_{jk} is always determined from the conditions of covariant constancy of the tensor g_{ij} relative to the connection (Γ, C) to within a choice of the skew-symmetric part $C^i_{[jk]}$. For example, Cartan [179] required that the tensor

$$C_{ijk} = g_{lj}C^l_{ik}$$

be symmetric relative to the last two indices and that $\overset{*}{\Gamma}{}^{i}_{jk} = \overset{*}{\Gamma}{}^{i}_{kj}$. Then the affine connection is determined uniquely from the conditions of covariant constancy of the tensor g_{ij}. The differential equations of the geodesic curves of space F_n do not depend upon the choice of the tensor C^i_{jk}. The geodesic lines of the Cartan connection coincide with the extremals of space F_n. Taylor [510] and Synge [501] determined affine connections in F_n by Christoffel symbols formed from the tensor $\overset{\circ}{g}_{ij}$ (in Finsler space the Christoffel symbols for the tensor g_{ij} do not form an object of affine connection) which defines the metric of the osculating Riemann space along a curve of the space F_n. The length of a vector and the angle between vectors change under a parallel displacement of vectors relative to a Berwald connection, i.e., the affine Berwald connection is not a metric connection.

If $l: x^i(t)$ $(0 \le t \le 1)$ is a curve in space V_n, then the curve $L: (x^i(t), \lambda(dx_i/dt))$ $(\lambda > 0)$ is called the tangent curve for the curve l. If a certain affine connection is given in F_n, then with its aid we can construct for each point of the curve l an evolute of the curve L, which is called the natural evolute L^* for l. The Cartan connection [179], and also the more general Chern connection [182, 183], possess the property that the natural evolutes of the geodesic curves of the connection are straight lines. Other authors have introduced affine connections such that the natural evolutes of the geodesic lines would be different from straight lines (see Kashiwabara [279]).

Vagner [24], Barthel [151, 152], and Laugwitz [336, 337] defined objects of affine connections for the space F_n with the aid of the intrinsic objects of the indicatrix

$$F(x, v) = 1 \qquad (x^i = \text{const})$$

of the local Minkowski space M_n (a centroaffine space with a fixed hypersurface), corresponding to the point $(x^i) \in V_n$. Varga found a whole series of geometric constructions [541, 542, 545] for obtaining the Cartan connection. Sulanke proposed a geometric method [496-498], different from Varga's method, for deriving other affine connections for the space F_n. Analogous questions have been treated by Laptev [82], Matsumoto [351, 355, 356], Kropina [78], Rund [445, 449, 450], and Ohkubo [409, 413]. Vasil'eva [36, 38, 39] treated the theory of connections in space F_n by the method of

Laptev (she obtained a number of interesting geometric interpretations for the subobjects of a sequence of fundamental differential-geometric objects of space F_n). Historical surveys of the theory of affine connections in space F_n have been presented by Kreter [327], Brown [174], Hashiguchi [238], and Laugwitz [336, 337].* Yano and Davies [575] introduced affine connection in F_n by means of the induced Riemann connection [they treated the space F_n as a surface in a $(2n-1)$-dimensional Riemann space].

The papers by Vagner [553], Kawaguchi [306], Otsuki [417, 418], and Rund [449] are devoted to nonlinear connections in space F_n.

The structure equations of the space F_n with the affine connection (Γ, C) have the form:

$$\begin{aligned} D\omega^i &= \omega^k \wedge \overset{*}{\omega}{}^i_k - R^i_{pq}\omega^p \wedge \omega^q + C^i_{pq}\overset{*}{\vartheta}{}^p \wedge \omega^q, \\ D\overset{*}{\vartheta}{}^i &= \overset{*}{\vartheta}{}^k \wedge \overset{*}{\omega}{}^i_k - R^i_{0pq}\omega^p \wedge \omega^q - P^i_{0pq}\omega^p \wedge \overset{*}{\vartheta}{}^q - S^i_{0pq}\overset{*}{\vartheta}{}^p \wedge \overset{*}{\vartheta}{}^q, \\ D\overset{*}{\omega}{}^i_j &= \overset{*}{\omega}{}^k_j \wedge \overset{*}{\omega}{}^i_k - R^i_{jpq}\omega^p \wedge \omega^q - P^i_{jpq}\omega^p \wedge \overset{*}{\vartheta}{}^q - S^i_{jpq}\overset{*}{\vartheta}{}^p \wedge \overset{*}{\vartheta}{}^q, \end{aligned} \tag{24}$$

where, for example, $R^i_{0pq} = R^i_{kpq}v^k$. If F_n has a Cartan connection, then $R^i_{pq} = 0$. Galvani [212, 215] proved that the structure equations of space F_n (in a holonomic frame) can be represented in the form (with a Cartan connection and by assuming that $v^1 = 1$, $v^\alpha = 0$, $\alpha = 2, \ldots, n$):

$$\begin{aligned} D\omega^i &= \omega^k \wedge \overset{*}{\omega}{}^i_k - A^i_{kl}\omega^k \wedge \overset{*}{\omega}{}^l_1, \\ D\overset{*}{\omega}{}^k_i &= \overset{*}{\omega}{}^l_i \wedge \overset{*}{\omega}{}^k_l - R^k_{ilm}\omega^l \wedge \omega^m - 2P^k_{ilm}\omega^l \wedge \overset{*}{\omega}{}^m_1 - S^k_{ilm}\overset{*}{\omega}{}^l_1 \wedge \overset{*}{\omega}{}^m_1, \end{aligned} \tag{25}$$

where $A^i_{kl} = A^i_{lk}, A^1_{kl} = 0$, while the tensors R and S are skew-symmetric in the indices l and m. The study of the identities which connect the torsion and curvature tensors in space F_n (analogs of the Bianchi identities), as well as of the Ricci identities which connect the second covariant derivatives of tensor fields, occupied Auslander [149], Berwald [165], Ispas [270], Misra [364-366], and Eisenhart [199], and the methods for obtaining all these identities

* Researches on Finsler geometry were initiated in the USSR by Vagner [17-19], Laptev [81-83], and Rashevskii [106, 107].

were spelled out in detail by Rund in his monograph [450] and in his papers [447, 451].

The theory of connections in space F_n from the global point of view was considered by Reeb [441], Kashiwabara [279], Matsumoto [351-356], Otsuki [417-419], Akbar-Zadeh [137-146], and Soós [477]. Aronszajn [148] and Laugwitz [339] considered infinite-dimensional Finsler spaces.

The foundations of the theory of curves in Finsler space were established by Finsler himself [203-205]. Another version of this theory was proposed by Cartan [179] and also by Nazim [404] (by a curve in space F_n we mean a set of line elements whose centers form a curve). I-P'ei Ch'en [124], Behari [159], Kaul [281], Mishra [367-370], Prakash [422], Prakash and Behari [426], Shamihoke [463], and Vinhelm [552] studied various curvatures of a curve in space F_n (and also special cases of curves). The variety of line elements of space F_n, whose centers form a surface (hypersurface) of the base space V_n for F_n, is called a surface (hypersurface) of space F_n. The surfaces and hypersurfaces of space F_n were first considered by Finsler [203-206], Cartan [179], Nazim [404, 405], and Wegener [564]. The rigging of a hypersurface of space F_n is determined by means of a line element (almost always it is assumed that the line element is orthogonal to the hypersurface). Wegener obtained the generalized Gauss — Peterson — Codazzi equations for the hypersurface, laid down the foundations of the theory of minimal surfaces, and introduced the notion of different curvatures of the hypersurface. The Gauss — Peterson — Codazzi equations, minimal hypersurfaces, and analogs of the different curvatures of a surface, as well as certain special classes of surfaces (or hypersurfaces) of a space F_n, have been studied, for example, by Su [115, 118, 486], Shapukov [126], Barthel [153-156], Bettingen [167], Debever [189], Dhawan [194, 195], Haimovici [231], Varga [543, 544], Laugwitz [340], Maebashi [350], Mishra [367, 368], Nagata [401], Prakash [425], Rapscak [433], Rund [448, 450, 452, 453], and Shamihoke [461, 464, 465]. Lehmann [342] investigated the geodesic lines of space F_n with the aid of Morse theory.

§ 3. Certain Classes of Finsler Space

If $F = \gamma_{ij}(x)v^i v^j$, then F_n is Riemannian (γ_{ij} is the Riemann metric tensor) and $g_{ij} = \gamma_{ij}$. If $A_i = 0$, where

$$A_i = A^k_{ik},\ A^k_{il} = FC^k_{il},$$

and if the metric function F is positive definite, then F_n is a Riemann space (see Deike [193]). If $R^i_{jpq} = 0$ and the covariant derivative of the first kind of the tensor A^i_{jk} equals zero, then F_n is a Minkowski space (see Cartan [179]). If the angle metric of space F_n is covariantly constant relative to a Cartan connection, then $S^i_{jkh} = 0$ and the Riemann curvature of this metric equals unity (see Cartan [179]). The Riemann — Christoffel tensor of the angle metric has the form (see Varga [539]):

$$\hat{R}_{ijkh} = \begin{vmatrix} g_{ik}-v_iv_k, & g_{ih}-v_iv_h \\ g_{jk}-v_jv_k, & g_{jh}-v_jv_h \end{vmatrix} + S_{ijkh},$$

where $v_i = g_{ik}v^k$, $S_{ijkh} = g_{jl}S^l_{ikh}$. If p^{ij} is a bivector transversal to the line element (x^i, v^i), then the scalar curvature

$$K(x, v, p) = \frac{R_{ijkh}p^{ij}p^{kh}}{(g_{ik}g_{jh}-g_{ih}g_{jk})\,p^{ij}p^{hk}}$$

is covariantly constant if and only if the covariant derivative of the first kind for the tensor S_{ijkh} equals zero (see Varga [539]). The tensor characteristics of spaces F_n for which the support line elements are parallel in the sense of a Cartan connection have been found by Varga [532, 533, 539, 540]. These characteristics have the form $R^i_{jkl}v^l = 0$.

If F_n is a Finsler space with absolute parallelism of the support line elements, then the osculating Riemann space along the vector field possesses a group translation whose trajectories coincide with the extremals of the field (see Varga [532]). The theory of spaces F_n based on the properties of the osculating Riemann spaces has been presented by Nazim [404], Varga [530, 541, 542], and Moór [376, 392].

Spaces F_n with an algebraic metric (the spaces $F^{(p)}_n$):

$$F^p = a_{i_1 \ldots i_p}(x)\, v^{i_1} \ldots v^{i_p}$$

have been thoroughly investigated. Vagner [17] investigated the spaces $F^{(3)}_2$, Liber [97] the spaces $F^{(p)}_2$ $(p > 3)$, Tonooka [518, 519] the spaces $F^{(3)}_3$ and $F^{(3)}_4$, Moór [374] the spaces $F^{(4)}_2$, Ermakov [50, 51] the spaces $F^{(p)}_n$ $(n \geq 4;\ p > 3)$, and Wegener [563] the spaces $F^{(3)}_n$ $(n \geq 2)$.

The scalar curvature of a Finsler space F_n is defined thus:

$$R(x, v, \xi) = \frac{R_{ijkh} l^i l^j \xi^j \xi^h}{(g_{ij} - l_i l_j) \xi^i \xi^j},$$

where $l_i = g_{ik} l^k$, $l^k = \frac{v^k}{F}$. If $R(x, v, \xi)$ is independent of the choice of the vector ξ^i, then F_n is called a Finsler space of scalar curvature (in the sense of Berwald [160, 166]). If R = const, then F_n is called a Finsler space of constant scalar curvature. If in F_n there exists a coordinate system in which the geodesic lines are defined by linear equations, then F_n is called a projectively-flat Finsler space (see Berwald [164]). Rapscak proved [434, 435] that in a space F_n we can pass a completely geodesic hypersurface (the locus of tangent line elements) through any line element in any orthogonal direction if and only if F_n is a projectively-flat Finsler space with scalar curvature. For example, if F_2 is projectively flat, then the metric function has the form (see Funk [211]):

$$F(x, y, \dot{x}, \dot{y}) = R \frac{d}{dt} \tan^{-1} \frac{\sqrt{(x-\xi)^2 + (y-\eta)^2}}{\sqrt{u^2 + v^2}},$$

where ξ, η, u, v are arbitrary constants and R is the curvature of the space. Berwald found all the projectively-flat F_2 for which

$$F = \frac{(\alpha \dot{x} + \beta \dot{y})^2}{\gamma \dot{x} + \delta \dot{y}},$$

where α, β, γ, δ are constants. Kropina determined all the projectively-flat spaces F_2 for which

$$F = \frac{a_{ij}(x) v^i v^j}{b_k(x) v^k} \quad (i, j = 1, 2).$$

For example, the spaces F_n with the metric function (see Kropina [76-79]):

$$v^1 F = v^1 \frac{\partial \psi}{\partial x^\alpha} v^\alpha + \sum_{i=1}^{n} K_i (v^i)^2 \quad (\alpha = 2, \ldots, n),$$

where $K_i = \pm 1$ and $\psi = \psi(x)$ is an arbitrary differentiable function, are projectively-flat Finsler spaces and among them there are no spaces of constant scalar curvature.

If F_n admits of two orthogonal families of completely extremal hypersurfaces of constant scalar curvatures R_1 and R_2, then

$R_1 = R_2$ and F_n is a Finsler space of constant scalar curvature $R = R_1$ (see Yuen-da Wang [555]). In order for an F_n admitting of one family of completely extremal hypersurfaces of constant scalar curvature to be a space of constant scalar curvature, it is necessary and sufficient that the curvature of the space and the curvature of the hypersurface be the same (see Yuen-da Wang [555]). If R_{ij} is the Ricci tensor for the tensor R^k_{ijh} and if

$$2Mg_{ik}v^i = \left(R_{ki} + R_{ik} + \frac{\partial R_{ij}}{\partial v^k} v^j\right)\frac{dv^i}{ds},$$

where $M = F^{-2}R_{ij}v^iv^j$, then F_n is called an Einstein—Finsler space. Any Finsler space of constant scalar curvature is an Einstein—Finsler space (see Rund [451]). A space F_n is called a space of quasiconstant curvature if

$$R_{0ijk} = R_{0i0k}l_j - R_{0i0j}l_k \quad (n \geqslant 3).$$

In such spaces we can pass a transversal hypersurface through any line element, such that its normal would coincide with the line element (see Rapscak [432]). The scalar

$$K_0 = 1/\omega_{n-1}\int_T R_0(x, v)\,d\mu,$$

where T is the unit tangent vector layer, R_0 is the scalar Ricci curvature, i.e., $R_0 = g^{ik}R_{ik}$ and ω_{n-1} is the volume of the layer relative to the natural metric, is called the total curvature of the space F_n. If F_n is without conjugate points, then always $K_0 \leq 0$ (see Moalla [373]). The lower bound of the arc lengths of the curves of class C^1 connecting two points of space F_n is called the distance d between these two points. The space F_n is called complete if every Cauchy sequence converges. For complete Finsler spaces there holds a Hopf—Rinow theorem analogous to the corresponding theorem for Riemann spaces. A space F_n is complete if and only if (see Moalla [372, 373]):

1) any geodesic arc can be extended in both directions;

2) any bounded point set $E \subset F_n$ is relatively compact (the closure is compact);

3) any two points can be connected by a geodesic arc whose length equals the distance between these points.

If $R_0 \geq e^2$ (e is a constant), then in F_n there does not exist a geodesic arc whose minimal length $\geq \pi (n-1)^{\frac{1}{2}} e^{-1}$. If F_n is complete and $R_0 \geq e^2$, then its universal space is compact and the fundamental group is finite. If $R_0 > 0$, then the first Betti number for F_n equals zero (see Moalla [371]).

If X^i and Y^i are vectors, then the quantity (λ is a bivector formed from X^i and Y^i)

$$K(x, v, \lambda) = \frac{R_{ijkh} X^i Y^j X^k Y^h}{\|X\|^2 \|Y\|^2 - (XY)^2},$$

where $\|X\|$ is the length of vector X^i and (XY) is the scalar product, is called the scalar curvature in the sense of Akbar-Zadeh [140]. If K does not depend on λ, then F_n is called a partially isotropic space. A space F_n is partially isotropic for a fixed line element (x^i, v^i) if and only if $(n > 2)$

$$R^l_{jkl} = R_0 (\delta^l_k g_{lj} - \delta^l_l g_{kj}),$$

where R_0 is the (constant) Ricci curvature, and in the case when $R_0 \neq 0$ it is a space of constant curvature in the sense of Berwald. If $\underset{a}{\xi}{}^{\lambda}$ are vectors ($a = 1, 2, \ldots, m-1$) defining an $(m-1)$-dimensional path and an m-dimensional path together with v^i, then the quantity

$$\underset{m}{R}(x, v, \xi) = \frac{K_{[i_1 i_2 [j_1 j_2} g_{i_3 j_3} \cdots g_{i_m] j_m]} l^{i_1} l^{j_1} \underset{1}{\xi}{}^{i_2} \underset{1}{\xi}{}^{j_2} \cdots \underset{m-1}{\xi}{}^{i_m} \underset{m-1}{\xi}{}^{j_m}}{2 g_{[i_1 [j_1} \cdots g_{i_m] j_m]} l^{i_1} l^{j_1} \underset{1}{\xi}{}^{i_2} \underset{1}{\xi}{}^{j_2} \cdots \underset{m-1}{\xi}{}^{i_m} \underset{m-1}{\xi}{}^{j_m}},$$

where K_{ijkh} is the curvature tensor of Rund connection [450], is called the scalar curvature at the point (x^i) and in the given m-dimensional path. When $\underset{m}{R}$ is independent of the $(m-1)$-dimensional path ξ, such F_n are called $(m-1)$-isotropic Finsler spaces (see Wrona [566]). Spaces F_n in which the Ricci tensor K_{ij} for the curvature tensor K^i_{jkh} of the Rund connection has the form

$$K_{ij} = \frac{1}{n-1} (g_{ij} - l_i l_j),$$

are $(m-1)$-isotropic and $(1 < m < n)$

$$\underset{m}{R} = \frac{m-2}{m(n-1)(n-2)} K,$$

where K is the Ricci curvature (for the tensor K^{i}_{jkh}). If F_n is a Finsler space with absolute parallelism of line elements, then it is an (m − 1)-isotropic space if and only if (2 < m < n − 1)

$$K_{ijkh} = \frac{2K}{(n-1)(n-2)} g_{[i[k}(g_{j]h]} - 2l_{j]}l_{h]}),$$

and if $\underset{m}{R}$ is independent of v^i, then $\underset{m}{R}$ is constant (see Wrona [566–568]).

Moór initiated the research of Finsler spaces with recurrent curvature [390], i.e., spaces F_n for which the covariant derivatives of one of the curvature tensors is proportional to the tensor itself (the coefficient of proportionality is a covector).

A space F_n with the metric function F is called the product of the spaces F_r and F_{n-r} if F_1 and F_2 are connected by the relation $F^2 = F_1^2 + F_2^2$. Such spaces were considered by Varga [538], and he found the necessary and sufficient conditions for the decomposability of the space F_n into two spaces. A space F_n is called the product of two Riemann spaces with quadratic forms $p_1 = a_{ab}dx^a dx^b$ and $p_2 = a_{pq}dx^p dx^q$, if $F = f(p_1, p_2)$. Such spaces were investigated by Hou-Sung Hu [262]. Symmetric Finsler spaces, i.e., spaces whose curvature and torsion tensors are covariantly constant, were investigated by Ting-chia Hsing [111]. Singular Finsler spaces have been considered by Vagner [28], Zhotikov [56], Kabanov [66], and Varga [538]. The papers by Varga [530, 531, 535, 543, 544] and by Barthel [151, 152] are devoted to Minkowski spaces. Varga [534] found the conditions for the space of line elements with affine connection to possess a Finsler metric (also see Rapscak [434,437]).

Papers have begun to appear recently on the geometry of tangent bundles of Riemann and Finsler spaces (and also of tangent bundles of arbitrary fiber spaces). The variety T(R) of tangent vectors to a Riemann space R_n is called the first-order tangent bundle of R_n (it is a 2n-dimensional variety). Higher-order tangent bundles are defined analogously. It turns out that the Finsler metric of space F_n induces certain algebraic structures of the tangent bundle for F_n (see Akbar-Zadeh [142, 144], Kurita [332], and Yano and Davies [576]). The papers by Heil [241], Prakash [423], and Rizza [442, 443] are devoted to complex structures of Finsler spaces.

Various classes of Finsler spaces (as well as certain other topics) are taken up by Grossman [229], Varga [529], Vasil'ev [34], Kawaguchi [282-284], Süss [499], Kikuchi [317], Laugwitz [336-340], Misra [362-366], Moór [379, 384-387, 391, 393, 396, 398], Nasu [402, 403], Sen [459, 460], Rashevskii [106, 107, 438-440], Shirokov [133], Soós [474, 476], Rapscak [435, 436], Sasayama [455], Shamihoke [466], Tachibana [503], Théodoresco [511-513], and Otsuki [416].

§ 4. Motions and Transformations

Transformations which preserve the metric are called the motions of the Finsler space F_n. The general theorem on motions was proven by Laptev [82] who was the first to apply the concept of the Lie derivative to the search for the motions and other transformations of Finsler spaces. If the motion group of a Finsler space F_n is of maximal possible order, then the space is Euclidean (see Yano [571]). If the space F_n admits of an r-parameter motion group and $r > (n(n-1)/2) + 1$, then F_n is a space of constant curvature (see Yuen-da Wang [555]). Chao-Hao Ku proved [330] that a space F_n ($n > 4$, $n \neq 8$) admits of a motion group of order $r = (n(n-1)/2) + 1$ if the metric function F is positive definite and has the form ($\alpha = 2, \ldots, n$):

$$F(x, v) = \mathfrak{F}\left(v^1, \sqrt{h_{\alpha\beta}(x)\, v^\alpha v^\beta}\right)$$

or

$$F(x, v) = \mathfrak{F}\left(v^1, e^{kx^1}\sqrt{(v^2)^2 + \ldots + (v)^{n2}}\right), \quad k = \text{const.}$$

Tashiro [509] found certain classes of spaces F_n admitting of a motion group of the same order. Maurin [357] considered the motion groups of a generalized Finsler space F_n (when the indicatrix is a star-shaped hypersurface). The space of variables (x^i, p_i), where $p_i = \partial \ln F / \partial v^i$, is called a phase space of space F_n. The one-term group of canonical transformations of a 2n-dimensional phase space generates a vector field to which there uniquely corresponds the Hamilton function $H(x, p)$ defining the figuratrix and the indicatrix of some Finsler space whose geodesic lines correspond to the trajectories of this group in the phase space. The Finsler metric also generates a one-term transformation group in the phase space.

Infinitesimal transformations defined by the vector field $\xi^i(x)$ are called homothetic transformations of a Finsler space if (L_ξ is the symbol for the Lie derivative)

$$L_\xi g_{ij} = c g_{ij} \ (c > 0, \ c = \text{const}).$$

If c = 0, the homothetic transformations coincide with motions. If the space F_n admits of an (r + 1)-parameter group H_{r+1} of homothetic transformations, then this group always has an r-parameter motion group G_r of space F_n as a normal divisor. This result is due to Soós [475]; he proved the nonexistence of Finsler spaces F_n ($n \neq 4$) with a homothety group H_r, where $\frac{1}{2}n(n+1) > r > \frac{1}{2}n \cdot (n-1) + 2$. Homothetic transformations always preserve the torsion and curvature tensors of the Finsler space F_n. Akbar-Zadeh proved [145] that if F_n is a locally-unreduced analytic variety, then every infinitesimal transformation which preserves the torsion and curvature tensors (as well as the covariant derivatives of these tensors) is also an infinitesimal homothety.

Transformations preserving affine connection are called affine motions of space F_n. Matsumoto proved [352, 353] that a space F_n admits of a maximal group of affine motions if and only if all torsion and curvature tensors of the given connection equal zero. Soós found the connection between homothetic and affine motions of a Finsler space. Homothety transformations do not exist in a Finsler space of constant curvature. In order for an r-parameter group G_r of affine motions of space F_n to have at least a one-dimensional homothety subgroup it is necessary and sufficient that

$$\text{rank} \, \| L_{\xi_\sigma} g_{ij}, \, g_{ij} \| = s < r + 1,$$

where ξ_σ is the basis of the Lie algebra of group G_r. If $s < r + 1$, then the subgroup of homothetic transformations has the dimension $r + 1 - s$ (see Soós [472, 473]). Homothetic transformations were studied by Shanks [468] and Yano [570, 571]. If the order of the homothetic transformations of space F_n is maximal, then the space F_n is Euclidean (see Hiramatsu [245, 246]). Soós [472] found the condition for the affine motion group of a Finsler space F_n to contain the motion group of space F_n as a nontrivial subgroup. Homothetic motions of two-dimensional Finsler spaces have been investigated in detail by Chetyrkina [125].

Finsler spaces for which the tensors A^i_{jk} and $R^i_{jkl} - A^i_{jm}R^m_{0kl}$ are covariantly constant are called quasisymmetric spaces (see Soós [474]). Any quasisymmetric space F_n admits of at least a one-parameter affine motion group. If the matrices $X_{kh} = \| R^i_{jkh} \|$ form an irreducible system of linear transformations, then the geodesics coincide with the autoparallel curves (see Teodorescu [514]).

Transformations which preserve the form $'F_i\omega^i$ are called dilations of the space F_n. Kurita [331, 332] proved that the volume metric of a Finsler space (the volume of the hyperflat element dual to a line element) is invariant relative to dilations.

Conformal transformations of Finsler metrics have been considered by Nazim [404], Hombu [251, 252], Misra [362, 363], and Rund [450]. Akbar-Zadeh [141] gave the name conformal motions of a Finsler space F_n also to the infinitesimal transformations for which

$$L_\xi\left(g^{-\frac{2}{n}} g_{ij}\right) = 0 \quad (g = \det \| g_{ij} \|),$$

and found the necessary and sufficient condition for the infinitesimal transformations of a compact space F_n to be conformal motions. Various transformations of Finsler spaces have been treated by Vagner [19], Kagan [70, 71], Akbar-Zadeh [144, 145], Kobayashi [322, 323], Hokari [247], Knebelmann [320, 321], Kondrat'ev [74, 75], Muto and Yano [400], and Ohkubo [414]. Earlier papers on motions of Finsler spaces have been covered in Yano's monograph [571].

In a Finsler space F_n there exist coordinate systems in which the components of the symmetric part of an object of affine connection equal zero (an analog of the normal coordinates in a Riemann space). Such coordinate systems (and also the theory of the invariants of space F_n) have been considered by Bucur [175], Busemann [176, 177], Davies [186], Moór [392, 394], Rapscak [427-430], and Varga [537].

Various transformations (such as projective, conformal, geodesic) of connection objects, as well as of the object H^i of space F_n, have been analyzed by Hsien-chu Wei [40], Golab [224, 225], Haimovici [232-237], Hombu [251-253], Misra [362, 363], Moór [389], Sinha [470].

§ 5. Immersions of Finsler Spaces

The space F_n is a fiber space (the layers are isomorphic Minkowski spaces). The set of local indicatrices of a Finsler space forms the global indicatrix of this space. We say that the space F_n has been embedded in an affine space A_N with a vector metric, if in A_N there exists an n-dimensional surface P_n such that the field, given in A_N, of hypersurfaces (they are defined by the vector metric) under an intersection with the tangent planes of surface P_n forms on the latter a field of local $(n-1)$-dimensional surfaces (they are taken as the local indicatrices of a new Finsler space with base P_n) relative to which the surface turns into a Finsler space isometric to the space F_n. The immersion is said to be regular if the surface P_n is tangentially nonsingular, and is said to be singular if the surface P_n is singular, i.e., the set of its tangent planes depends on p parameters $(p < n-1)$. In 1949, Vagner proved that an n-dimensional Finsler space can be regularly embedded (locally) in a 2n-dimensional Minkowski space, i.e., in a 2n-dimensional affine space with a vector metric. Other proofs of Vagner's theorem were obtained by Ingarden [267, 268] and by Chao-Hao Ku [42] who proved that every n-dimensional Finsler space of class C^r $(r \geq 2)$ is regularly immersed wholly in a 4n-dimensional Minkowski space. Singular immersions of two-dimensional Finsler spaces in three-dimensional affine spaces were considered by Kagan [70] who proved that F_2 admits of an imbedding in A_3 with a vector metric in the form of a (cyclic conical or tangential) surface if and only if the metric function F has the form (in special coordinates)

$$F = F(x^2, v^1, v^2), \quad F = e^{x^1} f(x^2, v^1, v^2), \quad F = F(x^2, v^1, x^1 - x^2, v^2).$$

By an immersion of a Finsler space in a Riemann space we mean a mapping under which to the line elements of the Finsler space there correspond points of the Riemann space R_N $(N \geq 2n-1)$ such that, in the tangent spaces E_N of the Riemann space R_N we can pick out an n-dimensional plane E_n and a path E in this plane such that the induced metric of this plane and the projection of an infinitely-close n-dimensional plane E_n' (the projection is established by means of the affine connection of the Riemann space R_N) coincide with the metric and with the parallel displacement of the immersed Finsler space. If the first n vectors E_i of the Euclidean tangent space E_{2n-1} for R_{2n-1} have been embedded in the plane E_n

and if the vector E_1 has been combined with the path E, then the immersion equations have the form ($i, j = 1, 2, \ldots, n$; $a = n + 1, \ldots, 2n - 1$; $I, J = 1, 2, \ldots, 2n - 1$)

$$\Omega^i \equiv \omega^i, \quad \Omega^i_j \equiv \omega^i_j \pmod{\Omega^a}, \tag{26}$$

where the Pfaffian forms Ω^I and Ω^I_J are connected by the structure equations of the Riemann space R_{2n-1} with torsion-free connection. Deike [191] proved that a space F_n with Cartan connection can be represented as an n-dimensional holonomic surface of a Riemann space if $A^i_{jk} \equiv 0 \pmod{\Omega^a}$. Thus, a Finsler space with Cartan connection can be immersed as a holonomic surface R_n of the space R_{2n-1} if and only if the system $\Pi^i = 0$, $D\Pi^i = 0$, where $\Pi^i = A^i_{jk}\omega^j \wedge \omega^k$, is in involution. Galvani [214] proved that this system is always in involution when $n = 2$. It turns out that any Finsler space can be immersed as an n-dimensional nonholonomic surface of a $(2n - 1)$-dimensional Riemann space with torsion (see Deike [192]). It should be noted that a Finsler space with Berwald connection can always be immersed in a $(2n - 1)$-dimensional Riemann space (as an n-dimensional holonomic surface). Any two-dimensional Finsler space with Cartan connection can be embedded (as a three-dimensional surface) in a five-dimensional Euclidean space (see Galvani [215]) (the arbitrariness of the embedding is four functions of one argument), while a two-dimensional Finsler space with Cartan connection and with absolute parallelism of the support line elements can be embedded in a three-dimensional Euclidean space (the arbitrariness of the embedding is two functions of one argument). The immersions of two-dimensional Finsler spaces were investigated in detail by Galvani [212-217].

§ 6. Metric Spaces of Line Elements

Various generalizations of Finsler spaces exist and, moreover, in some generalizations the metric function (taken as a fundamental differential-geometric object) is replaced by another differential-geometric object whose structure coincides with the structure of a certain differential-geometric object spanned by the differential continuations of the metric function F, while in other generalizations the fundamental object (the function F) does not change (in the sense of the structure) but another fiber space is taken. The generalizations of Finsler spaces (in the second sense) will be considered in other sections.

The space $L_n^{(1)}$ of line elements, with a nondegenerate symmetric tensor field g_{ij}:

$$\nabla g_{ij}=g_{ijk}\omega^k+g'_{ijk}\vartheta^k, \tag{27}$$

is called a metric space $\mathfrak{F}_n$ of line elements. It is clear that with $\mathfrak{F}_n$ we can always associate a Finsler space F_n with the metric function $F^2 = g_{ij}v^iv^j$. The fundamental difference between the space $\mathfrak{F}_n$ and the space F_n is that $g'_{ijk} \neq g_{ikj}$ (in the Finsler space the corresponding tensor is symmetric). The geometric meaning of this condition is that in the space $\mathfrak{F}_n$ the notions of transversality and orthogonality are different (in the Finsler space these notions are identical). The spaces $\mathfrak{F}_n$ under various names (for example, Finsler—Riemann systems, Finsler spaces with nonpotential metric, etc.), are considered by Ohkubo [408], Akbar-Zadeh [138, 140, 142, 143], Moór [377, 378, 380, 381], Freeman [208, 209], Bliznikas [2-4, 6, 8], Vanstone [528], Prakash [424], and Shamihoke [467]. The geometry of the spaces $\mathfrak{F}_n$ is fundamentally similar to the geometry of the spaces F_n, only the nonidentity of the notions of transversality and orthogonality (orthogonality in the sense of the metric of the associated Finsler space) shows up in the fact that the usual concepts of Finsler geometry are doubled in number (for example, for these spaces there are more analogs of the Schur theorem, the theory of curves and surfaces is enriched by new curvatures, etc.).

Spaces of line elements with an asymmetric tensor $h_{ij} \neq h_{ji}$, i.e., bilinear metric space of line elements, have also been considered (see Bliznikas [5, 7], Shapukov [127-129], Prakash [423], Moór [395], and Shamihoke [462]). Important results in the theory of surfaces of a bilinear metric space of line elements were obtained by Shapukov [127-129]. Kähler—Finsler spaces were analyzed by Prakash [423].

Finsler geometry is applicable not only in the calculus of variations (Vagner and his students) but also in physics. Not only Riemann spaces but also Finsler spaces are taken as a geometric model for some phenomenon or other in modern physics, in particular, in unified field theory (see Horvăth [254, 255]). For example, Horvăth proved that the Stephenson—Kilmister unified theory of gravitation and electromagnetism can be formulated in the language of Finsler space concepts if as the fundamental metric function of this space we take the Lagrangian in the Stephenson — Kilmister theory for the test particle which moves under the action of a unified electromagnetic and gravitational field. A pair of points is the fundamental element of the model space in the bilocal theory of Yukawa. Horváth and Moór [256, 257, 258] proved that the Yukawa space can be mapped onto a variety of line elements, whose metric structure is established by means of a twice covariant symmetric tensor which is not potential (the metric tensor of a Finsler space is potential). The nonpotentiality of the metric characterizes the anisotropy of the space, the concept of which is of significant value in physical theories.

Applications of Finsler geometry to physics, mechanics, and optics have been discussed by Vagner [23], Eliopoulos [200, 202], Fujinaka [210], Segre [458], Hosokawa [260], Horváth [256], Klein [319], Kondo [324], Lichnerowicz [343-346], Pihl [421], Rund [444], Suzuki [500], Uehara [525], and Wundheiler [569]. The fundamental aspects of the integral geometry of the space F_n are contained in Blaschke [168] and Owens [420].

§ 7. Cartan Space

The pair consisting of the point x^i and the covector u_i is called a hyperflat element. The structure equations of the space L_n^* of hyperflat elements have the form:

$$D\omega^i = \omega^k \wedge \omega_k^i,$$
$$D\theta_i = \omega_i^k \wedge \theta_k + \omega^k \wedge \theta_{ik}, \tag{28}$$

where $\theta_i = du_i - u_k \omega_i^k$ and $\theta_{ij} = -\omega_{ij}^l u_l$. If on L_n^* we are given a scalar field defined by the function F (the fundamental metric function):

$$dF - F(\omega_1^1 + \ldots + \omega_n^n) = F_i \omega^i + F^k \theta_k, \tag{29}$$

then the space L_n^* is called a Cartan space C_n (see Cartan [180]). This class of spaces, introduced by Cartan, were named Cartan spaces at Berwald's suggestion [162-164, 166].

The second differential extension of Eq. (29) includes the metric tensor and the formula for such an inclusion (in a holonomic frame) has the form:

$$g^{ij} = \frac{1}{2} \Delta^{-\frac{1}{n-1}} \frac{\partial^2 (F^2)}{\partial u_i \partial u_j},$$
$$\Delta = \det \|a^{ij}\|, \quad a^{ij} = F \frac{\partial^2 F}{\partial u_i \partial u_j} + \frac{\partial F}{\partial u_i} \frac{\partial F}{\partial u_j}.$$

In this case the angle metric is defined by the formula

$$d\varphi^2 = F^{-1} \frac{\partial^2 F}{\partial u_i \partial u_j} du_i du_j.$$

Cartan indicated a system of axioms [180] by means of which an affine connection can be uniquely introduced in the space C_n. Spaces C_n from the global point of view were considered by Reeb [441]. The motions of Cartan spaces were studied by Hiramatsu [243, 244] and Tashiro [509]. If a regular Cartan space admits of a motion group G_r of maximal order $n(n+1)/2$ $(n > 2)$ or of order $r > (n(n-1)/2) + 1$ $(n \neq 4)$, then it is a Riemann space of constant curvature. It turns out that a space C_n with a metric function F of the following form:

$$F = F\left((u_1)^2,\ w^*/\left(1 + \frac{K}{4} v\right)^{2n-4}\right),$$
$$F = F\left((u_1)^2,\ e^{2K(n-2)x^1} w^*\right),$$

where

$$v = \sum_{a=2}^{n} x^a x^a, \quad w^* = \sum_{a=2}^{n} u_a u_a, \quad K = \text{const},$$

admits of a motion group of order $(n(n-1)/2) + 1$ (see Tashiro [509]).

Schouten and Haantjes [456] considered a generalization of Cartan and Finsler spaces in the sense that they considered a support element in a covariant (or vector) density of specified weight. Moór developed the Schouten—Haantjes theory of generalized spaces, i.e., the metric in such spaces is given not by means of a scalar function but by a nonsingular symmetric tensor of a certain weight (see Moór [382, 383]). Medviedevaitė considered the connections of a generalized Cartan space (the metric in the space of hyperflat elements is given by means of a symmetric tensor) and constructed a theory of curves for such a space (see Medviedevaitė

105]). Various aspects of the geometry of Cartan spaces occupied Vagner [25, 29], Vasil'eva [37], Kabanov [65, 67, 68], Kaganov [72, 73], Khmelevskii [122, 123], Barthel [155, 158], Hölder [250], Moór [375, 376, 399], Ohkubo [411], Rashevskii [438], Rapscak [430], Tonowoka [521, 523], Berwald [162-164, 166], and Goldstine [228].

§ 8. Kawaguchi Spaces and Areal Spaces

Since the square of the differential of the projective arc of a curve is a function of the differentials of the coordinates of the point (up to the fifth order), A. Kawaguchi suggested that the length of an arc of a curve in a differentiable variety be introduced by the following formula:

$$s=\int_{t_1}^{t_2} F\left(x, \frac{dx}{dt}, \frac{d^2x}{dt^2}, \ldots, \frac{d^px}{dt^p}\right) dt. \tag{30}$$

The square of the length of an arc of a curve in Euclidean space is a (homogeneous) second-degree polynomial of the first differentials of the point coordinates and Finsler's generalization was that he took this function to be an arbitrary homogeneous function relative to the differentials of the point coordinates. Kawaguchi took it that this function depends also on the higher-order differentials.

The space $L_n^{(p)}$ of line elements with a metric function F:

$$dF=\sum_{a=0}^{p} F_{(a)i}\vartheta^{(a)i} \tag{31}$$

is called a Kawaguchi space (from the point of view of Laptev's method). The theory of these Kawaguchi spaces was first presented in 1927 at the Congress of the Japanese Physico-Mathematical Society (these spaces were named Kawaguchi spaces at the suggestion of Craig [184] and Synge [502]).

A sequential continuation of the differential equation (31) yields a fundamental sequence of differential-geometric objects (the fundamental sequence of extensors) of the Kawaguchi space K_n:

$$F_{(a)i},\ F_{(a_1)i_1(a_2)i_2}, \ldots,\ F_{(a_1)i_1(a_2)i_2\ldots(a_q)i_q},$$

the differential equations for which have the form:

$$dF_{(a)i} - F_{(c)k}\vartheta^{(c)k}_{(a)i} = F_{(a)i(c)k}\vartheta^{(c)k},$$

$$\cdots\cdots\cdots\cdots\cdots\cdots\cdots\cdots$$

$$dF_{(a_1)i_1\ldots(a_r)i_r} - \sum_{s=1}^{r}\frac{r!}{s!(r-s)!}\vartheta^{(e)l}_{\{(a_1)i_1\ldots(a_s)i_s}\times$$

$$F_{(a_{s+1})i_{s+1}\ldots(a_r)i_r\}(e)l} = F_{(a_1)i_1\ldots(a_r)i_r(e)l}\vartheta^{(e)l},$$

moreover, the quantities $F_{(a_1)i_1\ldots(a_r)i_r}$ are symmetric relative to the paired indices. It turns out that the third differential continuation of the metric function F includes the connection objects of the Kawaguchi space K_n. The geometry of spaces K_n has been studied by Losik [99-102], McKiernan [316], Pên-wang Sun [120], Katsurada [280], Kano [277, 278], A. Kawaguchi [282-292, 299], M. Kawaguchi [311], Ide [263-265], S. Kawaguchi [313-315], Okumura [415], Sharma [469], Srivastava [478], Watanabe [556-562], Tonowoka [522], Yoshida [577], Michihiro [358], and Mikami [360]. The investigations on the geometry of spaces K_n have been carried out mainly by means of the extensor calculus created by Craig [184, 185], A. Kawaguchi [293-296, 298, 303, 307], M. Kawaguchi [312], and Ohkubo [412]. In some of the papers it was assumed that

$$F = F\left(t,\ x,\ \frac{dx}{dt},\ \ldots,\ \frac{d^p x}{dt^p}\right).$$

The space $\overset{\circ}{K}{}^{(1)}_{n,m}$ with a fundamental differential-geometric object F $(\theta^i_\alpha = \nabla p^i_\alpha)$:

$$dF - F\theta^\alpha_\alpha = F_k\omega^k + F^\alpha_i\theta^i_\alpha, \tag{32}$$

is called an areal space. It is obvious that a partial continuation of Eq. (32) yields a sequence of tensors [relative to the group GL(n, R) × GL(m, R)]:

$$\nabla F^\alpha_i \equiv F^{\alpha\beta}_{ij}\,\theta^j_\beta,$$

$$\nabla F^{\alpha\beta}_{ij} \equiv F^{\alpha\beta\gamma}_{ijk}\,\theta^k_\gamma, \quad (\mathrm{mod}\ \omega^k)$$

$$\cdots\cdots\cdots\cdots\cdots$$

Areal spaces were first considered by A. Kawaguchi and his students. We remark that the geometry of areal spaces is a special case of the geometry of the integral:

$$\int \ldots \int F(x^i, p^i_\alpha, p^i_{\alpha\beta}, \ldots, p^i_{\alpha_1 \ldots \alpha_p}) du^1 \ldots du^m. \qquad (33)$$

The geometry of the space connected with the integral (33) is equivalent to the geometry of the space of m-dimensional flat elements of p-th order, $\overset{\circ}{K}{}^{(p)}_{n,m}$ with a fundamental differential-geometric object F:

$$dF - F\theta^\alpha_\alpha = F_i \omega^i + \sum_{a=1}^{p} F_i^{\alpha_1 \ldots \alpha_a} \theta^i_{\alpha_1 \ldots \alpha_a}. \qquad (34)$$

Sometimes these spaces are called higher-order areal spaces. The sequential continuation of Eq. (34) (we have in mind a partial continuation) yields a fundamental sequence of multidimensional extensors.

Fundamental results in the geometry of integral (33) have been obtained by Vagner [20-23, 25, 28, 30-32] (and also by his students Kabanov [64-69], Zhotikov [52-56], and Losik [99-103]), Vasil'eva [1, 35, 37], A. Kawaguchi [297, 299-302, 304, 305, 308, 310], Evtushik [47-49], Freeman [207], Hokari [248], and Iwamoto [271-275].

A simple m-vector $p^{i_1 \ldots i_m}$, is always associated with an m-dimensional line element of first order defined by the quantities p^i_α, i.e.,

$$p^{i_1 \ldots i_m} = m! p^{i_1}_{[1} \ldots p^{i_m}_{m]}.$$

Then, as the support element of the space $\overset{\circ}{K}{}^{(1)}_{n,m}$ we can take a simple m-vector and differential equation (34) can be written in the form

$$dF - F\theta^\alpha_\alpha = F_k \omega^k +' F_{i_1 \ldots i_m} \theta^{i_1 \ldots i_m},$$

where

$$\theta^{i_1 \ldots i_m} \equiv dp^{i_1 \ldots i_m} + \sum_{a=1}^{p} p^{i_1 \ldots k \ldots i_m} \omega^{i_a}_k.$$

The second differential continuation of this transformed differential equation includes the tensor

$$g_{i_1 \ldots i_m, j_1 \ldots j_m} = F'' F_{i_1 \ldots i_m, j_1 \ldots j_m} + 'F_{i_1 \ldots i_m} 'F_{j_1 \ldots j_m}, \tag{35}$$

by means of which the metric in the space of m-vectors is defined, while the tensor $''F_{i_1 \ldots i_m, j_1 \ldots j_m}$ is defined by the angle metric (see Aussem [1]). If an areal space is Riemannian, then with the aid of the metric tensor γ_{ij} (x) we can define a function $\mathring{F}$ in the following fashion:

$$\mathring{F} = \sqrt{\det \|a_{\alpha\beta}\|}, \quad a_{\alpha\beta}(x, p) = \gamma_{ij}(x) p^i_\alpha p^j_\beta,$$

i.e., to the metric tensor γ_{ij} of a Riemann space there always corresponds a metric function F of special form of some areal space. For this function the tensor $g_{i_1 \ldots i_m, j_1 \ldots j_m}$ is expressed only in terms of the metric tensor γ_{ij} in the following way ($\gamma_{ij} = g_{ij}$):

$$g_{i_1 \ldots i_m, j_1 \ldots j_m} = g_{[i_1[j_1} \cdots g_{i_m]j_m]}. \tag{36}$$

Iwamoto [274] proved that the metric tensors of Riemann, Finsler, and Cartan spaces satisfy relations (36). In a number of papers on areal spaces it was assumed that the differential continuations of the metric function F of an areal space include a nonsingular tensor $g_{ij} = g_{ji}$ depending on a support element of the space being considered. Areal spaces for which there exists a nonsingular tensor g_{ij} (included by differential continuations of the metric function F) satisfying relations (36), are called areal spaces of metric class (on a suggestion by Debever [190]). A. Kawaguchi [302] proved that in the case when the numbers n and m are relatively prime the determinant formed from the components of the tensor $g_{i_1 \ldots i_m, j_1 \ldots j_m}$, vanishes and the differential continuations of the metric function F include a nonsingular tensor g_{ij} which is not always related to the tensor $g_{i_1 \ldots i_m, j_1 \ldots j_m}$ by conditions (36). Tandai [504] proved a theorem according to which only Riemann, Finsler, and Cartan spaces are spaces of metric class. In other words, Tandai's result is that if in an areal space there exists a nonsingular tensor g_{ij} satisfying conditions (36), then the areal space is a Riemann, Finsler, or Cartan space.

If a tensor $g_{\alpha\beta}$ is given and is such that

$$\det \| g_{\alpha\beta} \| = F^2,$$

then the tensor

$$g_{ij} = g_{\alpha\beta}\left(\frac{1}{m} L_{ij}^{\alpha\beta} + p_i^\alpha p_j^\beta\right),$$

where

$$L_{ij}^{\alpha\beta} = F^{-2}\left(F''F_{ij}^{\alpha\beta} - 'F_i^\alpha\, 'F_j^\beta + 'F_j^\alpha\, 'F_i^\beta\right),$$

is called the normal metric tensor of an areal space. The tensor

$$^*L_{ij}^{\alpha\beta} = L_{ij}^{\alpha\beta} - \frac{1}{m} g^{\alpha\beta} g_{\nu\sigma} L_{ij}^{\nu\sigma}$$

is called the exmetric tensor of an areal space (see A. Kawaguchi and Tandai [310]). If $^*L_{ij}^{\alpha\beta} = 0$, then the areal space is a space of metric class (see Iwamoto [274]). Areal spaces for which $^*L_{ij}^{(\alpha\beta)} = 0$, are called areal spaces of semimetric class (see Tandai [506]), and these spaces are characterized by the fact that the differential equations of the extremal subspaces do not contain derivatives of the Christoffel symbols as in the case of Riemann, Finsler, or Cartan spaces (see Tandai [506]). An areal space is called a space of submetric class if there exists a tensor g_{ij}, as an algebraic inclusion of the metric function F, of the transversal vector $G_{i_1 \ldots i_m} = m!\, F^{1-m'} F_{i_1}^{[1} \ldots 'F_{i_m}^{m]}$ and of the tensor $g_{i_1 \ldots i_m,\, j_1 \ldots j_m}$.

The geometries of different types of areal spaces were looked at by Vagner [20–22], Alardin [147], Chao-Hao Ku and Buchin Su [46, 494], Buchin Su [114, 116, 117, 119, 489–493], Barthel [157, 158], Brickell [173], Davies [187, 188], Debever [189, 190], Gama [218–222], Kikuchi [318], Varga [536], Tandai [504–508], Tonooka [516], and Ueno [526]. Igarashi [266] has considered motions in areal spaces.

§ 9. Geometry of Differential Equation Systems

It is very well known that the geometry of Pfaffian systems having integral varieties of nonmaximal rank is the base of the geometry of nonholonomic subspaces of a Euclidean space (or of some other Klein space). The geometry of the Monge equations (the fundamentals of this geometry have been presented by Darboux,

Voss, Rodgers, and Sintsov) is closely connected with the problems of the line geometry of a projective space (connex theory). The majority of papers devoted to the geometry of Pfaffian systems or of Monge-type equations are concerned with the geometry of line or other subvarieties of a Euclidean space (or of some other Klein space), and we shall not consider them here (see A. Haimovici [230] and Mirodan [361]).

The equations of the geodesic curves of the shortest lines of a Finsler space have the form $\frac{d^2x^i}{ds^2}+H^i\left(x, \frac{dx}{ds}\right)=0$, i.e., have the very same structure as do the equations of the geodesics of a space of affine connection (in particular, of a Riemann space) except only that the functions H^i are not necessarily second-degree polynomials in the variables dx^i/ds. The differential equations of the minimal surfaces of a Riemann space have the form (m-dimensional surfaces):

$$\frac{\partial^2 x^i}{\partial u^\alpha \partial u^\beta}+H^{\ i}_{\alpha\beta}\left(x, \frac{\partial x}{\partial u}\right)=0, \tag{37}$$

where the $H^{\ i}_{\alpha\beta}$ are certain functions of p^i_α, of the components of the metric tensor and of its partial derivatives. Many problems of Finsler geometry are connected with the study of the structure of the differential equations of the geodesics of the Finsler space F_n.

Research started in 1920-1930 on the geometry of varieties in which the geodesic lines are the integral curves of some system or other of differential equations. Usually by the geometry of a system of differential equations $\frac{d^2x^i}{ds^2}+H^i\left(x, \frac{dx}{ds}\right)=0$ we mean the geometry of the space $L_n^{(1)}$ of line elements with a fundamental differential-geometric object H^i whose structure differential equations have the form (21). The geometry of such spaces has been treated by Bliznikas [14, 15], Bark [150], Cartan [178], Gaukhman [41], Berwald [161], G. F. Laptev [96], Matsumoto [354], Stepanov [112, 113], Kosambi [325, 326], Rapscak [437], Slebodzinski [471], Suguri [495], Douglas [196], Chern [181], and Veblen [546-551].

The geometry of systems of differential equations of form (37), i.e., the geometry of the space of m-dimensional line elements of first order, $\overset{\alpha}{K}{}^{(1)}_{n,m}$ with a fundamental differential-geometric ob-

object

$$\nabla H^{i}_{\alpha\beta}-\varphi^{i}_{\alpha\beta}=H^{\;i}_{\alpha\beta k}\,\omega^{k}+H^{\;i}_{\alpha\beta\gamma}\,\theta^{\gamma}+H^{i\cdot\gamma}_{\alpha\beta k}\,\theta^{k}_{\gamma}, \tag{38}$$

where $\varphi^{i}_{\alpha\beta}=\omega^{i}_{kh}\,p^{k}_{\alpha}\,p^{h}_{\beta}-\theta^{\gamma}_{\alpha\beta}\,p^{i}_{\gamma}$, is called the geometry of K-spread spaces (in Douglas' terminology). It is usually required that the structure equations of the fundamental object $H^{i}_{\alpha\beta}$ be invariant relative to linear transformations of the m-dimensional element. The geometry of K-spread spaces has been considered by Bliznikas [14, 16], Chao-Hao Ku [43-46, 328, 329], Buchin Su [479-481, 483, 484, 487, 488, 492, 494], Moór [388], B. L. Laptev [86], Bortolotti [171], Douglas [197], Mikami [359], Dumitras [198], and Yano [573, 574].

Motions in K-spread spaces were studied by Ishihara [269], Dumitras [198], and Yano [573, 574].

There has been very little development in the geometry of systems of higher-order differential equations, i.e., in the geometry of the space $L^{(p)}_{n}$ of line elements with a fundamental differential-geometric object H^{i} (see Bliznikas [15]):

$$d\overset{(p)}{H}{}^{i}+\overset{(p)}{H}{}^{k}\,\omega^{i}_{k}-\overset{(p)}{\varphi}{}^{i}=\overset{(p)}{H}{}^{i}_{0}\,\theta+\sum_{c=0}^{p}\overset{(p)}{H}{}^{i}_{(c)k}\,\vartheta^{(c)k}, \tag{39}$$

where

$$\overset{(p)}{\varphi}{}^{i}=(p+1)!\sum_{a=2}^{p+1}\frac{1}{a!}\left\{\omega_{k_1\ldots k_a}^{\;i}\sum_{(p_1+\ldots+p_a=p+1)}\frac{1}{p_1!\ldots p_a!}\,(v^{p_1})^{k_1}\ldots v^{(p_a)\,k_a}\right\}$$

and in the geometry of the spaces $\mathring{K}^{(p)}_{n,m}$ of m-dimensional line elements of higher order with a fundamental differential-geometric object $H^{i}_{\alpha_1\ldots\alpha_{p+1}}$:

$$\Delta H^{i}_{\alpha_1\ldots\alpha_{p+1}}-\varphi^{i}_{\alpha_1\ldots\alpha_{p+1}}=H^{i}_{\alpha_1\ldots\alpha_{p+1}k}\omega^{k}+\sum_{a=1}^{p}H^{i\cdot\beta_1\ldots\beta_a}_{\alpha_1\ldots\alpha_{p+1}k}\,\theta^{k}_{\beta_1\ldots\beta_a}, \tag{40}$$

where

$$\varphi^{i}_{\alpha_1\ldots\alpha_{p+1}}=\sum_{s=2}^{p+1}\frac{(p+1)!}{s!}\,\omega^{i}_{k_1\ldots k_s}\sum_{(p_1+\ldots+p_s=p+1)}\frac{1}{p_1!\ldots p_s!}\times$$

$$p^{k_1}_{(\alpha_1\ldots\alpha_{j_1}}\cdots p^{k_s}_{\alpha_{q+1}\ldots\alpha_{p+1})}-\sum_{s=2}^{p+1}\frac{(p+1)!}{s!(p-s+1)!}\,\theta^{\gamma}_{(\alpha_1\ldots\alpha_s}\,p^{i}_{\alpha_{s+1}\ldots\alpha_{p+1})\gamma}.$$

The geometries of systems of higher-order differential equations have been considered by Vagner [26, 27], Hashimoto [240], Izrailevich [59, 60], A. Kawaguchi [309], Hokari [249], Tonooka [517], Moór [397], and Ohkubo [410].

The geometry of various systems of differential equations has been considered from a new point of view by G. F. Laptev [96] and Vasil'ev [33] (and his students Stepanov [112, 113], Gaukhman [41], and Kuz'mina [80], and Shvartsburd [130]).

§ 10. Support Element Spaces

A support element space $V^{(p)}_{n,N}$ is defined by the following structure equations:

$$D\omega^i = \omega^k \wedge \omega^i_k, \quad D\theta^\alpha = \theta^\beta \wedge \theta^\alpha_\beta + \omega^k \wedge \theta^\alpha_k, \tag{41}$$

where

$$\theta^\alpha = dy^\alpha + \sum_{a=1}^{p} Y^{\alpha i_1 \ldots i_a}_k (Y)\, \omega^k_{i_1 \ldots i_a},$$

and the functions $Y^{\alpha i_1 \ldots i_a}_k (y)$ are such that the forms $\theta^\alpha|_{\omega^i=0}$ form a completely integrable system of forms ($\alpha = 1, 2, \ldots, N$). We note that the forms θ^α_β and θ^α_k are defined by the equalities

$$\theta^\alpha_\beta = \sum_{a=1}^{p} \partial_\beta Y^{\alpha i_1 \ldots i_a}_k \omega^k_{i_1 \ldots i_a}, \quad \theta^\alpha_k = \sum_{a=1}^{l} Y^{\alpha i_1 \ldots i_a}_h \omega^h_{i_1 \ldots i_a k}.$$

By extending the structure equations (41) we obtain the new forms

$$\theta^\alpha_{\beta_1 \ldots \beta_a}, \quad \theta^\alpha_{k_1 \ldots k_a}, \quad \theta^\alpha_{k_1 \ldots k_c \beta_1 \ldots \beta_e} \ (c + e = a),$$

which define fiber differential structures adjoined to the support element space $V^{(p)}_{n,N}$. The vertical differential group of space $V^{(p)}_{n,N}$ is determined by the forms

$$\vartheta^\alpha_{\beta_1 \ldots \beta_a} = \theta^\alpha_{\beta_1 \ldots \beta_a} \big|_{\omega^i=0,\ \theta^\alpha=0}.$$

The general theory of spaces $V^{(p)}_{n,N}$ was worked out by B. L. Laptev [82-86, 88-91]. The spaces $L^{(p)}_n$ and $K^{(p)}_{n,m}$ are special cases of a support element space. If the support object y^α is a tensor, then the space $V^{(p)}_{n,N}$ is called a space of tensor support elements. The spaces of tensor support elements were first considered by A. Kawaguchi (he took the support element to be a system of tensors of arbitrary valence and arbitrary weight) who called them hypergeneralized Finsler spaces (see A. Kawaguchi [283]).

Various versions of the connection theory of the space $V^{(p)}_{n,N}$ were proposed by Bliznikas [9-14]. Special cases of a support element space were considered by A. Kawaguchi [282-286], Maziliauskaitė [104], Rzhekhina [108-110], Urbonas [121], Sinkunas [131,132], and Yablokov [134-136].

It turns out that we can adjoin a space of tensor support elements to a uniform space (a Lie group with a fixed closed subgroup). If the orbits of the linear isotropy group of a uniform space are not tangentially degenerate, then in the uniform space there exists an invariant Finsler metric. If the transformation group of a uniform space is effective, then an invariant Finsler metric (in the corresponding support element space) exists if and only if the linear isotropy group is an exact representation of the stationary subgroup of points of the space. Concrete classes of uniform spaces in which invariant Finsler metrics exist are to be found in Jonušauskas [61-63].

At the present time support element spaces are being investigated by B. L. Laptev and his students. Great attention is being paid to the question of Lie differentiation in these spaces [B. L. Laptev, "Lie differentiation," in: Progress in Mathematics, Vol. 6, Plenum Press, New York (1970), pp. 229-269].

LITERATURE CITED

1. M. V. Aussem (Vasil'eva), "Metric spaces of n dimensions based on the concept of the area of m-dimensional surfaces," Uch. Zap. Mosk. Gor. Ped. Inst., Vol. 35, pp. 77-94 (1955).
2. V. I. Bliznikas, "On the theory of curves of a metric line-element space," Dokl. Akad. Nauk SSSR, 127(1): 9-12 (1959).
3. V. I. Bliznikas, "Congruence of centroidal geodesic curves of a metric line-element space," Dokl. Akad. Nauk SSSR, 132(4): 735-738 (1960).
4. V. I. Bliznikas, "On the differential geometry of metric line-element spaces," Uch. Zap. Vil'nyussk. Gos. Ped. Inst., Vol. 10, pp. 11-29 (1960).
5. V. I. Bliznikas, "On the differential geometry of bilinear metric line-element spaces," Vilniaus Univ. Mokslo Darbai, Matem., Fiz., 33(9): 97-106 (1960).

6. V. I. Bliznikas, "Certain geometric objects of a metric line-element space," Liet. Matem. Rinkinys, 1(1-2): 15-23 (1961).
7. V. I. Bliznikas, "Certain aspects of the differential geometry of bilinear metric line-element spaces," Liet. Matem. Rinkinys, 1(1-2): 372-373 (1961).
8. V. I. Bliznikas, "Euclidean connection of Cartan type in a metric line-element space," Liet. Matem. Rinkinys, 2(2): 33-37 (1963).
9. V. I. Bliznikas, "Certain varieties of support elements," Liet. Matem. Rinkinys, 3(2): 221-222 (1963).
10. V. I. Bliznikas, "The complete object of a central-projective connection and the torsion-curvature object of a space of central copunctors," Liet. Matem. Rinkinys, 4(4): 457-475 (1964).
11. V. I. Bliznikas, "Affine connection in a support element space," in: Reports Third Siberian Conf. Math. Mech., 1964, Tomsk Univ., Tomsk (1964), pp. 181-183.
12. V. I. Bliznikas, "On the theory of curvature of a support element space," Liet. Matem. Rinkinys, 5(1): 9-24 (1965).
13. V. I. Bliznikas, "Symmetric spaces of central copunctors," Liet. Matem. Rinkinys, 5(3): 381-389 (1965).
14. V. I. Bliznikas, "Nonholonomic Lie differentiation and linear connections in a support element space," Liet. Matem. Rinkinys, 6(2): 141-208 (1966).
15. V. I. Bliznikas, "The geometry of normal systems of ordinary higher-order differential equations," Liet. Matem. Rinkinys, 7(2): 51-68 (1967).
16. V. I. Bliznikas, "The geometry of systems of second-order partial differential equations," Liet. Matem. Rinkinys, 7(2): 69-84 (1967).
17. V. V. Vagner, "A two-dimensional space with a cubic metric," Uch. Zap. Saratovsk. Gos. Univ., Ser. Fiz.-Mat., Vol. 1(XIV), Issue 1, pp. 29-34 (1938).
18. V. V. Vagner, "Generalized Berwald spaces," Dokl. Akad. Nauk SSSR, Vol. 39, pp. 3-5 (1943).
19. V. V. Vagner, "Two-dimensional Finsler spaces with finite continuous holonomy groups," Dokl. Akad. Nauk SSSR, Vol. 39, pp. 99-102 (1943).
20. V. V. Vagner, "The geometry of a space with an areal metric and its application to the calculus of variations," Mat. Sb., Vol. 19, pp. 341-404 (1946).
21. V. V. Vagner, "The geometry of an n-dimensional space with an m-dimensional Riemann metric and its application to the calculus of variations," Mat. Sb., Vol. 20, pp. 3-25 (1947).
22. V. V. Vagner, "The geometrical interpretation of extremal surfaces in the Lagrange problem for multiple integrals," Dokl. Akad. Nauk SSSR, Vol. 55, pp. 91-94 (1947).
23. V. V. Vagner, "Field theory of local hyperstrips in X_n and its application to the mechanics of a system with nonlinear nonholonomic connections," Dokl. Akad. Nauk SSSR, 66(6): 1033-1036 (1949).
24. V. V. Vagner, "Finsler geometry as a field theory of local hypersurfaces in X_n," in: Proc. Seminar Vector Tensor Analysis, Issue 7, pp. 65-166 (1949).
25. V. V. Vagner, "The geometry of a space with a hyperareal metric as a field theory of local hypersurfaces in a composite variety," in: Proc. Seminar Vector Tensor Analysis, Issue 8, pp. 144-196 (1950).

26. V. V. Vagner, "The concept of an indicatrix in the theory of partial differential equations," Usp. Mat. Nauk, 18(2): 188-189 (1947).
27. V. V. Vagner, "The concept of an indicatrix in the theory of differential equations," Dokl. Akad. Nauk SSSR, Vol. 57, pp. 219-222 (1947).
28. V. V. Vagner, "Geometric theory of the simplest n-dimensional singular problem of the calculus of variations," Mat. Sb., Vol. 21, pp. 321-364 (1947).
29. V. V. Vagner, "The geometry of generalized Cartan spaces and the theory of geometric differential objects," Dokl. Akad. Nauk SSSR, Vol. 77, pp. 777-780 (1951).
30. V. V. Vagner, "Differential-geometric methods in the calculus of variations," Uch. Zap. Kazansk. Gos. Univ., 115(10): 4-7 (1955).
31. V. V. Vagner, "Field theory of local surfaces," in: Proc. Third All-Union Math. Congr., Vol. 2, Akad. Nauk SSSR, Moscow (1956), pp. 57-60.
32. V. V. Vagner, "The calculus of variations as a field theory of central semicones," in: Scientific Year Book of Saratov Univ., Mech.-Math. Faculty, 1955, Saratov (1959), pp. 27-34.
33. A. M. Vasil'ev, "A system of three first-order partial differential equations with three unknown functions and two independent variables (local theory)," Mat. Sb., 70(4): 457-480 (1966).
34. A. M. Vasil'ev, "Families of line elements enveloped by completely geodesic families," Izv. Vysshikh. Uchebn. Zavedenii, Matematika, No. 3, pp. 28-35 (1964).
35. M. V. Vasil'eva, "The geometry of an integral," Mat. Sb., 36(1): 57-92 (1955).
36. M. V. Vasil'eva, "Geometric characteristics of certain invariants of Finsler geometry," in: Proc. Third All-Union Math. Congr., Vol. 2, Akad. Nauk SSSR, Moscow (1956), p. 139.
37. M. V. Vasil'eva, "Invariant description of the Cartan geometry of an integral," Uch. Zap. Mosk. Gos. Ped. Inst. im. V. I. Lenina, No. 208, pp. 76-85 (1963).
38. M. V. Vasil'eva, "Finsler geometry in an invariant setting," Uch. Zap. Mosk. Gos. Ped. Inst., No. 243, pp. 38-54 (1965).
39. M. V. Vasil'eva, "Invariant description of certain Finsler geometries," Uch. Zap. Mosk. Gos. Ped. Inst., No. 243, pp. 55-68 (1965).
40. Hsien-chu Wei, "Veblen identities in Finsler spaces and in generalized Finsler spaces," Acta Sci. Nat. Univ. Amoiensis, 12(2): 23-31 (1965).
41. V. A. Gaukhman, "The geometry of an ordinary second-order differential equation relative to a conformal group of transformations of two variables," Dokl. Akad. Nauk SSSR, 140(1): 15-18 (1961).
42. Chao-hao Ku, "Embedding of a Finsler space in a Minkowski space," Acta Math. Sinica, 6(2): 215-232 (1956).
43. Chao-hao Ku, "Connection pairs and integral varieties of a system of second-order partial differential equations," Acta Math. Sinica, 6(3): 426-432 (1956).
44. Chao-hao Ku, "Connection pairs and integral varieties of a system of second-order partial differential equations. I, II," Acta Math. Sinica, 6(2): 153-162, 163-169 (1956).
45. Chao-hao Ku, "Embedding of Finsler varieties in a Minkowski space," Acta Math. Sinica, 8(2): 272-275 (1958).

46. Chao-hao Ku and Buchin Su, "First and second variations of a multiple integral in a space with a multiple areal metric," Acta Math. Sinica, 2(4): 231-245 (1953).

47. L. E. Evtushik, "On the geometry of a double integral," Mat. Sb., 37(1): 197-208 (1955).

48. L. E. Evtushik, "Geometry of the integral $\int F(x^{\alpha}, x^{n}, x^{n}_{\alpha}, x^{n}_{\alpha\beta})dx^{1}\dots dx^{n-1}$]," Nauchn. Dokl. Vysshei Shkoly, Fiz.-Mat. Nauk, No. 6, pp. 114-118 (1958).

49. L. E. Evtushik, "Lie derivative and differential equations of the field of a geometric object," Dokl. Akad. Nauk, SSSR, 132(5): 998-1001 (1960).

50. Yu. I. Ermakov, "A three-dimensional space with a cubic semimetric," Dokl. Akad. Nauk SSSR, 118(6): 1070-1073 (1958).

51. Yu. I. Ermakov, "Spaces X_n with an algebraic metric and a semimetric," Dokl. Akad. Nauk SSSR, 128(3): 460-463 (1959).

52. G. I. Zhotikov, "On the field theory of local conic surfaces in a first-order tangent composite variety $E_n(X_n)$, I," Izv. Vysshikh. Uchebn. Zavedenii, Matematika, No. 3, pp. 53-64 (1959).

53. G. I. Zhotikov, "On the field theory of local surfaces in a first-order tangent composite variety $E_n(X_n)$," Izv. Vysshikh. Uchebn. Zavedenii, Matematika, No. 2, pp. 69-79 (1959).

54. G. I. Zhotikov, "On the field theory of local conic surfaces in a first-order tangent composite variety $E_n(X_n)$. II," Izv. Vysshikh. Uchebn. Zavedenii, Matematika, No. 4, pp. 64-69 (1959).

55. G. I. Zhotikov, "On the field theory of local surfaces in a first-order tangent composite variety $E_n(X_n)$," in: Proc. Seminar Vector Tensor Analysis, Issue 11, Moscow University (1961), pp. 189-218.

56. G. I. Zhotikov, "Differential singular Finsler metric defined in X_n by the field of local singular hypersurfaces of singularity class $n - m - 1$," Uch. Zap. Bashkirsk. Univ., No. 20, pp. 32-45 (1965).

57. V. L. Zaguskii, "One form of Finsler space and motions in Minkowski space," Nauchn. Dokl. Vysshei Shkoly, Fiz.-Mat. Nauk, No. 3, pp. 50-52 (1958).

58. V. L. Zaguskii, "Certain aspects of Finsler geometry," Uch. Zap. Yaroslavsk. Gos. Ped. Inst., No. 34, pp. 83-110 (1960).

59. V. L. Izrailevich, "Invariant characteristics of hypercones in the space $W_{(2,n)}$," Tr. Saratovsk. Inst. Mekhaniz. S. Kh., No. 26, pp. 211-217 (1963).

60. V. L Izrailevich, "Theory of cones in the space $W_{(2,n)}$," in: Works of Young Scientists at Saratov Univ., Math. Sec., Saratov (1964), pp. 37-51.

61. A. Jonušauskas, "The existence of invariant Finsler metrics in uniform spaces," Liet. Matem. Rinkinys, 5(1): 45-55 (1965).

62. A. Jonušauskas, "The existence of invariant Finsler metrics in uniform spaces with a linear isotropy group of tensory type," Liet. Matem. Rinkinys, 6(1): 51-57 (1966).

63. A. Jonušauskas, "The existence of invariant Finsler metrics in certain uniform spaces," Liet. Matem. Rinkinys, 6(4): 621-622 (1966).

64. N. I. Kabanov, "Geometric theory of Caratheodory transformations in the Lagrange problem," in: Proc. Seminar Vector Tensor Analysis, Issue 11, Moscow University (1961), pp. 219-240.

65. N. I. Kabanov, "A Cartan space defined to within Caratheodory transformations," Uch. Zap. Balashovsk. Gos. Ped. Inst., Vol. 3, pp. 47-77 (1958).
66. N. I. Kabanov, "A singular Finsler space defined to within Caratheodory transformations," Sibirsk. Mat. Zh., 2(5):655-671 (1961).
67. N. I. Kabanov, "On the geometric theory of the simplest singular variational problem for an (n − 1)-fold integral," Dokl. Akad. Nauk SSSR, 140(1):37-40 (1961).
68. N. I. Kabanov, "On the geometric theory of the simplest singular variational problem for an (n − 1)-fold integral," in: Proc. Seminar Vector Tensor Analysis, Issue 12, Moscow University (1963), pp. 239-268.
69. N. I. Kabanov, "Certain aspects of the geometric theory of Caratheodory transformations in the calculus of variations," Liet. Matem. Rinkinys, 3(2):225 (1963).
70. F. I. Kagan, "Two-dimensional Finsler spaces admitting of a singular imbedding in a three-dimensional affine space with a vector metric," Izv. Vysshikh. Uchebn. Zavedenii, Matematika, No. 1, pp. 46-55 (1964).
71. F. I. Kagan, "An infinitesimal S-extension operation relative to a uniform S-distribution given on X_n," Volzhsk. Mat. Sb., No. 4, pp. 103-108 (1966).
72. S. A. Kaganov, "Geometry of a space with a singular hyperareal metric," Mat. Sb., 42(4):497-512 (1957).
73. S. A. Kaganov, "Linear connections in composite varieties defined by a regular hyperareal metric given in X_n," Sb. Tr. Ufimsk. Neft. Inst., No. 2, pp. 227-286 (1958).
74. A. T. Kondrat'ev, "Uniform line-element spaces of affine connection $A_3(x, \dot{x})$ of high mobility," Volzhsk. Mat. Sb., No. 4, pp. 82-92 (1966).
75. A. T. Kondrat'ev, "Line-element spaces of affine connection $A_3(x, \dot{x})$ admitting of an affine motion group G_r ($r \leq 3$)," Volzhsk. Mat. Sb., No. 5, pp. 152-157 (1966).
76. V. K. Kropina, "Projective Finsler spaces with a metric of a certain special form," Nauchn. Dokl. Vysshei. Shkoly, Fiz.-Mat. Nauk, No. 2, pp. 38-42 (1959).
77. V. K. Kropina, "Projective Finsler spaces," Uch. Zap. Arkhang. Gos. Ped. Inst., No. 4, pp. 111-118 (1959).
78. V. K. Kropina, "The introduction of absolute differentiation in Finsler space," Uch. Zap. Yaroslavsk. Gos. Ped. Inst., No. 34, pp. 113-123 (1960).
79. V. K. Kropina, "Projective two-dimensional Finsler spaces with a special metric," in: Proc. Seminar Vector Tensor Analysis, Issue 11, Moscow University (1961), pp. 277-292.
80. G. M. Kuz'mina, "The geometry of a system of two partial differential equations," Uch. Zap. Mosk. Gos. Ped. Inst., No. 243, pp. 99-108 (1965).
81. B. L. Laptev, "Covariant integration in a Finsler space of two and three dimensions," Izv. Fiz.-Mat. Obshch., Kazan', 3(9):61-76 (1937).
82. B. L. Laptev, "Lie derivative for objects which are functions of the point and the direction," Izv. Fiz.-Mat. Obshch., Kazan', 3(10):3-38 (1938).
83. B. L. Laptev, "Invariant form of the second variation obtained by Lie differentiation in Finsler space," Izv. Fiz.-Mat. Obshch., Kazan', 3(12):3-8 (1940).

84. B. L. Laptev, "Differential invariants of an affinely connected space of tensor support elements," Uch. Zap. Kazansk. Univ., 116(1):10-14 (1956).

85. B. L. Laptev, "Invariants of a space of tensor support elements," Uch. Zap. Kazansk. Univ., 115(10):12 (1955).

86. B. L. Laptev, "Covariant differential and theory of differential invariants in a space of tensor support elements," Uch. Zap. Kazansk. Univ., 118(4):75-147 (1958).

87. B. L. Laptev, "Application of Lie differentiation to the search for the geodesic displacement in a line-element space," Izv. Vysshikh. Uchebn. Zavedenii, Matematika, No. 2, pp. 173-181 (1958).

88. B. L. Laptev, "Lie derivative in a support element space," in: Proc. Seminar Vector Tensor Analysis, Issue 10, Moscow University, pp. 227-248 (1956).

89. B. L. Laptev, "Lie derivative of geometric objects in a support element space," Uch. Zap. Kazansk. Univ., 117(2):16-18 (1957).

90. B. L. Laptev, "Lie derivative of geometric objects in a support element space," in: Proc. Third All-Union Math. Congr., Vol. 1, Akad. Nauk SSSR, Moscow (1956), p. 157.

91. B. L. Laptev, "Support element space," in: Proc. Fourth All-Union Math. Congr., 1961, Vol. 2, "Nauka," Leningrad (1964), pp. 221-226.

92. G. F. Laptev, "Differential geometry of the immersions of varieties. Group-theoretical method of differential-geometric investigations," Tr. Mosk. Mat. Obshch., No. 2, pp. 275-382 (1953).

93. G. F. Laptev, "Group-theoretical method of differential-geometric investigations," in: Proc. Third All-Union Math. Congr., Vol. 2, Akad. Nauk SSSR, Moscow (1956), pp. 60-62.

94. G. F. Laptev, "Group-theoretical method of differential-geometric investigations," in: Proc. Third All-Union Math. Congr., Vol. 3, Akad. Nauk SSSR, Moscow (1958), pp. 409-418.

95. G. F. Laptev, "Varieties immersed in generalized spaces," in: Proc. Fourth All-Union Math. Congr., 1961, Vol. 2, "Nauka," Leningrad (1964), pp. 226-233.

96. G. F. Laptev, "Geometry of differential equations," in: First All-Union Geometric Conf., Kiev (1962), pp. 6-7.

97. A. E. Liber, "Two-dimensional spaces with algebraic metric," in: Proc. Seminar Vector Tensor Analysis, Issue 9, pp. 319-350 (1952).

98. A. M. Lopshits, "Parallel displacement in spaces with nonquadratic metric," in: Proc. All-Russian Math. Congr., (1927), pp. 241-242.

99. M. V. Losik, "Certain class of Kawaguchi spaces," Dokl. Akad. Nauk SSSR, 134(6):1299-1302 (1960).

100. M. V. Losik, "Geometric interpretation of certain conditions in an ordinary variational problem with higher derivatives," Sibirsk. Mat. Zh., 2(4):556-566 (1961).

101. M. V. Losik, "A Klein space as a Kawaguchi space," Dokl. Akad. Nauk SSSR, 139(6):1299-1301 (1961).

102. M. V. Losik, "Kawaguchi spaces connected with Klein spaces," in: Proc. Seminar Vector Tensor Analysis, Issue 12, Moscow University (1963), pp. 213-237.

103. M. V. Losik, "Connection of Klein spaces with spaces with higher-order areal metric," in: Works of Young Scientists at Saratov Univ., Math. Sec., Saratov (1964), pp. 55-59.
104. S. Maziliauskaitė, "Torsion and curvature tensors of the space of central punctors," Liet. Matem. Rinkinys, 5(3): 427-433 (1965).
105. I. H. Medviedevaitė, "Certain aspects of the geometry of a metric space of hyperflat elements," Liet. Matem. Rinkinys, 6(4): 533-539 (1966).
106. P. K Rashevskii, "Metric duality in a two-dimensional Finsler geometry, in particular, on an arbitrary surface," Dokl. Akad. Nauk SSSR, Vol. 3, pp. 147-150 (1935).
107. P. K. Rashevskii, "Polymetric geometry," in: Proc. Seminar Vector Tensor Analysis, Issue 5 (1941), pp. 21-147.
108. N. F. Rzhekhina, "On the field theory of local curves in X_n," Dokl. Akad. Nauk SSSR, Vol. 72, pp. 461-464 (1950).
109. N. F. Rzhekhina, "Field theory of local hypertorses in X_n," in: Proc. Seminar Vector Tensor Analysis, Issue 9 (1952), pp. 411-430.
110. N. F. Rzhekhina, "Curves and surfaces in higher-order tangent spaces," in: Scientific Year Book of Saratov Univ., Mech.-Math. Faculty, 1955, Saratov (1959), pp. 37-38.
111. Ting-chia Hsing, "Symmetry properties in certain Finsler spaces," Acta Math. Sinica, 9(2): 191-198 (1959).
112. N. V. Stepanov, "Geometry of two ordinary second-order differential equations," Dokl. Akad. Nauk SSSR, 140(1): 62-65 (1961).
113. N. V. Stepanov, "Classification of pairs of differential equations," Uch. Zap. Velikoluksk. Gos. Ped. Inst., No. 19, pp. 115-116 (1962).
114. Buchin Su, "Volume geometry of an affinely connected space with an areal metric," Acta Math. Sinica, 2(4): 246-257 (1953).
115. Buchin Su, "Isomorphic transformations of minimal hypersurfaces in Finsler space," Acta Math. Sinica, 5(4): 471-488 (1955).
116. Buchin Su, "What's new in the geometry of generalized spaces," Kesyue Tunabao, No. 8, pp. 29-32 (1955).
117. Buchin Su, "What's new in the geometry of generalized spaces," Shusyue Tszin'chzhan', 1(4): 615-637 (1955).
118. Buchin Su, "Koschmieder invariant and the associated differential equation of a minimal hypersurface in a regular Cartan space," Acta Math. Sinica, 6(3): 374-388 (1956).
119. Buchin Su, "Certain affinely connected spaces with an areal metric," Acta Math. Sinica, 7(2): 285-294 (1957).
120. Pên-wang Sun, "The equivalence problem for the integral $\int F(x, y, y', \ldots, y^{(n)}dx$," Acta Math. Sinica, 4(2): 223-224 (1954).
121. A. P. Urbonas, "Connections in a support element space," Liet. Matem. Rinkinys, 6(2): 279-290 (1966).
122. É. I. Khmelevskii, "Application of the Lie derivative to the search for the extremals in spaces of hyperflat elements," in: Collection of Aspirants' Papers at Kazan Univ., Math., Mech., Fiz., Kazan (1964), pp. 94-96.

123. É. I. Khmelevskii, "Invariant form of the second variation of an (n − 1)-fold integral, obtained by Lie differentiation in Cartan space," in: Collection of Aspirants' Papers at Kazan Univ., Math., Mech., Fiz., Kazan (1964), pp. 61-68.
124. I-P'ei Ch'en, "A special deformation of a curve in Finsler space," Acta Sci. Nat. Univ. Amoiensis, 12(2): 123-127 (1965).
125. Z. N. Chetyrkina, "Homotheties and motions in two-dimensional Finsler spaces," Volzhsk. Mat. Sb., No. 5, pp. 366-372 (1966).
126. B. N. Shapukov, "Extremal displacement of a minimal hypersurface in Riemann and Finsler spaces," Izv. Vysshikh. Uchebn. Zavedenii, Matematika, No. 5, pp. 112-116 (1961).
127. B. N. Shapukov, "On the theory of bilinear metric spaces," Uch. Zap. Kazansk. Univ., 123(1): 172-179 (1963).
128. B. N. Shapukov, "Surfaces and singular paths in a bilinear metric line-element space," Uch. Zap. Kazansk. Univ., 123(1): 180-195 (1963).
129. B. N. Shapukov, "On the geometry of bilinear metric line-element space," in: Collection of Aspirants' Papers at Kazan Univ., Exact Sciences, Kazan (1962), pp. 158-169.
130. É. M. Shvartsburd, "Structure equations of a system of four first-order partial differential equations," Uch. Zap. Mosk. Gos. Ped. Inst., No. 243, pp. 192-199 (1965).
131. J. Šinkūnas, "The space of support lineals," Liet. Matem. Rinkinys, 6(3): 449-455 (1966).
132. J. Šinkūnas, "Connections in spaces of special support elements," Liet. Matem. Rinkinys, 6(4): 622 (1966).
133. A. P. Shirokov, "Gonometric system in Finsler geometry," in: Proc. Seminar Vector Tensor Analysis, Issue 8 (1950), pp. 414-424.
134. D. M. Yablokov, "Euclidean connections in the space of pairs of line elements," in: Proc. Seminar Geom. Dept., Issue 2, Uch. Zap. Kazansk. Univ., Vol. 126, Book 1, pp. 90-102 (1966).
135. D. M. Yablokov, "Some applications of Lie differentiation in the space of line element pairs," in: Proc. Seminar Geom. Dept., Issue 2, Uch. Zap. Kazansk. Univ., Vol. 126, Book 1, pp. 103-116 (1966).
136. D. M. Yablokov, "Covariant differentiation and affine connection in a space of line element pairs," in: Collection of Aspirants' Papers at Kazan University, Mathematics, Kazan (1966), pp. 80-107.
137. H. Akbar-Zadeh, "Sur la réductibilité d'une variétés finslériennes Compt. Rend. Acad. Sci., 234(16): 945-947 (1954).
138. H. Akbar-Zadeh, "Sur les isométries infinitésimales d'une variété finslériennn," Compt. Rend. Acad. Sci., 242(5):608-610 (1956).
139. H. Akbar-Zadeh, "Sur une connexion euclidienne d'espace d'éléments linéaires," Compt. Rend. Acad. Sci., 245(1): 26-28 (1957).
140. H. Akbar-Zadeh, "Sur les espaces de Finsler isotropes," Compt. Rend. Acad. Sci., 252(14): 2061-2063 (1961).
141. H. Akbar-Zadeh, "Transformations infinitésimales conformes des variétés finslériennes compactes," Compt. Rend. Acad. Sci., 252(19): 2807-2809 (1961).

142. H. Akbar-Zadeh, "Les espaces de Finsler et certaines de leurs généralisations," Ann. Sci. Ecole Norn. Supér., 80(1): 1-79 (1963).

143. H. Akbar-Zadeh, "Une généralisation de la géometrie finslérienne," Cahiers Semin. Topol. et Géom. Différent. Ch. Ehresmann, Fac. Sci. Paris, Vol. 6, pp. 1-9 (1964).

144 H. Akbar-Zadeh, "Sur les automorphismes de certaines structures presque cosymplectiques," Can. Math. Bull., 8(1): 39-57 (1965).

145. H. Akbar-Zadeh, "Sur les homothétiew infinitésimales des variétés finslériennes," Compt. Rend. Acad. Sci., AB262(19): A1058-A1060 (1966).

146. H. Akbar-Zadeh and E. Bonan, "Structure presque kahlérienne naturelle sur le fibré tengent à une variété finslérienne," Compt. Rend. Acad. Sci., 258(23): 5581-5582 (1964).

147. F. Alardin, "L'autoparallélisme des courbes extrémales dans les espaces métriques fondés sur la notion d'aire," J. Math. Pures et Appl., 9(27): 255-336 (1948).

148. N. Aronszajn, "Sur quelques problèmes concernant les espaces de Minkowski et les espaces vectorièls généraux," Atti Accad. Naz. Lincei, Rend., 6(26): 374-376 (1937).

149. L. Auslander, "On curvature in Finsler Geometry," Trans. Am. Math. Soc., 79(2): 378-388 (1955).

150. O. Bark, "On the projective connection space and general projective geometry of paths," Kyungpook Math. J., Vol. 1, pp. 1-12 (1958).

151. W. Barthel, "Über eine Parallelverschiebung mit Längeninvarianz in lokal-Minkowskischen Räumen. I, II," Arch. Math., 4(4): 346-354 (1953).

152. W. Barthel, "Über Minkowskische und Finslersche Geometrie," in: Convegno Internationale di Geometria Differenziele, Italia, 1953, Ed. Gremonese, Rome (1954), pp. 71-76.

153. W. Barthel, "Zur Flächentheorie in Finslerschen Räumen," in: Proc. Internat. Congr. Math., 1954, Amsterdam (1954), pp. 194-196.

154. W. Barthel., "Über die Minimalflächen in gefäserten Finslerräumen," Ann. Mat. Pura ed Appl., Vol. 36, pp. 159-190 (1954).

155. W. Barthel, "Über das Verhältnis der Vektoräbertragung zu den Variationsproblemen in Cartanschen Raümen," Rend. Circolo Mat. Palarmo, 3(2): 270-3(2): 270-281 (1954).

156. W. Barthel, "Zur Flächentheorie in Finslerschen Räumen," in: Proc. Internat. Congr. Math., Amsterdam (1954), p. 2; Math. Z., 62(1): 23-26 (1955).

157. W. Barthel, "Über metrische Differentialgeometrie, begründet auf dem Begriff eines p-dimensionalen Areals," Math. Ann., 137(1): 42-63 (1959).

158. W. Barthel, "Zur Minkowski-Geometrie, begründet auf dem Flächeninhaltsbegriff," Monatsh. Math., 63(4): 217-343 (1959).

159. R. Behari and N. Prakash, "A study of normal curvature of a vector field in Minkowskian Finsler space," J. Indian Math. Soc., 24(3-4): 443-456 [1960 (1961)].

160. L. Berwald, "Untersuchung der Krümmung allgemeiner metrischer Räume auf Grund des in ihnen herrschenden Parallelismus," Math. Z., 25: 40-73 (1926).

161. L. Berwald, "On the projective geometry of paths," Ann. Math., 37:879-898 (1936).

162. L. Berwald, "Über Finslersche und Cartansche Geometrie. I. Geometriesche Erklärungen der Krümmung und des Hauptskalers im zweidimensionalen Finslerschen Raum," Mathematica, Timisoara, 17:34-55 (1941).

163. L. Berwald, "Über Finslersche und Cartansche Geometrie. II. Invarianten bei der Variation vielfachen Integrale und Parallelhyperflächen in Cartanschen Räumen," Compositio Math., 7:141-176 (1939).

164. L. Berwald, "On the Finsler and Cartan geometry. III. Two-dimensional Finsler spaces with rectilinear extremals," Ann. Math., Ser. 2, 42:84-112 (1941).

165. L. Berwald, "Über Beziehungen zwischen den Theorien der Parallelübertragung in Finslerschen Räumen," Nederl. Akad. Weten. Proc., 49:642-647 (1946); Indagationes Math., 8:401-406 (1946).

166. L. Berwald, "Über Finslerschen und Cartansche Geometrie. IV. Projektivkrümmung allgemeiner affiner Räume und Finslerscher Räume skalaler Krümmung," Ann. Math., Ser. 2, 48:755-781 (1949).

167. W. Bettingen, "Zum Satz von Gauss-Bonnet in der dreidimensionalen metrischen Differentialgeometrie," Rend. Circolo Mat. Palermo, 9(3):347-359 (1960).

168. W. Blaschke, "Integralgeometrie. XI. Zur Variationsrechnung," Abhandl. Math. Sem. Univ. Hamburg, 11:359-366 (1936).

169. G. A. Bliss, "A generalization of the notion of angle," Trans. Am. Math. Soc., 7:184-196 (1906).

170. E. Bompiani, "Enti geometrici definiti da sistemi differenziali," Atti Accad. Naz. Lincei. Rend. Cl. Sci. Fis. Mat. Nat., 8(1):187-194 (1946).

171. E. Bortolotti, "Geometry of a system of partial differential equations," Tensor, 4:25-34 (1941).

172. F. Brickell, "On the existence of metric differential geometries based on the notion of area," Proc. Cambridge Phil. Soc., 46:67-72 (1950).

173. F. Brickell, "On areal spaces," Tensor, 13:19-30 (1963).

174. G. M. Brown, "Metric differential geometry," Doctoral Dissertation, University of Toronto, 1965; Dissertation Abstr., B27(4):1211 (1966).

175. I. Bucur, "Asupra unei proprietati globale a linlilor geodezice ale unui spatiu," Commun. Acad. RPR, 5(6):965-968 (1955).

176. H. Busemann, "On normal coordinates in Finsler spaces," Math. Ann., 129(5):417-423 (1955).

177. H. Busemann, The Geometry of Geodesic, Academic Press, New York (1955).

178. E. Cartan, "Sur les variétés à connexion projective," Bull. Soc. Math. France, 52:205-241 (1924).

179. E. Cartan, "Les espaces de Finsler. (Actual. Scient. et Industr., No. 79), Paris (1934).

180. E. Cartan, "Las espaces métriques fondés sur la notion d'aire," Paris (1933) (Exposes de géometrie, 72).

181. S. Chern, "Sur la géometrie d'une équation différentielle du troisiéme ordre," Compt. Rend. Acad. Sci., 204:1227-1229 (1937).

182. S. Chern, "On the Euclidean connections in a Finsler space," Proc. Nat. Acad. Sci. USA, 29:33-37 (1943).

183. S. Chern, "Local equivalence and Euclidean connection in Finsler spaces," Sci. Rept. Nat. Tsing. Hua Univ., Ser. A, 5:95-121 (1948).

184. H. V. Craig, "On a generalized tangent vector," Am. J. Math, 57:457-462 (1935).

185. H. V. Craig, Vector and Tensor Analysis, McGraw-Hill, New York (1943).

186. E. T. Davies, "On the use of osculating spaces," Tensor, 14:86-98 (1963).

187. E. T. Davies, "The geometry of a multiple integral," J. London Math. Soc., 20:163-170 (1945).

188. E. T. Davies, "Areal spaces," Ann. Mat. Pura ed Appl., 55:63-76 (1961).

189. R. Debever, "Sur une structure infinitésimale reguliére associée aux intégrales d'hypersurfaces du calcul de variations," in: Convegno Internationale di Geometria Differenziale, Italy, 1953, Ed. Gremonese, Rome (1954), pp. 214-221.

190. R. Debever, "Sur une classe d'espaces," Thesis, Brussells (1947).

191. A. Deike, "Über die Darstellung von Finsler-Räumen durch nichtholonome Mannigfaltigkeiten in Riemannschen Räumen," Arch. Math., 4(3):234-238 (1953).

192. A. Deike, "Finsler spaces as nonholonomic subspaces of Riemannian spaces," J. London Math. Soc., 30(1):53-58 (1955).

193. A. Deike, "Über die Finsler-Räume mit $A_i = 0$," Arch. Math., 4:45-51 (1953). (1953).

194. M. Dhawan, "Curvature properties of a subspace embedded in a Finsler space," Ganita, 16(1):25-36 (1965).

195. M. Dhawan and N. Prakash, "Generalizations of Gauss — Codazzi equations in a subspace embedded in a Finsler manifold," Tensor, 15(2):159-167 (1964).

196. J. Douglas, "The general geometry of paths," Ann. Math., 29:143-168 (1928).

197. J. Douglas, "Systems of K-dimensional manifolds in an n-dimensional space," Math. Ann., 105:707-733 (1931).

198. V. Dumitras, "Sur le groupe de stabilité d'un espace H_n," Bull. Math. Soc. Sci. Math. et Phys. RPR, 3(1):17-20 (1959).

199. L. P. Eisenhart, "Finsler spaces derived from Riemann spaces by contact transformations," Ann. Math., Ser. 2, 49:227-254 (1948).

200. H. A. Eliopoulos, "Methods of generalised metric geometry with applications to mathematical physics," Thesis, Toronto (1956), 112 pp.

201. H. A. Eliopoulos, "Sur la définition de la courbure totale d'une hypersurface plongée dans un espace de Finsler localement Minkowskien," Bull. Cl. Sci. Acad. Roy. Belg., 45(3):205-214 (1959).

202. H. A. Eliopoulos, "Multi-particle theory derived from the geometry of a locally Minkowskian Finsler space," Bull. Cl. Sci. Acad. Roy. Belg., 52(1): 69-75 (1966).

203. P. Finsler, "Über Kurven und Flächen in allgemeinen Räumen," Dissertation, Göttingen (1918).

204. P. Finsler, Über Kurven und Flächen in allgemeinen Räumen, Basel (1951).

205. P. Finsler, "Über die Krümmungen der Kurven und Flächen," in: Reale Accad. Ital., Fondazione Alessandro Volta, IX Convegno Volta, Rome (1940).
206. P. Finsler, "Über eine Verallgemeinerung der Satzes von Meusnier," Viertel. Naturforsch. Ges. Zürich, Supplement 85, pp. 155-164 (1940).
207. J. G. Freeman, "First and second variations of the length integral in a generalized metric space," Quart. J. Math., Oxford, Ser. 15, pp. 70-83 (1944).
208. J. G. Freeman, "Finsler-Riemann systems," Quart. J. Math., 7(26): 100-109 (1956).
209. J. G. Freeman, "Complete Finsler-Riemann systems," Quart. J. Math., 8(31): 161-171 (1957).
210. M. Fujinaka, "On Finsler spaces and dynamics with special reference to the equations of hunting," in: Proc. 3rd Jap. Nat. Congress Appl. Mechanics, Tokyo (1954), pp. 433-436.
211. P. Funk, "Eine Kennzeichnung der zwei-dimensionalen elliptischen Geometrie," Sitzber. Österr. Akad. Wiss. Math.-Naturwiss. Kl., Abt. 2, 172(9-10): 251-269 (1963).
212. O. Galvani, "Sur la réalisation des espaces de Finsler," Compt. Rend. Acad. Sci., Paris, 222: 1067-1069 (1946).
213. O. Galvani, "Les connexions finsleriennes de congruences de droites," Compt. Rend. Acad. Sci. Paris, 222: 1200-1202 (1946).
214. O. Galvani, "Sur l'immersion du plan de Finsler dans certains espaces de Riemann a trois dimensions," Compt. Rend. Acad. Sci., 223: 1088-1090 (1946).
215. O. Galvani, "La réalization des connexions euclidiennes d'éléments linéaires et des espaces de Finsler," Ann. Inst. Fourier, Grenoble, 2: 123-146 (1950).
216. O. Galvani, "La réalisation des espaces de Finsler," Compt. Rend. Congrès. Soc. Savantes Paris et des Departements, Grenoble (1952), pp. 57-60.
217. O. Galvani, "Réalisations euclidiennes des plans de Finsler," Ann. Inst. Fourier, 5: 421-454 [1953-1954 (1955)].
218. M. Gama, "On areal spaces of submetric class," Tensor, 16(3): 262-268 (1965).
219. M. Gama, "On areal spaces of the submetric class. II," Tensor, 16(3): 291-293 (1965).
220. M. Gama, "On areal spaces of the submetric class. III," Tensor, 17(1): 79-85 (1966).
221. M. Gama, "On areal spaces of the submetric class. IV," Tensor, 18(1): 49-53 (1967).
222. M. Gama, "Theory of subspaces in areal spaces of the submetric class," Tensor, 18(2): 168-180 (1967).
223. St. Gołab, "Einige Bemerkungen über Winkelmetrik in Finslerschen Räumen," Verhandl. Intern. Math. Kongresses Zürich, 11: 178-179 (1932).
224. St. Gołab, "Sur la representations conforme de l'espace euclidien," Compt. Rend. Acad. Sci. Paris, 196: 25-27 (1933).
225. St. Gołab, "Sur la representation conforme de deux espaces de Finsler," Compt. Rend. Acad. Sci., Paris, 196: 986-988 (1933).
226. St. Gołab, "Sur la mesure des aires dans les espaces de Finsler," Compt. Rend. Acad. Sci., Paris, 200: 197-199 (1935).

227. St. Gołab, "On Finsler's measurement of angle," Ann. Soc. Polon. Math., 24: 78-84 (1954).
228. H. H. Goldstine, "The calculus of variations in abstract spaces," Duke Math. J., 9: 811-822 (1942).
229. N. Grossman, "On real projective spaces as Finsler manifolds," Proc. Am. Math. Soc., 18(2): 325-326 (1967).
230. A. Haimovici, "Geometria unei ecutii Monge de tip particular," Studii Si Cercetări Stint., 5(1-2): 17-27 (1954).
231. M. Haimovici, "Formules fondamentales dans la theorie des hypersurfaces d'un espace de Finsler," Compt. Rend. Acad. Sci., Paris, 198: 426-427 (1934).
232. M. Haimovici, "Sur les espaces généraux qui se correspondent point par point avec conservation du parallélisme de Cartan," Compt. Rend. Acad. Sci., Paris, 198: 1105-1108 (1934).
233. M. Haimovici, "Sur quelques types de metriques de Finsler," Compt. Rend. Acad. Sci., Paris, 199: 1091-1093 (1934).
234. M. Haimovici, "Sur les espaces de Finsler à connexion affine," Compt. Rend. Acad. Sci., Paris, 204: 837-839 (1937).
235. M. Haimovici, "La parallélisme dans les espaces de Finsler et la differentiation invariante de M. Levi-Civita," Ann. Sci. Univ. Jassy, 24: 214-218 (1938).
236. M. Haimovici, "Sulle superficie totalemente geodetiche negli spazi di Finsler," Rend. Lincei, 27: 633-641 (1938).
237. M. Haimovici, "Variétés totalement extremales et variétés totalement géodésiques dans les espaces de Finsler," Ann. Sci. Univ. Jassy, 25: 559-644 (1939).
238. M. Hashiguchi, "On parallel displacements in Finsler spaces," J. Math. Soc. Japan, 10(4): 365-379 (1958).
239. H. Hashimoto, "On the geometry of a system of partial differential equations of third order," Tensor, 4: 55-59 (1941).
240. H. Hashimoto, "On the geometry of a system of partial differential equations of third order," J. Fac. Sci. Hokkaido Univ., Ser. Math., 8(14): 163-172 (1940).
241. E. Heil, "A relation between Finslerian and Hermitian metrics," Tensor, 16(1): 1-3 (1965).
242. E. Heil, "Eine Charakterisierung lokal-Minkowskischer Räume," Math. Ann., 167(1): 64-70 (1966).
243. H. Hiramatsu, "On affine collineations in a space of hyperplanes," Kumamoto J. Sci., 1: 1-7 (1952).
244. H. Hiramatsu, "On projective collineations in a space of hyperplanes," Tensor, New Series, 2: 1-14 (1952).
245. H. Hiramatsu, "On some properties of groups of homothetic transformations in Riemannian and Finslerian spaces," Tensor, 4: 28-29 (1954).
246. H. Hiramatsu, "Groups of homothetic transformations in a Finsler space," Tensor, 3(3): 131-143 (1954).
247. S. Hokari, "Winkeltreue Transformationen und Bewegungen im Finslerschen Räume," J. Fac. Sci. Hokkaido Univ., 5: 1-8 (1936).
248. S. Hokari, "Geometry in an n-dimensional space based on the idea of K-dimensional volume," Tensor, 4: 72-77 (1941).
249. S. Hokari, "On a geometrical treatment of a system of higher partial differential equations," Tensor, 5: 89-103 (1942).

250. E. Hölder, "Über die auf Extremalintegrale gegründeten metrischen Räume," Schr. Forschungsinst. Math., 1:178-193 (1957).

251. H. Hombu, "Konforme Invarianten im Finslerschen Raum," J. Fac. Sci. Hokkaido Univ., Ser. I, Math., 2:157-168 (1934).

252. H. Hombu, "Konforme Invarianten im Finslerschen Raum," J. Fac. Sci. Hokkaido Univ., Ser. I, Math., 4:51-66 (1935).

253. H. Hombu, "Die Krümmungstheorie im Finslerschen Raum," J. Fac. Sci. Hokkaido Univ., Ser. I., Math., 5:67-94 (1936).

254. J. I. Horváth, "Un model geometric pentru teoria unitaria a campurilor fizice. Studdii si cercetări fiz.," Acad. RPR, 4(1-2):109-111 (1953).

255. J. I. Horváth, "Contribution to Stephenson-Kilmister's unified theory of gravitation and electromagnetism," Nuovo Cimento, 4(3):571-576 (1956).

256. J. I. Horváth, "New geometrical methods of the theory of physical fields," Nuovo Cimento, 9, Suppl., No. 2:444-496 (1958).

257. J. I. Horváth and A. Moór, "Entwicklung einer Feldtheorie begründet auf einen allgemeinen metrischen Linienelementraum. I, II," Proc. Koninkl. Ned. Akad. Wetenschap., A58(4):421-430; A58(5):581-587 (1955).

258. J. I. Horváth, "A italános metricus vonalelementérre alapozott térelmélet," Magy. Tud. Akad. Mat. Fiz. Tud. Oszt. Közl., 6(1):53-72 (1956).

259. T. Hosokawa, "On the various linear displacements in the Berwald-Finsler's manifold," Sci. Rept. Tokyo, 19:37-51 (1930).

260. T. Hosokawa, "Finslerian wave geometry and Milne's world structure," J. Sci. Hiroshima Univ., A8:249-270 (1938).

261. Hou-Sung Hu, "A new geometry of a space of K-spreads," Sci. Rec., 3(3):107-111 (1959).

262. Hou-Sung Hu, "A Finslerian product of two Riemannian spaces," Sci. Rec., 3(10):446-448 (1959).

263. S. Ide, "On the connections in higher order spaces," Tensor, 4(3):135-140 (1955).

264. S. Ide, "On the theory of curves in an n-dimensional space with the metrics $s=\int \{A_i(x,x')x''^i+B(x,x')\}^{1/p}dt$," Tensor, 9:25-29 (1949).

265. S. Ide, "On the geometrical meanings of Wirtinger's connections based on Kawaguchi's," Tensor, 14:216-218 (1963).

266. T. Igarashi, "On Lie derivaties in areal spaces," Tensor, 18(2):205-211 (1967).

267. R. S. Ingarden, "Über die Einbettung eines Finslerschen Raumes in einen Minkowskischen Raum," Bull. Acad. Polon. Sci., Cl. 3, 2(7):305-308 (1954).

268. R. S. Ingarden, "Über die Einbettung eines Finslerschen Raumes in einen Minkowskischen Raum," Bull. Acad. Polon. Sci., Cl. 3, 2(7):309-311 (1954).

269. S. Ishihara and T. Fukami, "Groups of projective transformations in a space of K-spreads," Japan J. Math., 26:79-93 (1956).

270. C. I. Ispas, "Identités de type Ricci dans l'espace de Finsler," Comun. Acad. RPR, 2:13-18 (1952).

271. H. Iwamoto, "On the conformal theory of metric geometry of higher order," Tensor, 7:50-57 (1944).

272. H. Iwamoto, "La géométrie des espaces métriques fondés sur la notion d'aire. I," Proc. Japan. Acad., 21:119-123 (1945).

273. H. Iwamoto, "La géométrie des espaces métriques fondés sur la notion d'aire. II," Proc. Japan. Acad., 21: 223-226 (1945).
274. H. Iwamoto, "On geometries associated with multiple integrals," Jap. J. Math., 1(1): 74-91 (1948).
275. H. Iwamoto, "Über eine geometrische Theorie der mehrfachen Integrale," Jap. J. Math., 19: 479-512 (1948).
276. H. Iwamoto, "On the geometry in a space based on the notion of area. I, II," Tensor, 9: 7-12 (1949); 9: 13-17 (1949).
277. C. Kano, "Conformal geometry in an n-dimensional space with the arc length $s=\int\{(A_i(x, x')x''^i+B(x, x')\}^{1/p}dt$," Tensor, 5(3): 187-196 (1956).
278. C. Kano, "Conformal geometry in an n-dimensional space with the arc length $s=\int\{A_i(x, x')x''^i+B(x, x')\}^{1/p}dt$, II," Tensor, 10(3): 210-217 (1960).
279. S. Kashiwabara, "On Euclidean connections in a Finsler manifold," Tohoku Math. J., 10(1): 69-80 (1958).
280. Y. Katsurada, "On the theory of curves in a higher order space with some special metrics," Tensor, 7: 58-64 (1944).
281. R. N. Kaul, "Curvatures in Finsler space," Bull. Calcutta Math. Soc., 50(4): 189-192 (1958).
282. A. Kawaguchi, "Theory of connections in the generalized Finsler manifold," Proc. Imp. Acad. Tokyo, 7(6): 211-214 (1931).
283. A. Kawaguchi, "Theory of connections in generalized Finsler manifold. II," Proc. Imp. Acad. Tokyo, 8(8): 340-343 (1932).
284. A. Kawaguchi, "Theory of connections in the generalized Finsler manifold," Proc. Imp. Acad. Tokyo, 9(6): 347-350 (1933).
285. A. Kawaguchi, "Some intrinsic derivations in a generalized space," Proc. Imp. Acad. Tokyo, 12(6): 149-152 (1936).
286. A. Kawaguchi, "Certain identities in a generalized space," Proc. Imp. Acad. Tokyo, 12(6): 152-155 (1936).
287. A. Kawaguchi, Die Geometrie des Integrals $\int\{A_i x''^i+B\}^{1/p}dt.$," Proc. Imp. Acad. Tokyo, 12: 205-208 (1936).
288. A. Kawaguchi, "Ein metrischer Raum, der eine Verallgemeinerung des Finslerschen Raumes ist," Monatsh. Math. Phys., 43: 289-297 (1936).
289. A. Kawaguchi, "Theorie des Raumes mit dem Zusammenhang, der von abhänggig ist," Monatsh. Math. Phys., 44: 131-152 (1936).
290. A. Kawaguchi, "Theory of connections in a Kawaguchi space of order two," Proc. Imp. Acad. Tokyo, 13: 183-186 (1937).
291. A. Kawaguchi, "Theory of connections in a Kawaguchi space of higher order," Proc. Imp. Acad. Tokyo, 13: 237-240 (1937).
292. A. Kawaguchi, "Views on higher order geometry of connections. I, II, III," Tensor, 1: 13-18 (1938); 2: 39-45 (1939); 3: 68-70 (1940).
293. A. Kawaguchi, "Die Differentialgeometrie höherer Ordnung. I. Erweiterte Koordinatentransformationen und Extensoren," J. Fac. Sci. Hokkaido Univ., Ser. I., 9: 1-152 (1940).
294. A. Kawaguchi, "Die Differentialgeometrie höherer Ordnung. II. Über die n-dimensionalen Flächenelement abhängigem Zusammenhang," J. Fac. Sci. Hokkaido Imp. Univ., Ser. I, 9: 153-188 (1940).

295. A. Kawaguchi, "Die Differentialgeometrie höherer Ordnung. III. Erweiterte Parametertransformationen und P-Tensoren," J. Fac. Sci. Hokkaido Univ., 10: 77-156 (1941).
296. A. Kawaguchi, "Views on higher order geometry of connections. IV," Tensor, 4: 66-68 (1941).
297. A. Kawaguchi, "Determination of the fundamental tensor in a five-dimensional space based on two-dimensional area," Tensor, 6: 49-61 (1943).
298. A. Kawaguchi, "On various tensors appearing in the higher order geometry of connection," Tensor, 6: 1-26 (1943).
299. A. Kawaguchi, "On certain metric space of higher order," Tensor, 7: 73-77 (1944).
300. A. Kawaguchi, "On areal spaces. I," Tensor, 1: 14-45 (1950).
301. A. Kawaguchi, "On areal spaces. II," Tensor, 1: 67-88 (1951).
302. A. Kawaguchi, "On areal spaces. III," Tensor, 1: 89-101 (1951).
303. A. Kawaguchi, "Generalizzationi del calcolo tenzoriale e delle sue applicazioni," Atti Accad. Lincei. Rend. Cl. Sci., Fis., Mat. e Natur., 15(5): 255-261 (1953).
304. A. Kawaguchi, "Theory of areal spaces," Rend. Mat. e Applic., 12(3-4): 373-386 (1953).
305. A. Kawaguchi, "A remark to the theory of areal spaces," Nieuw Arc. Wiskunde, 2(2-3): 115-117 (1954).
306. A. Kawaguchi, "On the theory of nonlinear connections. II. Theory of nonlinear connections in a Finsler space," Tensor, 6(3): 165-199 (1956).
307. A. Kawaguchi, "Die Differentialgeometrie höherer Ordnung. IV. Erweiterung der verallgemeinerten Rheonomtransformation von Flächenelementen höherer Ordnung und R_k-Extensoren," Publ. Math., 7(1-4): 256-276 (1960).
308. A. Kawaguchi, "On functional form of integral invariants," Math. Notae, 18(1): 109-116 (1962).
309. A. Kawaguchi and H. Hombu, "Die Geometrie des Systems der partiellen Differentialgleichungen," J. Fac. Sci. Hokkaido Univ., Ser. I, 6(1): 21-62 (1937).
310. A. Kawaguchi and K. Tandai, "On areal spaces. IV," Tensor, 2: 47-58 (1952).
311. M. Kawaguchi, "An introduction to the theory of higher order spaces. I. The theory of Kawaguchi spaces," in: RAAG Mem. Unifying Study Basic Prob. Engng. and Phys. Sci. Means Geom., Vol. 3, Tokyo, Gakujutsu Bunken Fukyukai (1962), pp. 718-734.
312. M. Kawaguchi, "Une observation sur le calcul des calottes," Tensor, 14: 182-190 (1963).
313. S. Kawaguchi, "On some properties of projective curvature tensor in a special Kawaguchi space," Tensor, 13: 83-88 (1963).
314. S. Kawaguchi, "On a special Kawaguchi space of recurrent curvature," Tensor, 15(2): 145-158 (1964).
315. S. Kawaguchi and T. Nobuhara, "On extremal curves in a special Kawaguchi space," Tensor, 5(3): 197-200 (1956).
316. M. A. McKiernan, "Sufficiency of parameter invariance conditions in areal and higher order Kawaguchi spaces," Publ. Math., 13(1-4): 77-85 (1966).

317. Shigetaka Kikuchi, "Theory of Minkowski space and of nonlinear connections in a Finsler space," Tensor, 12(1): 47-60 (1962).
318. Shigetaka Kikuchi, "Some remarks on areal spaces of the submetric class," Tensor, 17(1): 44-48 (1966).
319. J. Klein, "Sur les trajectoires d'un système dynamique dans un espace finslérien ou variationnel généralisé," Compt. Rend. Acad. Sci., 238(22): 2144-2146 (1954).
320. M. S. Knebelmann, "Motions and collineations in general space," Proc. Nat. Acad. Sci. USA, 13: 607-611 (1927).
321. M. S. Knebelmann, "Collineations and motions in generalized spaces," Am. J. Math., 51: 527-564 (1929).
322. S. Kobayashi, "Groupe de transformations qui laissent invariante une connexion infinitésimale," Compt. Rend. Acad. Sci. Paris, 238: 644-645 (1954).
323. S. Kobayashi, "Le groupe des transformationes qui laissent invariant le parallélisme," Coll. de Topologie, Strasbourg (1954).
324. K. Kondo, "On the theoretical investigation based on abstract geometry of dynamical systems appearing in engineering," in: Proc. 3rd. Jap. Nat. Congress Appl. Mechanics, Tokyo (1954), pp. 425-432.
325. D. D. Kosambi, "Parallelism and path-space," Math. Z., 37: 608-618 (1933).
326. D. D. Kosambi, "Lie rings in path space," Proc. Nat. Acad. Sci. USA, 35: 389-394 (1949).
327. R. Kreter, "Zusammenhänge in Finslerschen Räumen," Wiss. Z. Humboldt Univ., Berlin, Math.-Naturwiss. Reihe, 6(4): 353-365 (1956-1957).
328. Chao-Hao Ku, "On the descriptive geometry of a space of K-spreads," Acad. Sinica Sci. Record, 3: 53-59 (1950).
329. Chao-Hao Ku, "New treatment of geometry in a space of K-spreads," Acad. Sinica Sci. Record, 3: 41-51 (1950).
330. Chao-Hao Ku, "On Finsler spaces admitting a group of motions of the greatest order," Sci. Rec., 1(4): 215-218 (1957).
331. M. Kurita, "On the dilatation in Finsler spaces," Osaka Math. J., 15(1): 87-98 (1963).
332. M. Kurita, "Theory of Finsler spaces based on the contact structure," J. Math. Soc. Japan, 18(2): 119-134 (1966).
333. G. Landsberg, "Über die Totalkrümmung," Jahresberichte Deut. Math.-Ver., 16: 36-46 (1907).
334. G. Landsberg, "Krümmungstheorie und Variationsrechnung," Jahresberichte Deut. Math.-Ver., 16: 547-551 (1907).
335. G. Landsberg, "Über die Krümmung in der Variationsrechnung," Math. Ann., 65: 313-349 (1908).
336. D. Laugwitz, "Zur geometrischen Begründungen der Parallelverschiebung in Finslerschen Räumen," Arch. Math., 6(6): 448-453 (1955).
337. D. Laugwitz, "Die Vektorübertragungen in der Finslerschen Geometrie und der Wegegeometrie," Proc. Koninkl. Ned. Akad. Wetenschap., A59(1): 21-28 (1956); Indagationes Math., 18(1): 21-28 (1956).
338. D. Laugwitz, "Zur projectiven und konformen Geometrie der Finsler Räume," Arch. Math., 7(1): 74-77 (1956).

339. D. Laugwitz, "Grundlagen für die Geometrie der unendlichdimensionalen Finsler-Räume," Ann. Mat. Pura ed Appl., 41: 21-41 (1956).

340. D. Laugwitz, "Zur Differentialgeometrie der Hyperflächen in Vektorräumen und zur affingeometrischen Deutung der Theorie der Finsler-Räume," Math. Z., 67(1): 63-74 (1957).

341. D. Laugwitz, "Eine Beziehung zwischen affiner und Minkowskischer Differentialgeometrie," Publ. Math., 5(1-2): 72-76 (1957).

342. D. Lehmann, "Théorie de Morse en géométrie finslérienne," Cahiers Semin. Topol. et Géom. Différent. Ch. Ehresmann, Fac. Sci., Paris, Vol. 6 (1964).

343. A. Lichnerowicz, "Les espaces à connexion semi-symétrique et la mécanique," Compt. Rend. Acad. Sci. Paris, 212: 328-331 (1941).

344. A. Lichnerowicz, "Sur une généralisation des espaces de Finsler," Compt. Rend. Acad. Sci., Paris, 214: 599-601 (1942).

345. A. Lichnerowicz, "Sur une extension de la formule d'Allendorerfer-Weil à certaines variétés finslériennes," Compt. Rend. Acad. Sci., Paris, 223: 12-14 (1946).

346. A. Lichnerowicz and Y. Thiry, "Problémes de calcul des variations liés à la dynamique classique et à la théory unitaire du champ," Compt. Rend. Acad. Sci. Paris, 224: 529-531 (1947).

347. H. Lippmann, "Zur Winkeltheorie in zweidimensionalen Minkowski- und Finsler-Räumen," Proc. Koninkl. Ned. Akad. Wetenschap., A60(2): 162-170 (1957); Indagationes Math., 19(2): 162-170 (1957).

348. H. Lippmann, "Metrische Eigenschaften verschiedener Winkelmasse in Minkowski und Finsler-Räumen. II," Proc. Koninkl. Akad. Wetenschap., A61(2): 223-230 (1958); Indagationes Math., 20(2): 223-230 (1958).

349. H. Lippmann, "Metrische Eigenschaften verschiedener Winkelmasse in Minkowski- und Finsler-Räumen. II," Proc. Koninkl. Akad. Wetenschap., A61(2): 231-238 (1958); Indagationes Math., 20(2): 231-238 (1958).

350. T. Maebashi, "A weakly osculating Riemann space of the Finsler space and its application to a theory of subspaces in the Finsler space," Tensor, 9(1): 62-72 (1959).

351. M. Matsumoto, "A global fundation of Finsler geometry," Mem. Coll. Sci., Univ. Kyoto, A33(1): 171-208 (1960).

352. M. Matsumoto, "Affine transformations of Finsler spaces," J. Math. Kyoto Univ., 3(1): 1-35 (1963).

353. M. Matsumoto, "Linear transformations of Finsler connections," J. Math. Kyoto Univ., 3(2): 145-167 (1964).

354. M. Matsumoto, "Paths in a Finsler space," J. Math. Kyoto Univ., 3(3): 305-318 (1964).

355. M. Matsumoto, "On R. Sulanke's method deriving H. Rund's connection in a Finsler space," J. Math. Kyoto Univ., 4(2): 355-368 (1965).

356. M. Matsumoto, "A Finsler connection with many torsions," Tensor, 17(3): 217-226 (1966).

357. K. Maurin, "Eingliedrige Gruppen der homogenen kanonischen Transformationen und Finslersche Räume," Ann. Polon. Math., 2(1): 97-102 (1955).

358. T. Michihiro, "Theory of curves in a two-dimensional space with arc length $s = \int (A_i x''^i + B)^{1/p} dt$," Tensor, 4: 63-66 (1941).
359. M. Mikami, "Projective theory of a system of paths of higher order," Tensor, 6: 86-94 (1943).
360. M. Mikami, "Geometry of the integral $\int (A_i x^{(m)i} + B)^{1/p} dt$," Jap. J. Math., 18: 663-673 (1943).
361. R. Mirodan, "Geometrizarea ecuatiilor cu derivate partiale linaire si omogene," An. Univ. "C. I. Parhon," Ser. Stint. Natur., No. 14, pp. 35-39 (1957).
362. R. B. Misra, "The projective transformation in a Finsler space," Ann. Soc. Sci. Bruxelles, Ser. 1, 80(3): 227-239 (1966).
363. R. B. Misra, "Projective tensors in a conformal Finsler space," Bull. Cl. Sci. Acad. Roy. Belg., 52(10): 1275-1279 (1966).
364. R. B. Misra, "The commutation formulae in a Finsler space. I," Ann. Mat. Pura ed Appl., 75: 363-370 (1967).
365. R. B. Misra, "The commutation formulae in a Finsler space. II," Ann. Mat. Pura ed Appl., 75: 371-383 (1967).
366. R. B. Misra, "The Bianchi identities satisfied by curvature tensors in a conformal Finsler space," Tensor, 18(2): 187-190 (1967).
367. R. S. Mishra and R. S. Sinha, "Relative Frenet formulae for curves in a subspace and a hypersurface of a Finsler space," Tensor, 16(2): 114-132 (1956).
368. R. S. Mishra and R. S. Sinha, "Union and hyperasymptotic curves of a Finsler subspace and hypersurface," Rend. Circolo Mat. Palermo, 14(1): 119-128 (1965).
369. R. S. Mishra and R. S. Sinha, "Union curvature of a curve in a Finsler space," Tensor, 16(2): 160-168 (1965).
370. R. S. Mishra and U. P. Singh, "On the union curvature of a curve of a Finsler space," Tensor, 17(2): 205-211 (1966).
371. F. Moalla, "Espaces de Finsler complets," Compt. Rend. Acad. Sci., 258(8): 2251-2254 (1964).
372. F. Moalla, "Espaces de Finsler complets a courbure de Ricci positive," Compt. Rend. Acad. Sci., 258(10): 2734-2737 (1964).
373. F. Moalla, "Espaces de Finsler sans points conjugués," Compt. Rend. Acad. Sci., 260(25): 6510-6512 (1965).
374. A. Moór, "Finslersche Räume mit algebraischen Grundfunktionen," Publ. Math., 2(3-4): 178-180 (1952).
375. A. Moór, "Ergänzung zu meiner Arbeit: 'Über die Dualität von Finslerschen und Cartanschen Räumen'," Acta Math., 91(3-4): 187-188 (1954).
376. A. Moór, "Die oskulierenden Riemannschen Räume regulärer Cartanscher Räume," Acta Math. Acad. Sci. Hung., 5(1-2): 59-72 (1954).
377. A. Moór, "Allgemeine metrische Räume von skalarer Krümmung," Publ. Math. 4(3-4): 207-228 (1956).
378. A. Moór, "Entwicklung einer Geometrie der allgemeinen metrischen Linienelementräume," Acta Sci. Math., 17(1-2): 85-120 (1956).
379. A. Moór, "Über die Torsions- und Krümmungsinvarianten der dreidimensionalen Finslerschen Räume," Math. Nachr., 16(2): 85-99 (1957).
380. A. Moór, "Über den Schurschen Satz in allgemeinen metrischen Linienelementräumen," Proc. Koninkl. Ned. Akad. Wetenschap., A60(3): 290-301 (1957); Indagationes Math., 19(3): 290-301 (1957).

381. A. Moór, "Über die autoparallele Abweichung in allgemeinen metrischen Linienelementräumen," Publ. Math., 5(1-2): 102-118 (1957).

382. A. Moór, "Konformgeometrie der verallgemeinerten Schouten-Haantjesschen Räume. I," Proc. Koninkl. Ned. Akad. Wetenschap., A61(1): 94-102 (1958); Indagationes Math., 20(1): 94-102 (1958).

383. A. Moór, "Konformgeometrie der verallgemeinerten Schouten-Haantjesschen Räume. II," Proc. Koninkl. Ned. Akad. Wetenschap., A61(1): 103-113 (1958); Indagationes Math., , 40(1): 103-113 (1958).

384. A. Moór, "Über nichtholonome allgemeine metrische Linienelementräume," Acta Math., 101(3-4): 201-233 (1959).

385. A. Moór, "Erweiterung des Begriffs der Räume skalarer und konstanter Krümmung," Acta Sci. Math., 21(1-2): 53-77 (1960).

386. A. Moór, "Untersuchungen über die kovariante Ableitung in Linienelementenräumen," Publ. Math., 7(1-4): 41-53 (1960).

387. A. Moór, "Über affine Finslerräume von skalarer Krümmung," Acta Sci. Math., 22(3-4): 157-189 (1961).

388. A. Moór, "Über die Form der Fundamentalgrössen gewisser affiner Räume," Publ. Math., 9(3-4): 289-297 (1962).

389. A. Moór, "Über projektive Veränderung der Übertragung in Linienelementmannigfaltigkeiten," Acta Sci. Math., 24(1-2): 119-128 (1963).

390. A. Moór, "Untersuchungen über Finslerräume von rekurrenter Krümmung," Tensor, 13: 1-8 (1963).

391. A. Moór, "Eine Verallgemeinerung der metrischen Übertragung in allgemeinen metrischen Räumen," Publ. Math., 10(1-4): 145-150 (1963).

392. A. Moór, "Untersuchungen über die oskulierenden Punkträume der metrischen Linienelementräume," Liet. Matem. Rinkinys., 3(2): 212-213 (1963).

393. A. Moór, "Gleichung der autoparallelen Abweichung in n-dimensionalen Linienelementräumen," Acta Sci. Math., 25(3-4): 266-282 (1964).

394. A. Moór, "Untersuchungen über die oskulierenden Punkträume der metrischen Linienelementräume," Acta Math. Acad. Sci. Hung., 16(1-2): 57-74 (1956).

395. A. Moór, "Linienelementräume mit nicht-symmetrischem Fundamentaltensor," Publ. Math., 11(1-4): 245-256 (1964).

396. A. Moór, "Über eine skalare Form der Gleichung der autoparallelen Abweichung im affinem Raum," Publ. Math., 12(1-4): 281-291 (1965).

397. A. Moór, "Begründung einer affinen Geometrie der Bahnen dritter Ordnung," Tensor, 16(1): 37-55 (1965).

398. A. Moór, "Übertragungstheorie bezüglich der allgemeinen Linienelemententransformationen," Publ. Math., 13(1-4): 263-287 (1966).

399. A. Moór and Gy. Soos "Über affinzusammenhängende Mannigfaltigkeiten von Hyperflächenelementen, insbesondere deren Aquivalenz," Acta Sci. Math., 16(1-2): 29-42 (1955).

400. Y. Muto and K. Yano, "Sur les transformations de contact et les espaces de Finsler," Tohoku Math. J., 45: 295-307 (1939).

401. Y. Nagata, "Normal curvature of a vector field in a hypersurface in a Finsler space," Tensor, 5(1): 17-22 (1955).

402. Y. Nasu, "On the normality in Minkowskian spaces," Kumamoto J. Sci., A2(1): 11-17 (1954).

403. Y. Nasu, "On similarities in a Finsler space," Tensor, 9(3) : 175-189 (1959).
404. A. Nazim, "Uber Finslersche Räume," Dissertation, Munich (1936).
405. A. Nazim, "Über den Satz von Gauss-Bonnet im Finslerschen Raum," University of Instanbul (1948), pp. 26-32.
406. E. Noether, "Invarianten beliebiger Differentialausdrücke," Gött. Nachr., 25: 37-44 (1918).
407. E. Noether, "Invarianten beliebiger Differentialausdrücke," Jahresber. Deutsch. Mach.-Ver., 32: 177-184 (1923).
408. T. Ohkubo, "Geometry in a space with generalized metrics. II," J. Fac. Sci. Hokkaido Univ., Ser. I, 10: 157-178 (1941).
409. T. Ohkubo, "On a symmetric displacement in a Finsler space," Tensor, 4: 53-55 (1941).
410. T. Ohkubo, "Descriptive geometry of paths," Tensor, 5: 81-86 (1942).
411. T. Ohkubo, "A generalization of Cartan space," Tensor, 6: 45-48 (1943).
412. T. Ohkubo, "Über die Extensorrechnung in den verallgemeinerten Räumen von Flächenelementen höherer Ordnung," J. Fac. Sci. Hokkaido Univ., Ser. I, 11: 1-37 (1946).
413. T. Ohkubo, "On relations among various connections in Finslerian space," Kumamoto J. Sci., A1(3): 1-6 (1954).
414. T. Ohkubo, "On the order of the groups of affine collineations in the generalized spaces of paths. I, II, III," Tensor, 6(3): 141-158 (1956); 7(1): 1-17 (1957); 7(1): 18-33 (1957).
415. M. Okumura, "On some remarks of special Kawaguchi spaces," Tensor 11(2): 154-160 (1961).
416. T. Otsuki, "On geodesic coordinates in Finsler spaces," Math. J. Okayama Univ., 6(2): 135-145 (1957).
417. T. Otsuki, "Theory of affine connections of the space of tangent directions of a differentiable manifold. I, II," Math. J. Okayama Univ., 7(1): 1-74 (1957).
418. T. Otsuki, "Theory of affine connections of the space of tangent directions of a differentiable manifold. III," Math. J. Okayama Univ., 7(2): 95-122 (1957).
419. T. Otsuki, "Note on curvature of Finsler manifolds," Math. J. Okayama Univ., 8(2): 102-116 (1958).
420. O. G. Owens, "The integral geometry definition of arc-length for two-dimensional Finsler spaces," Trans. Am. Math. Soc., 73: 198-210 (1952).
421. M. Pihl, "Classical mechanics in a geometrical description," Danske Vid. Selsk. Math.-Fys. Medd., Vol. 30, No. 12 (1955).
422. N. Prakash, "Generalised normal curvature of a curve and generalised principal directions in Finsler space," Tensor, 11(1): 51-56 (1961).
423. N. Prakash, "Kählerian Finsler manifolds," Math. Student, 30(1-2) : 1-12 (1962).
424. N. Prakash and Ram Behari, "Deviations from parallelism and equidistance in Finsler space," Proc. Indian Acad. Sci., A52(5): 209-227 (1960).
425. N. Prakash and Ram Behari, "Generalizations of Codazzi's equations in a subspace imbedded in a Finsler manifold," Proc. Nat. Inst. Sci., India, A26(5): 532-540 (1960).

426. N. Prakash and R. Behari, "Union curves and union curvature in Finsler space," Proc. Nat. Inst. Sci., India, A26, Suppl., No. 2, pp. 21-30 (1960).
427. A. Rapscak, "A normálkoordináták egy uj ertelmezése a finsler terben," Acta Univ. Debrecen, 1: 109-116 (1954).
428. A. Rapscak, "Theorie der Bahnen in Linienelementmannigfaltigkeiten und eine Verallgemeinerung ihrer affinen Theorie," Acta Sci. Math., 16(3-4): 251-265 (1955).
429. A. Rapscak, "Invariante Taylorsche Reihe in einem Finslerschen Raum," Publ. Math., 4(1-2): 49-60 (1955).
430. A. Rapscak, "Über das vollständige System von Differentielinvarianten im regulärer Cartanschen Raum," Publ. Math., 4(3-4): 276-293 (1956).
431. A. Rapscak, "Eine neue Charakterisierung Finslerscher Räume skalarer und konstanter Krümmung und projectiv-ebene Räume," Acta Math. Acad. Sci. Hung., 8(1-2) : 1-18 (1957).
432. A. Rapscak, "Metrische Charakterisierung der Finslerschen Räume mit verschwindender projectiver Krümmung," Acta Sci. Math., 18(3-4):192-204 (1957).
433. A. Rapscak, "Hypersikok a Finsler-féle terben," Acta Univ. Debrecen, 4: 85-87 [1957 (1959)].
434. A. Rapscak, "Metricus és affinösszefüggö pályaterek pályatartó leképezései," Magy. Tud. Akad. Mat. Fiz. Tud. Oszt. Kozleman., 11(4): 339-369 (1961).
435. A. Rapscak, "Über die bahntreuen Abbildungen metrischer Räume," Publ. Math., 8(3 -4): 285-290 (1961).
436. A. Rapscak, "Die Bestimmung der Grundfunktionen projectiv-ebener metrischer Räume," Publ. Math., 9(1-2): 164-167 (1962).
437. A. Rapscak, "Über die Metrisierbarkeit affinzusammenhängender Bahnräume," Ann. Mat. Pura ed Appl., 57: 233-238 (1962).
438. P. Rashevskii (Rashevsky), "Une géometrie métrique quale, fondée sur les espaces de Cartan généralisés," Compt. Rend. Acad. Sci., Paris, (1935), pp. 921-923.
439. P. Rashevskii (Rashevsky), "Système bimétrique dual," Compt. Rend. Acad. Sci. Paris, 201: 1088-1090 (1935).
440. P. Rashevskii (Rashevsky), "Systèmes trimétriques et la métrique de Finsler Généralisée," Compt. Rend. Acad. Sci., Paris, 202: 1237-1239 (1936).
441. Georges Reeb, "Sur les espaces de Finsler et les espaces de Cartan," in: Colloq. Intern. Centre Nat. Rech. Sci., 52, Strasbourg, 1953, Paris (1953), pp. 35-40.
442. G. B. Rizza, "Strutture de Finsler sulle varieta quasi complesse," Atti Accad. Naz. Lincei Rend. Cl. Sci. Fiz. Mat. Nat., 33(5): 271-275 [1962 (1963)]. (1963)].
443. G. B. Rizza, "Strutture di Finsler di tipo quasi hermitiano," Riv. Mat. Univ. Parma, 4: 83-106 (1963).
444. H. Rund, "Die Hamiltonsche Funktion bei allgemeinen dynamischen Systemen," Arch. Math., 3: 207-215 (1952).
445. H. Rund, "On the geometry of generalized metric spaces," Convegno Internazionale di Geometria Differenziale, Italy, 1953, Ed. Cremonese, Rome (1954), pp. 114-121.
446. H. Rund, "On the analytical properties of curvature tensors in Finsler spaces," Math. Ann., 127(1): 82-104 (1954).

447. H. Rund, "Über nicht-holonome allgemeine metrische Geometrie," Math. Nachr., 11(1-2): 61-80 (1954).
448. H. Rund, "Hypersurfaces of a Finsler space," Can. J. Math., 8(4): 487-503 (1956).
449. H. Rund, "Some remarks concerning the theory of nonlinear connections," Proc. Koninkl. Ned. Akad. Wetenschap., A61(3): 341-347 (1958); Indagationes Math., 20(3): 341-347 (1958).
450. H. Rund, The Differential Geometry of Finsler Spaces, Spinger, Berlin (1959), XIII, p. 283.
451. H. Rund, "Über Finslersche Räume mit speziellen Krümmungseigenschaften," Monatsh. Math., 66(3): 241-251 (1962).
452. H. Rund, "Curvature properties of hypersurfaces of Finsler and Minkowskian spaces," Tensor, 14: 226-244 (1963).
453. H. Rund, "The intrinsic and induced curvature theories of subspaces of a Finsler space," Tensor, 16(3): 294-312 (1965).
454. S. Sasaki, "On some properties in the large in the geometry of paths," Tensor, 8: 41-53 (1948).
455. H. Sasayama, "On the nonholonomic quasi-Euclidean space of line-elements," J. Spat. Math. Sasayama Res. Room, 4(2-3): 62-88 (1961).
456. J. A. Schouten and J. Haantjes, "Über die Festlegung von allgemeinen Massbestimmungen und Übertragungen in Bezug auf k_0 und kontravariante Vektordichten," Monatsh. Math. Phys., 43: 161-176 (1936).
457. V. Seetharaman, "Differential invariants for path-space of order 2," Proc. Indian Acad. Sci., A5: 161-165 (1937).
458. B. Segre, "Geometria non euclidea ed ottica geometrica. I, II," Atti Accad. Naz. Lincei, Rend. Cl. Sci. Fis., Mat. Nat., 8(7): 16-19 (1949); 8(7): 20-26 (1949).
459. R. N. Sen, "Application of an algebraic system in Finsler geometry," Tensor, 18(2): 191-195 (1967).
460. R. N. Sen, "On curvature tensors in Finsler geometry," Tensor, 18(2): 217-226 (1967).
461. A. C. Shamihoke, "On the subspaces of a Finsler space," J. Indian Math. Soc., 25(3-4): 215-220 (1961).
462. A. C. Shamihoke, "Some properties of curvature tensors in a generalised Finsler space," Tensor, 12(2): 97-109 (1962).
463. A. C. Shamihoke, "Normal curvature of a vector field in a hypersurface of a generalised Finsler space," Istanbul Univ. Fen. Fak. Mecmuasi, A27: 9-14 (1962).
464. A. C. Shamihoke, "Subspaces of a generalised Finsler space. I," Ganita, 14(1): 43-59 (1963).
465. A. C. Shamihoke, "Hypersurfaces of a generalised Finsler space," Tensor, 13: 129-144 (1963).
466. A. C. Shamihoke, "A note on a curvature tensor in a Finsler space," Tensor, 15(1): 20-22 (1964).
467. A. C. Shamihoke, "Parallelism and covariant differentiation in a generalized Finsler space of n dimensions," Riv. Mat. Univ. Parma, 5: 189-200 (1964).

468. E. B. Shanks, "Homothetic correspondences between Riemannian spaces," Duke Math. J., 17: 299-311 (1950).

469. N. K. Sharma and R. Behari, "Some properties of spaces with the arc length $s=\int \{A_i (x^i, \dot{x}^i) \ddot{x}^i\}^{1/3} dt$," Ganita, 15(1): 1-8 (1964).

470. B. B. Sinha, "Projective invariants," Math. Student, 33(2-3): 121-127 (1965).

471. W. Slebodzinski, "Sur deux connexions affines généralisées," Prace Mat.-Fizycz., 43: 167-205 (1935).

472. Gy. Soós, "Über Gruppen von Affintäten und Bewegungen in Finslerschen Räumen," Acta Math. Acad. Sci. Hung., 5(1-2): 73-84 (1954).

473. Gy. Soós, "Über Gruppen von Automorphismen in affinzusammenhängenden Räumen von Linienelementen," Publ. Math., 4(3-4): 294-302 (1956).

474. Gy. Soós, "Über eine spezielle Klasse von Finslerschen Räumen," Publ. Math., 5(1-2): 150-153 (1957).

475. Gy. Soós, "Über die homothetische Gruppe von Finslerschen Räumen," Acta Math. Acad. Sci. Hung., 10(3-4): 391-394 (1959).

476. Gy. Soós, "Über einfache Finslersche Räume," Publ. Math., 7(1-4): 364-373 (1960).

477. Gy. Soós, "A Finsler-féle fibrált terek elméletéhez," Magy. Tud. Akad. Mat. Fiz. Tud. Oszt. Közlemen., 13(1): 17-64 (1963).

478. T. N. Srivastava, "A few remarks on special Kawaguchi spaces," Tensor, 15(1): 12-19 (1964).

479. Buchin Su, "On the isomorphic transformations of K-spreads in a Douglas space," Acad. Sinica Sci. Record, 2: 11-19 (1947).

480. Buchin Su, "On the isomorphic transformations of K-spreads in a Douglas space. II," Acad. Sinica Sci. Record, 2: 139-146 (1948).

481. Buchin Su, "A characteristic property of affine collineations in a space of K-spreads," Bull. Am. Math. Soc., 54: 136-138 (1948).

482. Buchin Su, "Geodesic deviation in generalized metric spaces," Acad. Sinica Sci. Record, 2: 220-226 (1949).

483. Buchin Su, "Axiom of the plane in a space of K-spreads," Acad. Sinica Sci. Record, 3: 7-16 (1950).

484. Buchin Su, "A generalization of descriptive collineations in a space of K-spreads," J. London Math. Soc., 25: 236-238 (1950).

485. Buchin Su, "Extremal deviation in a geometry based on the notion of area," Acta Math., 85: 99-116 (1951).

486. Buchin Su, "Koschmider invariant and the associate differential equation of a minimal hypersurface in a regular Cartan space," Math. Nachr., 16: 117-129 (1957).

487. Buchin Su, "Axiom of the plane in a descriptive geometry of K-spreads," Math. Nachr., 16(3-4): 215-226 (1957).

488. Buchin Su, "A generalization of descriptive collineations in a space of K-spreads," Math. Nachr., 16(3-4): 227-232 (1957).

489. Buchin Su, "On the determination of certain affine connections in an areal space," Sci. Rec., 1(4): 195-198 (1957).

490. Buchin Su, "The geometry of spaces with areal metrics," Math. Nachr., 16(5-6): 281-287 (1957).

491. Buchin Su, "Certain affinely connected spaces with areal metrics," Scientia Sinica, 6(6): 967-975 (1957).

492. Buchin Su, "Recent progress in the differential geometry of a space of K-spreads," in: Lucrǎrile Conf. Geometrie Si Topol., 1958, Acad. RPR, Bucurest (1962), pp. 21-30.

493. Buchin Su, "On the theory of affine connections in an areal space," Bull. Math. Soc., Sci. Math. et Phys. (RPR), 2(2): 185-190 (1958).

494. Buchin Su and Chao-hao Ku, "The developments of differential geometry in China for the past ten years," Scientia Sinica, 8(11): 1238-1242 (1959).

495. T. Suguri, "Theory of invariants in the geometry of paths," J. Math. Soc. Japan, 4: 231-268 (1952).

496. R. Sulanke, "Die eindeutige Bestimmtkeit des von Hanno Rund eigeführten Zusammenhangs in Finsler-Räumen," Wiss. Z. Humboldt-Univ. Berlin, Math.-Naturwiss. Reihe, 4(4): 229-233 (1954).

497. R. Sulanke, "Anmerkung zu der Arbeit: 'Die eindeutige Bestimmtheit des von Hanno Rund eigeführten Zusammenhangs in Finsler-Räumen'," Wiss. Z. Humboldt-Univ. Berlin, Math.-Naturwiss. Reihe, 5(3): 269 (1955/56).

498. R. Sulanke, "Eine Ableitung des Cartanschen Zusammenhangs eines Finslerschen Raumes," Publ. Math., 5(1-2): 197-203 (1957).

499. W. Süss, "Affine und Minkowskische Geometrie eines ebenen Variationsproblems," Arch. Math., No. 4-6, pp. 441-446 (1954).

500. Y. Suzuki, "Finsler geometry in classical physics," J. Coll. Arts Sci. Chiba Univ. Natur. Sci. Ser., 2(1): 12-16 (1956).

501. S. L. Synge, "A generalization of the Riemannian line element," Trans. Am. Math. Soc., 27: 61-67 (1925).

502. S. L. Synge, "Some intrinsic and derived vectors in a Kawaguchi space," Am. J. Math., 57: 679-691 (1935).

503. Syun-ichi Tachibana, "On Finsler spaces which admit a concurrent vector field," Tensor, 1: 1-5 (1950).

504. K. Tandai, "On areal spaces. VI. On the characterization of metric areal spaces," Tensor, 3: 40-45 (1953).

505. K. Tandau, "On areal spaces. VII. The theory of the canonical connection and m-dimensional subspaces," Tensor, 4(2): 78-90 (1954).

506. K. Tandau, "On areal spaces. VIII. Theory of a space of the semi-metric class," Tensor, 10: 161-166 (1960).

507. K. Tandau, "On general connections in an areal space. I. General connection on a fibre of the tangent m-frame bundle," Tensor, 13: 277-291 (1963).

508. K. Tandau, "On general connections in an areal space. II. On general connections on the tangent m-frame bundle," Tensor, 14: 26-46 (1963).

509. Y. Tashiro, "A theory of transformation groups on generalized spaces and its applications to Finsler and Cartan spaces," J. Math. Soc. Japan, 11(1): 42-71 (1959).

510. J. H. Taylor, "A generalization of Levi-Civita's parallelism and the Frenet formulas," Trans. Am. Math. Soc., 27: 246-264 (1925).

511. N. Théodoresci, "Sur les géodésiques de longeur nulle de certains éléments lineaires finslériens," Bull. École Polytech. Bucarest, 12: 9-16 (1941).

512. N. Théodoresci, "Géométrie finslérienne et propagation des ondes," Acad. Roumaine Bull. Sec. Sci., 23:138-144 (1942).

513. N. Théodoresci. "Introduction physico-mathématique à la théorie invariante de la propagation des ondes," Rev. Univ. si Politechn. Bucuresti, Ser. Stint. Natur., 1:25-51 (1952).

514. N. N. Teodorescu, "Asupra unor spatii Finsler," Studii Si Cercetari Mat. Acad. RPR, 13(3):499-510 (1962).

515. K. Tonooka, "Generalized rheonomic geometry of K-spreads," Tensor, 3:26-39 (1953).

516. K. Tonooka, "Theory of subspaces in a geometry based on a multiple integral. I. Metric tensor and theory of connections," Tensor, 4:75-83 (1954).

517. K. Tonooka, "On intrinsic theories in the manifold of surface elements of higher order. II. Intrinsic geometry of a system of partial differential equations," Tensor, 4(2):67-77 (1954).

518. K. Tonooka, "On a geometry of three-dimensional space with an algebraic metric," Tensor, 6(1):60-68 (1956).

519. K. Tonooka, "On three- and four-dimensional Finsler spaces with the fundamental form $\sqrt[3]{a_{\alpha\beta\gamma}x'^{\alpha}x'^{\beta}x'^{\gamma}}$," Tensor, 9(3):209-216 (1959).

520. K. Tonowoka, "On a metric displacement along a curve in a special Kawaguchi space," Tensor, 4:60-62 (1941).

521. K. Tonowoka, "On a geometrical treatment of an (n − 1)-ple integral of some kind," Tensor, 7:16-23 (1944).

522. K. Tonowoka, "On invariants of $\int_{(n-1)}\{A_{ij}^{\alpha(2)\beta(3)}p^i_{\alpha(2)}p^j_{\beta(3)}+B_j^{\beta(3)}p^j_{\beta(3)}+C\}^{1/p}$ $du^1 \ldots du^{n-1}$. ," Tensor, 9:18-24 (1949).

523. K. Tonowoka, "A problem on the generalization of Cartan space," Sugeku Jap. Math. Soc., 2:47-50 (1950).

524. T. N. Srivastava, "Les identités généralisées de Bianchi et de Veblen dans les espaces de Kawaguchi spéciaux," Compt. Rend. Acad. Sci., Paris, 254(15):2706-2708 (1962).

525. T. Uehara, "The discontinuous character of the three-dimensional motion of a compressible fluid in terms of the geometry of Cartan space," RAAG Mem. Unifying Study Basic Probl. Engng. and Phys., Sci. Means Geom., Vol. 3, Tokyo Gakujutsu Bunken Fukyu-kai (1962), pp. 735-742.

526. S. Ueno, "On the densities in a two-dimensional generalized space," Mem. Fac. Sci. Kyusyu Univ., A9(1):65-77 (1955).

527. A. Urban, "On the geometry of a system of partial differential equations of the second order," Proc. Koninkl. Ned. Akad. Wetenschap., A52:855-867 (1949); Indagationes Math., 11:303-315 (1949).

528. J. R. Vanstone, "A generalization of Finsler geometry," Can. J. Math., 14(1): 87-112 (1962).

529. O. Varga, "Bestimmung des invarianten Differentials in Finslerschen Räumen," Math.-Fiz. Lapok, 48:423-435 (1941).

530. O. Varga, "Aufbau der Finslerschen Geometrie mit Hilfe einer oskulierenden Minkowskischen Massbestimmung," Math. Nat. Anz. Ungarn. Acad. Wiss., 61:14-22 (1942).

531. O. Varga, "Zur Begründung der Minkowskischen Geometrie," Acta Univ. Szeged. Sec. Ser. Math., 10:149-163 (1943).

532. O. Varga, "Olyan finsler-terek jellemzése, amelyekben létezik a vonalemek abszolut parhuzamossaga," Acta Univ. Debrecen, 1:105-108 (1954).
533. O. Varga, "Eine Charakterisierung der Finslerschen Räume mit absolutem Parallelismus der Linienelemente," Arch. Math., 5(1-3):128-131 (1954).
534. O. Varga, "Bedingungen für die Metrisierbarkeit von affinzusammenhängenden Linienelementmannigfaltigkeiten," Acta Math. Acad. Sci. Hung., 5(1-2): 7-16 (1954).
535. O. Varga, "Die Krümmung der Eichfläche des Minkowskischen Raumes und die geometrische Deutung des einen Krümmungstensors des Finslerschen Raumes," Abhandl. Math. Seminar Univ. Hamburg, 20(1-2): 41-59 (1955).
536. O. Varga, "Eine Charakterisierung der Kawaguchischen Räume metrischer Klasse mittels eines Satzes über derivierte Matrizen," Publ. Math., 4(3-4): 418-430 (1956).
537. O. Varga, "Normalkoordinaten in Kawaguchischen Räumen und seine affinen Verallgemeinerungen sowie eine Anwendung derselben zur Bestimmung von Differentialinvarianten," Math. Nachr., 18(1-6):141-151 (1958).
538. O. Varga, "Über die Zerlegbarkeit von Finslerschen Räumen," Acta math. Acad. Sci. Hung., 11(1-2):197-203 (1960).
539. O. Varga, "Bemerkung zur Winkelmetrik in Finslerschen Räumen," Ann. Univ. Sci. Budapest, Sec. Math., 3-4:379-382 (1960-1961).
540. O. Varga, "Über eine Charakterisierung der Finslerschen Räume konstanter Krümmung," Monatsh. Math., 65(3): 277-286 (1961).
541. O. Varga, "Herleitung des Cartanschen euklidischen Zusammenhanges in Finslerräumen mit Hilfe der Riemannschen Geometrie," Acta Univ. Debrecen, Ser. Phys. et Chim., 8:121-124 (1962).
542. O. Varga, "Eine einfache Herleitung der Cartanschen Übertragung der Finslergeometrie," Math. Notae, 18(1):185-196 (1962).
543. O. Varga, "Über Hyperflächen konstanter Normalkümmung in Minkowskischen Räumen," Tensor, 13:246-250 (1963).
544. O. Varga, "Hyperflächen mit Minkowskischer Massbestimmung in Finslerräumen," Publ. Math., 11(1-4):301-309 (1964).
545. O. Varga, "Die Methode des beweglichen n-Beines in der Finsler-Geometrie," Acta Math. Acad. Sci. Hung., 18(1-2):207-215 (1967).
546. O. Veblen, "Normal coordinates for the geometry of paths," Proc. Nat. Acad. Sci. USA, 8:192-197 (1922).
547. O. Veblen, "Equiaffine geometry of paths," Proc. Nat. Acad. Sci. USA, 9:3-4 (1923).
548. O. Veblen, "Generalized projective geometry," J. Lond. Math. Soc., 4:140-160 (1929).
549. O. Veblen and T. Y. Thomas, "The geometry of paths," Trans. Am. Math. Soc., 25: 551-608 (1923).
550. O. Veblen and T. Y. Thomas, "Projective normal coordinates for the geometry of paths," Proc. Nat. Acad. Soc., 11: 204-207 (1925).
551. O. Veblen and T. Y. Thomas, "Projective invariants of affine geometry of paths," Ann. Math., 27:278-296 (1926).
552. Vaclav Vinhelm, "Krivky v prostorèch Minkowského," Casop. Pestov. Mat., 82(3): 283-300 (1957).

553. V. V. Vagner (Wagner), "Geometria del calcolo delle variationi," Centro Intern. Mat. Estivo (1965).
554. Hsièn-Chung Wang, "On Finsler spaces with completely integrable equations of Killing," J. Lond. Math. Soc., 22: 5-9 (1947).
555. Yuen-da Wang, "Finsler spaces of constant curvature and totally extremal hypersurfaces," Sci. Rec., 2(7): 211-214 (1958).
556. S. Watanabe, "On special Kawaguchi spaces," Tensor, 7(2):130-136 (1957).
557. Shoji Watanabe, "On special Kawaguchi spaces. II," Tensor, 8(3): 69-176 (1958).
558. S. Watanabe, "On special Kawaguchi spaces. III. Generalizations of affine spaces and Finsler spaces," Tensor, 11(2): 144-153 (1961).
559. S. Watanabe, "On special Kawaguchi spaces. IV. Extremal curves in the generalized affine spaces," Tensor, 11(3): 254-262 (1961).
560. S. Watanabe, "On special Kawaguchi spaces. V. Some remarks on the special Kawaguchi spaces," Tensor, 11(3): 279-284 (1961).
561. S. Watanabe, "On special Kawaguchi spaces. VI. Some transformations in certain special Kawaguchi spaces," Tensor, 12(3): 244-253 (1962).
562. S. Watanabe and M. Yoshida, "On special Kawaguchi spaces. VII. Some transformations in certain special Kawaguchi spaces H," Tensor, 13:31-41 (1963).
563. J. M. Wegener, "Untersuchung der zwei- und dreidimensionalen Räume mit der Grundfunktion $L=(a_{ijk}v^i v^j v^k)^{1/3}$," Proc. Koninkl. Ned. Akad. Wetenschap., Ser. A, 38: 949-955 (1935).
564. J. M. Wegener, "Hyperflächen in Finslerschen Räumen als Transversalflächen einer Schar von Extremalen," Monatsh. Math. Phys., 44: 115-130 (1936).
565. A. Winternitz, "Über die affine Grundlage der Metrik eines Variationsproblems," Sitzber. Acad. Wiss. Berlin (1930), pp. 457-469.
566. W. Wrona, "On multi-isotropic Finsler spaces," Bull. Acad. Polon. Sci. Ser. Sci. Math., Astron. Phys., 11(5): 285-288 (1963).
567. W. Wrona, "A necessary and sufficient condition for the Finsler space with distant parallelism of line elements to be multi-isotropic," Bull. Acad. Polon. Sci. Ser. Sci. Math., Astron. Phys., 11(5): 289-292 (1963).
568. W. Wrona, "Generalized F. Schur theorem in Finsler spaces," Bull. Acad. Polon. Sci. Ser. Sci. Math., Astron. Phys., 11(5): 293-295 (1963).
569. A. Wundheiler, "Über Variationsgleichungen für affine geodätische Linien und nicht-holonome, nicht-konservative dynamische Systeme," Prace Mat. Fiz., (1931), pp. 129-147.
570. K. Yano, Groups of Transformations in Generalized Spaces, Akad. Press, Tokyo (1949).
571. K. Yano, The Theory of Lie Derivatives and Its Applications, Amsterdam (1957), 299 pp.
572. K. Yano and H. Hiramatsu, "Affine and projective geometries of systems of hypersurfaces," J. Math. Soc. Japan, 3: 116-136 (1951).
573. K. Yano and H. Hiramatsu, "On projective geometry of K-spreads," Compositio Math., 10: 286-296 (1952).
574. K. Yano and H. Hiramatsu, "On groups of projective collineations in a space of K-spreads," J. Math. Soc. Japan, 6(2): 131-150 (1954).

575. K. Yano and E. T. Davies, "On the connection in Finsler space as an induced connection," Rend. Circolo Mat. Palermo, 3(3): 409-417 (1954).
576. K. Yano and E. T. Davies, "On the tangent bundles of Finsler and Riemannian manifolds," Rend. Circolo Mat. Palermo, 12(2): 221-228 (1963).
577. M. Yoshida, "On the connections in a subspace of the special Kawaguch space," Tensor, 17(1): 49-52 (1966).

Structures On Differentiable Manifolds

A. P. Shirokov

Introduction

Structures on differentiable manifolds have been very intensively studied in recent years both in differential topology as well as in differential geometry. Along with almost complex structures a systematic examination was begun of almost contact, affinor, symplectic, quaternion, and many other structures, a simple listing of which would take up considerable space. Not all of these structures are of equal value, but there is no doubt that the study of manifolds with various structures is of interest from both the global as well as the local point of view; moreover, for certain structures the possibility was outlined for applying them to various aspects of geometry and mechanics.

The majority of the structures being considered are included in the general scheme of G-structures. Therefore, the present survey starts with a brief description of the papers devoted to G-structures. However, a systematic consideration of all the particular types of G-structures known to date would take us too far afield, since here we could include all the research on Riemann geometry. Therefore, the object of this survey will be principally those structures whose systematic study started during the past ten to twelve years. Here we shall almost completely ignore the papers specially devoted to the study of the topological properties of manifolds with structures. We shall also leave aside the investigations on complex structures because the initial period of the study of such structures has been covered in detail in the survey of Beklemishev [4], while so many papers have been devoted to this theme in the past five years that after some time a special survey will be needed. Of course, it would be impossible to avoid completely points of contact with complex structures. They appear in §6 and also in §10.

In connection with the large role played by the field of area element in the theory of affinor and other structures, we have specially put together §3, in which we consider papers on distributions in the spaces V_n and L_n.

§1. General Aspects of the Theory of G-Structures

The concept of a G-structure, arising in the work of Cartan on infinite Lie groups, was explicitly introduced in 1953 by Chern [192] (also see [193, 194]). However, earlier still the problem of studying such structures (called subordinate structures) was posed by Ehresmann [205, 212] and served as a research topic to Libermann [315]. After this Bernard devoted a number of papers and a dissertation [165-168] to this question.

Let X_n be a differential manifold of dimension n, for which there has been constructed a principal fiber space $E(X_n)$ of all its linear frames. The general linear group $GL(n, R)$ is the structure group of the space $E(X_n)$. By taking an n-dimensional vector space W_n in which the group $GL(n, R)$ acts, we select some closed subgroup $G \subset GL(n, R)$. By T_x we denote the tangent space to X_n at a point x. A G-structure on X_n is given at each point $x \in X_n$ by a subset $S_x \subset \mathrm{Hom}[W_n, T_x]$ such that the following conditions are fulfilled:

1) linear mappings of S_x are invertible;

2) if $g \in G$ and $\alpha \in S_x$, then $\alpha \circ g \in S_x$;

3) if $\alpha \in S_x$ and $\beta \in S_x$, then we find a transformation $g \in G$ such that $\alpha \circ g = \beta$;

4) the set $E(X_n, G) = \bigcup_{x \in X_n} S_x$ forms a principal fiber space with base X_n and with structure group G.

Having chosen a certain frame e in W_n we can represent each element $\alpha \in S_x$ by a frame z of T_x, serving as the image of the frame e under a linear mapping $\alpha: W_n \to T_x$. Consequently, $E(X_n, G)$ can be treated as a principal fiber subspace in the space $E(X_n)$. The frames z forming this subspace are said to be adapted for the given

G-structure. By p we denote the natural projection $p: E(X_n, G) \to X_n$.

Let $\underline{G}$ be the Lie algebra of group G. By treating $\underline{G}$ as a subspace in $W_n \otimes W_n^*$, we construct a sequence of mappings $\underline{G} \otimes W_n^* \xrightarrow{i} W_n \otimes W_n^* \otimes W_n^* \xrightarrow{A} W_n \otimes W_n^* \wedge W_n^*$, where i is the inclusion mapping and A is the alternation operation. By ∂ we denote the resulting mapping

$$\partial = A \cdot i.$$

On the space $E(X_n, G)$ we define a 1-form ω with values in W_n, the so-called fundamental form of the G-structure. The value of this form at the vector v, which is tangent to the space $E(X_n, G)$ at the point z, is obtained as follows: we project the vector v onto the base, we refer the vector $p_* v$ obtained to the frame z represented as the point $z \in E(X_n, G)$, and then we find the image $\omega(v) = \alpha^{-1} p_* v$, where α is the mapping considered above taking e into z. Denoting the tangent space to $E(X_n, G)$ at point z by T_z, we choose a linear mapping $\Phi: W_n \to T_z$ such that $\omega \circ \Phi$ would be an identity transformation onto W_n. The mapping $d\omega \circ \Phi \wedge \Phi$ is an element of $\mathrm{Hom}(W_n \wedge W_n, W_n)$, and hence to it there uniquely corresponds an element of the space $W_n \otimes W_n^* \wedge W_n^*$. It turns out that the image of this element in the factor space $K = (W_n \otimes W_n^* \wedge W_n^*)/\partial(\underline{G} \otimes W_n^*)$ is independent of the choice of Φ. The function S, referring an element of K to each point of $E(X_n, G)$, is called the structure tensor of the G-structure.

The structure tensor of a G-structure can also be defined by means of the torsion tensor of the G-connection on X_n, by which we mean the infinitesimal connection on the principal fiber space $E(X_n, G)$. This connection is defined with the aid of a differential 1-form ω_1 with values in $\underline{G}$. If ρ is a representation of group G in the space W_n, while $\bar{\rho}$ is the corresponding representation in W_n of the algebra $\underline{G}$, then the torsion form of the G-connection ω_1 is determined by the formula

$$\Omega = d\omega + \bar{\rho}(\omega_1)\,\omega.$$

Having set $\omega = \omega^s \otimes e_s$, where $\{e_i\}$ is a frame in W_n and ω^i are forms on $E(X_n, G)$, we can represent the form Ω as

$$\Omega = T^k_{ij}\omega^i \wedge \omega^j \otimes e_k$$

and in this way introduce the torsion tensor T. If by q we denote the natural projection of the space $W_n \otimes W_n^* \wedge W_n^*$ into the factor space K, then it turns out that

$$q \circ T = S$$

for any G-connection ω_1. If the G-structure is integrable, i.e., if in the neighborhood of each point of manifold X_n we can choose adapted natural frames, then the structure tensor vanishes. In the general case the converse is false, and G-structures with zero structure tensors are said to be almost integrable.

Having constructed the complexification of the tangent vector spaces of manifold X_n, we can examine the principal fiber space of all complex frames of this manifold with structure group GL(n, C). After this it is natural to introduce the concept of a complex G-structure on the manifold X_n, defined by giving a fiber space of complex frames with structure group $G \subset GL(n, C)$.

If two G-structures are given by two different spaces H and H' of frames, going one into the other by some right shift, then these two G-structures are said to be equivalent: the structure groups of such G-structures are connected to each other by a similarity transformation. However, if the inclusion $H \subset H'$ holds, then we say that the first G-structure is subordinate to the second; in this case we have also $G \subset G'$ where G and G' are the structure groups of the two given G-structures. Bernard [166, 168] studied the question of the conditions under which a complex structure is equivalent to a real one, and also of when there does exist a real G-structure subordinate to a given complex one (in this case the complex G-structure is called an extension of the real structure).

Many G-structures may be defined with the aid of a tensor [168]. Let $\mathfrak{K}$ be a linear representation of the group GL(n, R) [or GL(n, C)] in some vector space W, let G be a subgroup of the group being considered, taking the vector $u \subset W$, into itself, and let W_u be the orbit described by the vector u under arbitrary transformations of the group being considered. Then the giving of a G-structure on X_n is equivalent to the giving, on a principal fiber space of real (complex) frames of manifold X_n, of a tensor of type $\mathfrak{K}$ with values in W_u, realizing a differentiable mapping in W_u. Here the adapted frames of the G-structure are distinguished as being those on which the tensor being considered takes a constant value (for example, u).

In this way we may define a Riemann structure (by picking out the orthoframes of manifold X_n), an almost complex structure, and a general tensor structure, studied by Clark and Bruckheimer [198] and also by Kobayashi [290].

For every G-connection on X_n we can construct its extension up to a connection on $E(X_n)$, i.e., up to a linear connection on X_n. If the G-structure $E(X_n, G)$ on X_n can be defined by a tensor t on $E(X_n)$, then the linear connection on X_n will be an extension of the G-connection if and only if the absolute differential of tensor t in this connection equals zero (Bernard [168], Clark and Bruckheimer [198]).

A number of authors ([165, 168] and others) have studied the automorphisms of G-structures. If $E(X_n)$ and $E'(X_n')$ are two principal fiber spaces of linear frames of manifolds X_n and X_n', then the regular diffeomorphism $\mu : X_n \to X_n'$ induces the prolonged mapping $\tilde{\mu} : E(X_n) \to E'(X_n')$. If a G-structure $E(X_n, G)$ is given on X_n, then $\tilde{\mu}$ defines a G-structure $E'(X_n', G)$ on X_n', being the image of the original G-structure under the diffeomorphism $\mu : E'(X_n', G) = \mu E(X_n, G)$. Such two G-structures $E(X_n, G)$ and $E'(X_n', G)$ are said to be isomorphic. The isomorphism of a G-structure onto itself is called its automorphism. If we consider the diffeomorphisms not of the manifolds X_n and X_n' themselves, but only of the two open sets $U \subset X_n$ and $V \subset X_n'$, then there arises the idea of a local isomorphism of two G-structures induced on U and V by the original G-structures. Two G-structures on X_n and X_n' are called locally isomorphic if for every pair of points $x \in X_n$ and $x' \in X_n'$ there exists the isomorphism of the G-structures induced in some neighborhoods of these points. The local automorphism of a G-structure is defined analogously. The set of local automorphisms of a G-structure forms a transformation pseudogroup Γ of manifold X_n. A G-structure is called homogeneous if Γ acts transitively on X_n; it is called isotropic if the prolongation $\tilde{\Gamma}$ of the pseudogroup Γ acts transitively on each layer of the space of adapted frames. If $\tilde{\Gamma}$ acts transitively on the whole space of admissible frames, then the G-structure is called transitive. If a G-structure is isotropic at a point $x_0 \in X_n$, then the structure tensor is constant on the layer of adapted frames over the point x_0; if a G-structure is transitive, then the structure tensor is constant on the whole space of adapted frames (see [168]).

Bernard has considered in detail analytic almost-transitive G-structures, i.e., analytic G-structures with a constant structure tensor. This analysis is closely related with the researches of Cartan [180, 181]. Infinitesimal pseudogroups and their application to G-structures were studied by Libermann [318, 320]. Let us note his results on such: if a G-structure is given on X_n and if in the neighborhood $\tilde{U}\subset X_n$ a cross section of the space $E(X_n, G)$ is indicated in the form of a field of adapted coframes $\{\omega^i\}$, then the local infinitesimal transformation v in the region $U\subset\tilde{U}$ defines a local infinitesimal automorphism of the G-structure if and only if the Lie derivative $L_v\omega^i$ has the form

$$L_v\omega^i = u^i_s(x)\,\omega^s,$$

where u^i_j defines at each point an element of the Lie algebra of group G.

Infinitesimal automorphisms of a G-structure were studied in Bernard's dissertation. He showed, for example, that if a G-structure is defined by a tensor t on $E(X_n)$, then the infinitesimal transformation v defines an infinitesimal automorphism of the G-structure if and only if the Lie derivative of tensor t with respect to the vector field v [or rather with respect to its prolongation v onto $E(X_n, G)$] vanishes. Fujimoto [220, 221] stated the conditions under which an automorphism of a G-structure is an automorphism of a G-connection and, conversely, when the infinitesimal automorphisms of a G-connection are infinitesimal automorphisms of G-structures. Analogous analysis had been carried out earlier by Hermann [258].

The papers by Guillemin, Sternberg, Singer, and Libermann [24, 235-237, 323, 324, 406], which preceded the investigations of Ehresmann [206-211], considered k-th order G-structures connected with the prolongations of the structure group (also see Cenkl [191], Clark [196], Kobayashi [292], Lewis [314], Ogiue [364], and Ruh [388]). These general constructions already brought completely concrete results. It will suffice only to mention Ruh's paper [388] in which simple criteria are obtained on when the automorphism group of a G-structure is a Lie group, on the basis of a consideration of prolonged G-structures and Lie algebras.*

* Chu and Kobayashi [195] have several theorems on when the transformation group of a manifold with structure is a Lie group; Ochiai's work [356] closely abuts the researches of Ruh.

To this same group of questions are closely related the fundamental investigations of G. F. Laptev on higher-order fibers over differentiable manifolds [68, 69].

Frölicher and Nijenhuis [218] and Nickerson [349] have constructed a theory of derivations for concrete and general G-structures.

Problems of the deformation of pseudogroup structures and G-structures have occupied Kodaira [294], Spencer [408-410], and Griffiths [232-234]. However, these investigations relate largely to the area of differential topology.

We mention a number of other researches in the theory of G-structures.

Ngo-van-Que [348] and Fujimoto [221] considered invariant G-structures on a uniform space. Also see [422].

Fujimoto [222, 223] studied the conformal geometry of G-structures. Let R_+ be a one-term subgroup in $GL(n, R)$, consisting of homotheties with a positive coefficient of similarity, and let the subgroup $G\subset GL(n, R)$ not contain the group R_+ in its composition. By $\hat{G}$ we denote the direct product of groups G and R_+. If a G-structure $E(X_n, G)$ is given on the manifold X_n, then we can construct a $\hat{G}$-structure $\hat{E}(X_n, \hat{G})$ obtained by extending the structure group G up to $\hat{G}$ and called the conformal extension of the original G-structure. Two G-structures on X_n are called conformally equivalent if their conformal extensions coincide. We can also define the homothetic equivalence of the G-structures $E(X_n, G)$ and $E'(X_n, G)$, which holds when there exists a fixed element $\lambda \in R_+$ such that

$$R_\lambda \cdot E(X_n, G) = E'(X_n, G),$$

where R_λ denotes a right shift in $E(X_n)$, determined by the element λ. Fujimoto indicated tests for the homothetic equivalence of two G-structures and analyzed conformal and homothetic transformations of G-structures.

The concept of a derived G-structure was introduced by Dazord [200].

Kobayashi and Nagano [293] found those Lie algebras G for which the corresponding G-structure admits of the existence of a connection with an arbitrarily given torsion tensor field on X_n.

Properties of G-connections have been investigated also by West [465], Libermann [320, 323, 324], and Nickerson [349]. The application of higher-order G-structures to the theory of conformal connections have been mentioned by Ogiue [363] (also see [176, 177]).

Srinivasacharyulu [411] has studied differentiable families of G-structures.

§2. Structure Defined by a Vector-Valued 1-Form and by a System of Such Forms

Among the different G-structures defined by means of tensors, particularly distinguished are those for which this tensor is an element of the space $W_n \otimes W_n^*$, i.e., is an affinor. In this case there arises a field of the affinor h on the manifold X_n. This field will be a special one (or, as we say, a 0-deformable one), i.e., will be such that at each point the matrix of affinor h can be reduced in a suitable frame to one and the same fixed form (for example, to the canonical Jordan form). Adapted frames will be those in which this matrix retains its form. Together with these it is interesting to study more general structures defined, for example, by fields of affinors with variable eigenvalues but with one and the same characteristic at each point of the manifold. One of the first papers relating to such a general affinor structure was that by Nijenhuis [350], which studied the conditions for the integrability (holonomicity) of the distributions defined by the contravariant or covariant eigenfunctions of affinor h. In these investigations a fundamental role is played by the tensor N defined in the natural frame by the formula

$$N^k_{ij} = 2h^s_{[i}\partial_{|s|}h^k_{j]} - 2h^k_s\partial_{[i}h^s_{j]}.$$

Subsequently this tensor was named the Nijenhuis tensor associated with the affinor field h. If we treat the affinor field h as a vector-valued 1-form $h=h^i_j dx^j \otimes e_i$, where $\{e_i\}$ is the frame dual to the coframe $\{dx^i\}$, then we can construct a vector-values 2-form [hh] whose value on any two vector fields u and v is computed by the formula $\frac{1}{2}[h, h](u, v) = [hu, hv] - h[hu, v] - h[u, hv] + hh[u, v]$. It is easy to be convinced that $\frac{1}{2}[h, h]=N^k_{ij}dx^i\otimes dx^j\otimes e_k$. In general, if two vector-valued 1-forms h and k are given, then we can construct their bracket [h, k] which is a vector-valued 2-form defined by the equality

$$[h, k](u, v) = [hu, kv]-h[u, kv]-k[hu, v]+hk[u, v]+[ku, hv]$$
$$-k[u, hv]-h[ku, v]+kh[u, v]$$

(in this regard see [351, 352, 217-219, 288]).

A number of authors have devoted themselves to the question of the integrability of a G-structure defined by a 0-deformable vector-valued 1-form. First of all we note Kobayashi [288, 289]. He studied in detail a wide class of 0-deformable vector-valued 1-forms h for which the conditions [h, h] = 0 ensured the integrability of the structure. (Since the Nijenhuis tensor vanishes for every integrable structure, the condition [h, h] = 0 is necessary for integrability.) In this connection he discovered that in the general case the condition [h, h] = 0 is not sufficient for the integrability of the structure. Let us cite an example he presented: in a four-dimensional manifold with local coordinates x^1, x^2, x^3, x^4 we consider the four vector fields $v_1=\frac{\partial}{\partial x^1}$, $v_2=\frac{\partial}{\partial x^2}$, $v_3=\frac{\partial}{\partial x^3}$, $v_4=(1+x^3)\frac{\partial}{\partial x^1}+\frac{\partial}{\partial x^4}$ and we construct the affinor h whose action on the vectors v_i is determined by the formulas

$$hv_1=v_2,\ hv_a=0 \quad (a=2, 3, 4).$$

Then we have [h, h] = 0, but by virtue of $[v_3, v_4]=v_1$ the distribution ker h is not integrable, so that the structure constructed is not integrable.

The equivalence of the condition [h, h] = 0 to the condition of integrability of a 0-deformable structure for affinors with simple roots of the minimal polynomial was proved by Muto [338] starting from analogous investigations by Kurita [297].

Hangan [238, 240] studied the field of the affinor h with variable eigenvalues but with constant characteristic. Having shown that the equality [h, h] = 0 is necessary and sufficient for the integrability of the distributions of the root subspaces, he next went on to an affinor h with a unique eigenvalue $\lambda(x)$, for which he constructed an auxiliary affinor $k = h - \lambda \cdot 1$, where 1 is the unit affinor. With the aid of this affinor he indicated the conditions under which the vanishing of the Nijenhuis tensor of affinor h is necessary and sufficient for the reducibility of the field h to the canonical Jordan form in local coordinates on X_n.

Nasu [347] studied the interrelations between the structure tensor of a G-structure defined by a 0-deformable vector-valued 1-form h and the Nijenhuis tensor of affinor h. Lehmann—Lejeune [309-311, 313] took up the question of the integrability of G-structures defined by 0-deformable affinor fields. She showed that the necessary and sufficient condition for the integrability of such a structure is the vanishing of the structure tensor, i.e., the existence of a torsion-free G-connection. (An analogous result was noted in [142].)

A number of authors have studied structures defined by a system of 0-deformable vector-valued 1-forms. Thus, Houh [261] considered simultaneously two affinor fields h and k satisfying the conditions $h^2 = \lambda \cdot 1$, $k^2 = \mu \cdot 1$, $hk = kh$, where $\lambda = \pm 1$, $\mu = \pm 1$. He showed that the structure defined by the pair of these fields will be integrable if and only if the vector-valued 2-forms $[h, h]$, $[k, k]$, and $[h, k]$ vanish identically. By generalizing this result, Hatakeyama [254] treated the case when the commuting affinors h and k have minimal polynomials without multiple roots and define 0-deformable 1-forms on X_n. In this case we can construct a 0-deformable vector-valued 1-form l such that h and k are expressed by polynomials in l with constant coefficients. The condition $[h, h] = [k, k] = [h, k] = 0$ guarantees the integrability of the structure determined by the 1-form l and, hence, in the neighborhood of each point on X_n there exist coordinates relative to which the affinors h and k have constant components. Thus the stated conditions are necessary and sufficient for the integrability of the structure defined by the considered pair of 1-forms h and k. This result carries over in a natural way to the case of a structure defined by a system of p intercommuting 0-deformable vector-valued 1-forms $h_1, h_2, \ldots, h_p$ possessing minimal polynomials with simple roots: the necessary and sufficient conditions for the integrability of such a structure have the form $[h_i, h_j] = 0$ for all possible pairs of indices i and j. This same circle of questions is taken up also by Sasaki [398], Hsu [262], Hsu and Houh [268]; also considered are the investigations on the so-called f-structures, almost-product structures, etc., which will be examined in the following sections.

Let us dwell further on the investigations of structures defined by general vector-valued 1-forms (i.e., not assuming 0-deformability). We have already noted above the papers by Frölicher and Nijenhuis [218] and by Hangan [240] pertinent here. Kurita [297] and Muto [338] studied the existence and the properties of a linear connection (with torsion, in general) relative to which the structure affinor is covariantly constant; here it was assumed that the minimal polynomial $f(\lambda)$ of the structure affinor does not have multiple roots; in this case, for example, Kurita proved that the above-mentioned connection exists if and only if the coefficients of the polynomial $f(\lambda)$ are constants.

Gaukhman [23] also studied the affinor structures $F^i_j(x)$ with simple roots of the minimal polynomial, calling them diagonalizable almost-tensor structures. He constructed a theory of connections on a manifold with a diagonalizable structure.

Tsalenko [137], generalizing the investigations of Gaukhman, analyzed the case of general tensor (affinor) structures given by the field of an arbitrary affinor $F^i_j(x)$ of class C^ω, whose characteristics are independent of the point. She introduced the notion of weak and strong integrability of a structure, as well as the weak and strong flag integrability of a manifold with tensor structure. She developed the theory of connections on a manifold with a general tensor structure.

Frölicher and Nijenhuis [217, 218], Hangan [240], and Lehmann — Lejeune [312] have studied derivations in the space of exterior forms defined by giving a vector-valued 1-form on X_n.

Slebodzinski [407] considered tensors and differential comitants, defined by the field of affinor h of class C^1, on a differentiable manifold of class C^2.

Tachibana and Koto [416] also have studied aspects of the differential geometry of manifolds X_n with a structure defined by a general affinor field φ^i_j. By analogy with the case of almost-complex structures (see [4]) they introduced the concepts of pure and hybrid tensors on such a manifold, and also examined certain properties of the so-called invariant subspaces. A submanifold $X_m \subset X_n$ $(m < n)$, given in local coordinates by the equation $x^i = x^i(\xi^a)$, is called an invariant subspace if the tangent m-plane to X_m goes into itself under the endomorphism $v^i \to \varphi^i_s v^s$. If we denote $B^i_a = \partial x^i / \partial \xi^a$, then the relation

$$\varphi^i_s B^s_a = \varphi^c_a B^i_c$$

is satisfied for an invariant subspace, whence it follows that a general affinor structure φ^a_b is induced on such a submanifold. It turns out here that if the Nijenhuis tensor of the affinor φ^i_j equals zero, then the Nijenhuis tensor of the affinor φ^a_b vanishes. Precisely similar results were obtained by Yano [475] for the invariant subspaces of a manifold with almost-complex structure. We further note Muto's paper [339] which considers a manifold with a general affinor structure φ^i_j, where, moreover, the minimal polynomial of affinor φ^i_j has simple roots. By generalizing an analogous concept from the theory of manifolds with almost-complex structure, Muto called a vector field v almost analytic if $L_v\varphi = 0$, where L_v is the Lie derivative defined by field v. He showed that the system $L_v\varphi = 0$ will be complete if and only if the structure φ is 0-deformable and integrable; in addition, he also studied the completeness conditions for the system $L_v\varphi^i_j = a^i_{js}v^s$, where a^i_{jk} is a tensor of class C^∞. Tachibana [413] also has constructed generalized analytic tensors.

§ 3. Distributions in Riemann Spaces and in Affinely-Connected Spaces

Distributions on differentiable manifolds, playing an important role both in the general theory of manifolds as well as in the area of applications (for example, see [2]), have been studied by many authors. The giving of a distribution, i.e., of a differentiable field of p-dimensional area elements,* defines on a manifold X_n the so-called T-structure, especially investigated by Legrand [304] (see § 9 below). Independently of this point of view, various distributions in Riemann spaces and in affinely-connected spaces were made the subject of study even in the very first stages of the de-

* Translator's note: The Russian word here is "ploshchadka" which means "small area" or "area element." However, the titles of the papers cited use the word "plane" in this connection.

velopment of multidimensional geometry. For example, here we could refer to vector field theory and to the theory of nonholonomic manifolds. Therefore, the literature on the various aspects of the theory of distributions cannot be surveyed completely, and here we shall chiefly mention only the papers of the last decade, which have a direct bearing on the theme of the present survey. We also shall not touch upon the topological investigations relating to manifolds with differentiable distributions (for example, see [467, 190]).

Let us start with the earlier papers of Walker [457-459] which studied parallel fields of area elements in Riemann spaces with undeterminate metrization. A field (distribution) of p-dimensional area elements is said to be parallel if a parallel displacement of an area element of this field along any curve leads it to an area element of this same field. However, if in the area elements of the field we can choose a basis formed by p covariantly constant vectors, then such a field is called strictly parallel. Walker showed that parallel fields of area elements are connected in a natural way with recurrent tensor fields and indicated the canonical form of the matrix of the metric tensor of a Riemann space admitting of fields of parallel or strictly parallel null area elements (all vectors of which are isotropic) and also of fields of parallel partially null area elements. Ruse [389] classified four-dimensional Riemann spaces admitting of parallel fields of null or partially null area elements and also considered partially null parallel fields in a general space V_n [390].

Wong [471], generalizing Walker's investigations, studied parallel and strictly parallel fields of area elements in a torsion-free affinely-connected space A_n.

In subsequent papers [461-463], Walker considered the problem of seeking connections in which a distribution given globally on a manifold turns out to be parallel. In fact, Walker gives on a manifold X_n two distributions D and $\overline{D}$ complementary to each other, of which the distribution $\overline{D}$ plays a fundamental role while the second distribution $\overline{D}$ is used only for the construction of two affinors $\underline{a}_i^i$ and $\overline{a}_j^i$, effecting projections onto the area elements $\grave{D}_x$ and $\overline{D}_x$, respectively, at each point $x \in X_n$. By choosing on X_n a global torsion-free connection $\overset{\circ}{\Gamma}$ he seeks a global connection $L_{jk}^i = \overset{\circ}{\Gamma}{}_{jk}^i + T_{jk}^i$, such that, in this connection, the distribution D would be parallel and such that, in the case when the distribution D is totally integrable, the connection L would have zero torsion. This problem can always be solved and one of the possible solutions is given by the formula

$$T^{i}_{jk} = \overset{\circ}{\nabla}_j a^i_s \cdot a^s_k - a^s_j \overset{\circ}{\nabla}_k a^i_s + a^t_j a^s_k \overset{\circ}{\nabla}_t a^i_s,$$

where $\overset{\circ}{\nabla}$ is the symbol for the covariant derivative in the connection $\overset{\circ}{\Gamma}$. This result can be generalized to any system of nonintersecting distributions $D_1, \ldots, D_m$. Such a system is called integrable if every partial direct sum of these distributions is totally integrable. Walker showed that there exists a global linear connection in which all the distributions D_α are parallel and which has zero torsion if the distribution system is integrable.

Willmore [466] proved this same theorem with the use of exterior forms (also see [467]). Global parallel distributions have been considered by Walker also in other papers [468-470].

Yano [473] carried out a similar analysis, studying the existence of global connections in which a globally given distribution is flat (semiparallel in Walker's sense), geodesic, or parallel. (A distribution is termed geodesic if every geodesic belonging to this distribution at one of its own points retains this property at all of its own points; it is termed flat if its area elements are parallel along any integral curve of this distribution.)

The fields of parallel area elements and paths in Riemann spaces were studied also by Debever [201], Ivanova [51], Kashiwabara [282-285], Ichijyo [272], etc.

Adati and Kandatu [155, 156, 280] generalized the above-mentioned investigations of Walker by studying Riemann spaces with fields of partially null area elements, closely related by properties with parallel fields, in particular, possessing the property that the area elements of the field intersect under a development along any curve. Similar fields of area elements (called developable) in torsion-free affinely-connected spaces were examined by Temirov [134]; he studied the motion groups of such spaces (see [32]).

Shapiro [138] investigated Riemann spaces with geodesic fields of m-paths, i.e., holonomic fields of area elements such that every (m + 1)-dimensional surface, formed by the integral surfaces of this field, passing through the points of any geodesic line, is a completely geodesic surface.

Special Riemann spaces, possessing a system of completely geodesic surfaces, namely "mirrors," were studied by Sabinin under the name subsymmetric spaces [121-123].

Shapukov [139] considered completely geodesic nonholonomic manifolds, arising naturally in spaces with a bilinear metric.

Distributions of $(n-1)$-dimensional area elements in Riemann spaces V_n also were the subject of Tanno's analysis [432, 433] who with their help defined generalized conformal transformations (which he called partially conformal). These are point transformations of Riemann space, which preserve the angles between paths belonging to the area elements of the field.

Legrand [307] and Tondeur [447] studied the properties of invariant distributions on uniform spaces.

§ 4. Almost Product Structures

The topological product $X_n = X_p \times X_q$ of two differentiable manifolds X_p and X_q possesses the property that in it there exist two mutually complementary area elements of dimensions p and q touching the submanifolds $X_p \times y$ and $x \times X_q$, where $x \in X_p$ and $y \in X_q$. In other words, in such a manifold there are two complementary totally-integrable distributions. If in some differentiable manifold X_n there are given two mutually-complementary distributions, their total integrability not being required, then we obtain a manifold with an almost-product structure. A complex almost-product structure on X_n is defined analogously if in a complexified tangent space we give two mutually-complementary complex subspaces at each point $x \in X_n$. For the sake of brevity almost-product structures are also called π-structures. Every π-structure is uniquely determined by a field of affinor f satisfying the equation $f^2 = 1$, in which the eigenvectors corresponding to the root +1 (respectively, −1) constitute area elements of the first (respectively, second) distribution. We call such an affinor a structure affinor. By taking as the adapted frames those in which the first p vectors belong to area elements of the first distribution and the last q to area elements of the second, we obtain a G-structure on X_n with structure group $G = KL(p, q)$, where K is the field of real or complex numbers; this group is represented by the matrices

$$\begin{pmatrix} A & 0 \\ 0 & B \end{pmatrix},$$

where A and B are nonsingular blocks of p-th and q-th orders, respectively, real or complex. For simplicity let us consider real π-structures. We denote the adapted frame by $\{e_\alpha, e_{\bar\alpha}\}$, where $\alpha = 1, \ldots, p$ and $\bar\alpha = p+1, \ldots, n$, and we let $\{\omega^\alpha, \omega^{\bar\alpha}\}$ be the coframe dual to it. Then the Lie algebra $\underline{G}$ of the structure group G has the basis $\{e_\alpha \otimes \omega^\beta, e_{\bar\alpha} \otimes \omega^{\bar\beta}\}$. The factor space $K = (W_n \otimes W_n^* \wedge W_n^*)/\partial(\underline{G} \otimes W_n^*)$ considered in § 1 can be identified here with the subspace in $T_x \otimes T_x^* \wedge T_x^*$ constructed on the elements

$$\{e_\alpha \otimes \omega^{\bar\beta} \wedge \omega^{\bar\gamma}\ (\bar\beta < \bar\gamma),\ e_{\bar\alpha} \otimes \omega^\beta \wedge \omega^\gamma\ (\beta < \gamma)\}.$$

Every G-connection (almost-product connection) can in our case be canonically identified with a linear connection whose form in an adapted frame can have only ω^α_β and $\omega^{\bar\alpha}_{\bar\beta}$ as nonzero components. This connection is characterized by the fact that relative to it the structure affinor is covariantly constant. Taking into account what we said in § 1, we see that the structure tensor S of the G-structure being considered is defined by the components $t^\alpha_{\bar\beta\bar\gamma}$ and $t^{\bar\alpha}_{\beta\gamma}$ of the torsion tensor of an arbitrary G-connection. In the G-connection in which the torsion tensor coincides with S there hold the structure equations

$$d\omega^\alpha = \omega^\sigma \wedge \omega^\alpha_\sigma + \frac{1}{2} t^\alpha_{\bar\beta\bar\gamma} \omega^{\bar\beta} \wedge \omega^{\bar\gamma},$$

$$d\omega^{\bar\alpha} = \omega^{\bar\sigma} \wedge \omega^{\bar\alpha}_{\bar\sigma} + \frac{1}{2} t^{\bar\alpha}_{\beta\gamma} \omega^\beta \wedge \omega^\gamma.$$

We have

$$d\omega^\alpha \equiv \frac{1}{2} t^\alpha_{\bar\beta\bar\gamma} \omega^{\bar\beta} \wedge \omega^{\bar\gamma} \pmod{\omega^\beta},$$
$$d\omega^{\bar\alpha} \equiv \frac{1}{2} t^{\bar\alpha}_{\beta\gamma} \omega^\beta \wedge \omega^\gamma \pmod{\omega^{\bar\beta}}. \qquad (1)$$

The tensor t^i_{jk}, having the nonzero components $t^\alpha_{\bar\beta\bar\gamma}$ and $t^{\bar\alpha}_{\beta\gamma}$, in the adapted frame, defined from Eqs. (1), was considered by Legrand [301, 303, 306] (also see Kerbrat [286]), who called it the torsion tensor of the π-structure. Thus, in the case being considered the structure tensor S is identified with the torsion tensor of a π-structure. This result is valid both for real as well as for complex π-structures. The difference between the complex and the real cases is apparent in the fact that only in a real π-structure is the vanishing of its torsion tensor equivalent to the integrability of this structure. In this connection we remark that the torsion tensor of a π-structure differs by only a constant factor from the Nijenhius tensor of a structure affinor. By introducing into consideration the torsion form of a π-structure, i.e., a vector 2-form defined by the torsion tensor of this structure, Legrand established the simple relation of such a form with the curvature form of an infinitesimal connection on a principal fiber space [305]: the fields of vertical and horizontal area elements in a fiber space determine a π-structure and the torsion form of this π-structure is identified in a natural way with the curvature form of an infinitesimal connection. Hsu [264] and Walker [464] considered manifolds with $r - \pi$-structures which differ from π-structures by the presence of r nonintersecting real or complex distributions, the sum of whose dimensions equals the dimension of the space; analogously derived for them was the concept of the torsion tensor of the structure, whose vanishing in the real case characterizes the integrability of the structure.

The main attention of the investigators concerned with manifolds with an almost-product structure was drawn to the study of different connections in such spaces. The almost-product connection has already been noted above; using this connection Legrand and Hsu determined the characteristic differential forms of a π-structure and of an $r - \pi$-structure [303, 264]; other types of connections are mentioned by Legrand [302, 303], Fukami [226], and a number of other authors.

The existence of certain remarkable global connections in almost-product spaces was proven by Walker [461, 462] and Willmore [466], of which we have already spoken in the preceding section. Walker [462] studied a manifold with a pair of complementary distributions D and $\overline{D}$ which define a pair of fields of projecting affinors. Walker proved the existence of global connections in which D and $\overline{D}$ are geodesic, parallel, or even parallel one along the other (a distribution D is said to be parallel along $\overline{D}$ if the area elements of distribution D still belong to it after a parallel displacement along any integral curve of distribution $\overline{D}$). These connections were described in terms of the projecting affinors. Yano's work [473], mentioned in the preceding section, immediately abuts these investigations of Walker and is devoted to the study of linear connections in manifolds with a real almost-product structure defined by the field of the structure affinor f^i_j (also see [472]). We denote by D and $\overline{\mathrm{D}}$ the distributions having the π-structure being considered; assume that distribution D corresponds to the eigenvalue +1 of the structure affinor, while distribution D corresponds to the eigenvalue −1. We denote the Nijenhuis tensor of the structure affinor by N^k_{ij}. Then the condition for the total integrability of distribution D (respectively, $\overline{\mathrm{D}}$) acquires the form $N^k_{ij} - N^s_{ij} f^k_s = 0$ (respectively, $N^k_{ij} + N^s_{ij} f^k_s = 0$), while the condition $N^k_{ij} = 0$, naturally, characterizes the integrability of the π-structure. Let $\overset{\circ}{\Gamma}$ denote an arbitrary torsion-free connection, globally given on the manifold being considered, and let us introduce the tensor

$$M^k_{ij} = f^s_i \overset{\circ}{\nabla}_s f^k_j + f^s_j \overset{\circ}{\nabla}_i f^k_s,$$

where $\overset{\circ}{\nabla}_i$ is the covariant derivative in this connection. Every globally defined connection can now be defined in the form $\Gamma^k_{ij} = \overset{\circ}{\Gamma}{}^k_{ij} + T^k_{ij}$. In this case, for example, the distribution D will be geodesic relative to connection Γ if

$$T^k_{ij} = -\frac{1}{8}\left[M^k_{(ij)} - M^s_{(ij)} f^k_s\right].$$

In just the same manner Yano found globally defined connections in which the field D is parallel or flat, in which both fields D and $\overline{\mathrm{D}}$ are flat, parallel, or geodesic, in which both fields are geodesic and one is parallel along the other, etc.

Houh [260] generalized these results to the case of $r - \pi$-structures.

Yano [472] and Hsu [266] investigated the analogs of Walker differentiation (see [4]) in manifolds with a π-structure and with an $r - \pi$-structure.

Spaces with Riemann metrics, in which the fundamental distributions of the π-structure are mutually orthogonal, were examined by Walker [460], Aragnol [160], Chern [193, 194], Fukami [226], Hsu [267], Reinhart [385-387], Tachibana [412], Cattaneo-Gasparini [185-189], etc.

Connections in manifolds with complex almost products were also studied by Cattaneo-Gasparini [185, 186, 189]. The physical applications of certain Riemann structures, almost products, were studied by Cattaneo [182-184].

Saeki [391] and Kerbrat [286] were occupied with affine transformations of an affinely-connected space with a π-structure. We say that a linear almost-product connection (see above) or a π-connection has a homogeneous π-irreducible holonomy group if this group leaves invariant the fundamental area elements of the π-structure but is irreducible in each of them. A manifold with such a connection is called π-irreducible. Saeki proved that if in a π-irreducible manifold there does not exist an isomorphism between the area elements of the π-structure, permuting with all transformations of the homogeneous holonomy group, then the affine transformation group of such a manifold preserves the π-structure. Kerbrat studied infinitesimal affine transformations in a special connection, i.e., in a linear torsion-free connection $\{\omega^i, \omega^i_j\}$, for which the connection $\hat{\omega}^i_j = \omega^i_j - \frac{1}{2} t^i_{js}\omega^s$ is a π-connection (here t^i_{jk} is the torsion tensor of the π-structure). Infinite transformations in metric spaces, in almost products, were considered by Hsu [267] and Tachibana [412].

In his survey [32], Egorov has considered the papers which study the reducibility of spaces admitting of motion groups.

Reinhart [383, 384] has investigated the decomposition of the curvature and of the torsion of an affinely-connected space, arising as the result of the expansion of the tangent space into a direct sum of subspaces; obviously, the results of these investigations may be used in the theory of spaces, namely, of almost products.

We now dwell on the papers by Norden and his students, which have a direct bearing to spaces, to almost products. In [97], Norden examined a topological product of r independent differential manifolds and arrived in this manner at a differentiable manifold X_n with an integrable $r-\pi$-structure which he called the composition of the original r base manifolds. Through each point of the composition X_n there pass r positions of the base manifolds, being the integral manifolds of the $r-\pi$-structure. A composition is called Cartesian if in the space X_n there is given a linear torsion-free connection in which all the distributions of the $r-\pi$-structure being considered are parallel. Examples of Cartesian composition spaces are reducible Riemann spaces as well as the Riemann spaces V_{2n} which are the Cartesian composition of two n-dimensional manifolds with completely isotropic positions (the spaces of Shirokov and Keller [67] and the spaces of Rashevskii [117]). By giving, in a projective space P_n, a system S_m^r of r copies of m-dimensional surfaces $\overset{a}{X}_m$ ($a = 1, \ldots, r$; $n = m + r - 1$) Norden constructed the composition Π_{rm} of these surfaces. For each position of the surface $\overset{a}{X}_m$ in a natural way there is defined its normalization (positional normalization), inducing an equiprojective connection on the position being considered. As a result, a connection arises in the composition Π_{rm} relative to which it turns out to be a Cartesian composition. When $r = 2$ the system S_m^2 represents a collection of two hypersurfaces in P_{m+1} and determines a Cartesian composition for which Norden has studied the parallel displacement of vectors and the geodesic images. In particular, if the system S_m^2 has been formed two regions of points of one and the same hyperquadric $(x, x) = 0$, then Π_{2m} is called the autocomposition of this hyperquadric; the intrinsic Riemann geometry of the autocomposition Π_{2m}, of the Shirokov — Rashevskii type, is determined by the kernel function $\theta = \ln(x, \bar{x})$, where x and $\bar{x}$ are points of both regions being considered.

The researches of Norden were continued by his students: Pisareva [105-107], who considered almost-reducible Weyl spaces and almost-reducible affinely-connected spaces; Leont'ev [72-74], who studied Chebyshev compositions in affinely-connected spaces, i.e., those compositions in which the area elements of the positions of one family are displaced parallel along the positions of a second family (multidimensional analogs of Chebyshev nets); Kolyan [62], who investigated the geometry of hypercircles in an n-dimensional conformal space with the aid of the autocomposition of a hyperquadric, representing a conformal point space; Matevosyan [77, 78], who examined the differential geometry of submanifolds in reducible Riemann spaces.

In this connection we note that a paper [373] by Okumura has appeared recently, in which hypersurfaces in a reducible Riemann space are studied.

§ 5. Almost Contact Surfaces and Their Generalizations

1. The intensive investigation of contact and almost-contact structures on manifolds started roughly at the end of the Fifties when Gray's articles [229-231] appeared, as well as that by Boothby and Wang [175], devoted to contact structures on manifolds. On a (2n + 1)-dimensional differentiable manifold a contact structure is given by a Pfaffian form η possessing the property that $\eta\wedge(d\eta)^n\equiv\eta\wedge d\eta\wedge\ldots\wedge d\eta\neq 0$. A manifold M equipped with a contact structure is called a contact manifold. The condition $\eta\wedge(d\eta)^n\neq 0$ signifies that the exterior 2-form $d\eta$ has rank 2n. At each point of a contact manifold there appears a 2n-dimensional area element formed by the vectors which transform the form η to zero, and there also appears a one-dimensional subspace, nonincident on this area element, of vectors X for which $d\eta$ (X, Y) = 0 for all vectors Y tangent to M. By normalizing the vector X so that the equality η (X) = 1 holds, we obtain a nonzero vector field on the manifold M. The contact structure (and the form η are called regular if every point of the manifold possesses a coordinate neighborhood in which the integral curves of the vector field mentioned are given by the equations $x^2=c^2, \ldots, x^{2n+1}=c^{2n+1}$ (c^i=const).

The terms "contact manifold" and "contact structure" are justified by the fact that if we represent the form η locally in the form $\eta=dz-\sum_{\alpha=1}^{n} y_\alpha dx^\alpha$ (see [116]) and we consider the (n + 1)-dimensional Euclidean space E_{n+1} with Cartesian coordinates $(x^1, \ldots, x^n, z)$ of a point, in which the y_α are interpreted as the tangential coordinates of the hyperplane passing through the point being considered, then the contact transformations of a certain region of the space E_{n+1} are represented by local diffeomorphisms of the contact manifold, preserving the form η to within a factor.

A distribution on a contact manifold, defined by the equation η = 0, is called a contact distribution. The greatest dimension of an integral submanifold of a contact distribution equals n [401].

The first investigators of contact manifolds were primarily interested in the topological aspects. Thus, for example, Boothby and Wang [175] paid particular attention to a compact manifold M with a regular contact form η and showed that such a manifold is a principal fiber space with a fibration by a circle over a symplectic manifold, in which the form η yields a connection, while the fundamental bilinear form of the symplectic manifold is determined by the curvature form of this connection. They studied homogeneous contact manifolds and established the cases in which these manifolds are homeomorphic to the bundle of unit tangent vectors of some manifold. Gray [231] also studied the global properties of contact manifolds. He cited an example of an odd-dimensional manifold which does not admit of a contact structure, and he also studied the deformations of contact structures. He also introduced the concept of an almost-contact manifold, i.e., a (2n + 1)-dimensional manifold the structure group of whose tangent bundle can be shortened (reduced) to the group $1 \times U(n)$, where $U(n)$ is a unitary group [in such a case the principal fiber space of all linear frames of the manifold admit of the existence of a principal fiber subspace with a structure group $1 \times U(n)$].

2. Almost-contact structures soon became the object of numerous differential-geometric investigations. Thus, Sasaki [397] showed that an almost-contact structure is induced on a manifold X_{2+1} by the so-called (φ, ξ, η, g)-structure defined by the fields of the affinor φ, the vector ξ, the covector (Pfaffian form) η, and the symplectic positive-definite bivalent tensor g, where this tensor satisfies the following conditions:

$$\operatorname{rank}(\varphi_j^i)=2n,\quad \eta_i\xi^i=1,\quad \varphi_s^i\xi^s=0,\quad \varphi_j^s\varphi_s^i=-\delta_j^i+\eta_j\xi^i, \tag{1}$$

$$\eta_i=g_{is}\xi^s,\quad g_{pq}\varphi_i^p\varphi_j^q=g_{ij}-\eta_i\eta_j. \tag{2}$$

If on the manifold X_{2n+1} there are given only the tensor fields φ, ξ, η satisfying conditions (1), then we say that a (φ, ξ, η)-structure has been defined on X_{2n+1}. To each such structure we can always (nonuniquely) add a positive-definite metric g satisfying conditions (2), i.e., we can convert it to a (φ, ξ, η, g)-structure. Sasaki showed that on a contact manifold with a contact form $\eta = \eta_i dx^i$ we can construct a (φ, ξ, η, g)-structure such that in local coordinates $\varphi_{ij}=\frac{\partial\eta_j}{\partial x^i}-\frac{\partial\eta_i}{\partial x^j}$, where $\varphi_{ij}=\varphi_i^s g_{sj}$. The existence of this structure follows also from the more general results of Hatekeyama [252]. Subsequently, a (φ, ξ, η)-structure came to be called an almost-contact structure [since the tangent bundle group of the manifold with such a structure can always be reduced to $1 \times U(n)$], while (φ, ξ, η, g)-structures were increasingly named almost-contact metric structures as well as almost-Gray structures. However, if η is a contact form, i.e., satisfies the condition $\eta \wedge (d\eta)^n \neq 0$ and if in addition the relation $\frac{1}{2}\varphi_{ij} = \nabla_{[i}\eta_{j]}$, is fulfilled, then the (φ, ξ, η, g)-structure is called a contact metric structure. However, Yano and Ishihara [480] suggested that the name contact metric structure be given to the contact (φ, ξ, η, g)-structure for which the condition $\nabla_{[i}\eta_{j]} = a\varphi_{ij}$ $(a = \text{const})$ is fulfilled (compare below with the definition of almost-Sasakian and Sasakian structures). An almost-contact structure is called integrable if the Nijenhuis tensor of the affinor φ^i_j vanishes.

3. If X_{2n+1} is a manifold with a (φ, ξ, η)-structure, then we can analyze the topological product $M = X_{2n+1} \times R$, where R is the real number line [403, 251]. We choose local coordinates x^α $(\alpha, \beta, \ldots = 1, \ldots, 2n + 2)$ on M is such a way that the first $2n + 1$ of them coincide with the local coordinates x^i on X_{2n+1}, while the last one, x^{2n+2}, with the coordinate on R. Then on the manifold M there arises in a natural way an affinor φ^α_β of almost-complex structure, whose matrix in the local coordinates mentioned has the form

$$(\varphi^\alpha_\beta) = \begin{pmatrix} \varphi^i_j & \eta_j \\ -\xi^i & 0 \end{pmatrix}. \tag{3}$$

The presence of a (φ, ξ, η, g)-structure in X_{2n+1} permits us to construct in the manifold M a metric tensor $g_{\alpha\beta}$ with the matrix, in local coordinates

$$(g_{\alpha\beta}) = \begin{pmatrix} g_{ij} & 0 \\ 0 & 1 \end{pmatrix}. \tag{4}$$

From (1) and (2) it follows that $\varphi^\sigma_\alpha g_{\sigma\beta} = -\varphi^\sigma_\beta g_{\sigma\alpha}$, i.e., that relative to the almost-complex structure (3) the metric (4) is almost Hermitian (see [4]). For an affinor of almost-complex structure (3) we can construct the Nijenhuis tensor $N^\alpha_{\beta\gamma}$; its components, pertaining to certain indices of the submanifold X_{2n+1}, define four tensors on X_{2n+1}, expressed in local coordinates by the formulas

$$\begin{gathered} N^i_{kj} = \varphi^s_k(\partial_s\varphi^i_j - \partial_j\varphi^i_s) - \varphi^s_j(\partial_s\varphi^i_k - \partial_k\varphi^i_s) + \eta_k\partial_j\xi^i - \eta_j\partial_k\xi^i, \\ N_{kj} = \varphi^s_k(\partial_j\eta_s - \partial_s\eta_j) - \varphi^s_j(\partial_k\eta_s - \partial_s\eta_k), \\ N^i_j = \xi^s(\partial_s\varphi^i_j - \partial_j\varphi^i_s) - \varphi^s_j\partial_s\xi^i, \\ N_j = \xi^s(\partial_j\eta_s - \partial_s\eta_j). \end{gathered} \tag{5}$$

Here the partial derivatives can be replaced by covariant derivatives in any torsion-free connection. The last three tensors can be expressed linearly in terms of the first and, therefore, when $N^i_{kj} = 0$ all the tensors in (5) vanish. In this case the (φ, ξ, η)-structure is called a normal almost-contact structure.* Morimoto [336] gave another interpretation of the concept of the normality of an almost-contact structure. A geometric interpretation of the normality condition for a (φ, ξ, η)-structure from the point of view of a $3-\pi$-structure defined on X_{2n+1} by the field of the affinor φ^i_j, was proposed by Sasaki and Hsu [405] (also see [402]). If a (φ, ξ, η, g)-structure (an almost-Gray structure) turns out to be normal, then it is called a Gray structure. Let us now assume that we have a contact (φ, ξ, η, g)-structure; then on the manifold there arise two 2-forms of rank 2n: the 2-form $d\eta$ and the so-called fundamental 2-form Φ having the local representation $\Phi = \frac{1}{2}\varphi_{ij}dx^i \wedge dx^j$, where $\varphi_{ij} = \varphi^s_i g_{sj} = -\varphi_{ji}$ If it turns out that $\Phi = d\eta$, then the tensors N_i and N_{ij} vanish and the original (φ, ξ, η, g)-structure is called almost-Sasakian (or, as noted above, a contact metric structure); if, however, it is also normal, then it is called a Sasakian structure.

For a Sasakian structure there is fulfilled the relation $\varphi^i_j = 2\nabla_j\xi^i$, or $\varphi_{ji} = 2\nabla_j\eta_i$, where the covariant differentiation is carried out in the metric g_{ij}; hence it follows that in this case the unit vector field ξ^i will be a Killing field (see [404]).

If a contact metric structure is not normal but is such that ξ is a Killing vector, or such that equivalently $N^i_j = 0$, then it is called a K-contact metric structure. Manifolds with a K-contact metric structure possess the property that their curvature in the direction of any two-dimensional area element containing the vector ξ is constant at each point [256]. The properties of the curvature tensor of a Sasakian space have been studied by Okumura [366], Ogiue [361], Ogawa [359, 360], and others. An odd-dimensional sphere is an example of a Sasakian manifold [404]; Goldberg [227] proved that a simply-connected complete manifold with a regular Sasakian structure, possessing a positive curvature in two-dimensional directions and a constant scalar curvature, is globally isometric to the hypersphere S^{2n+1} of the Euclidean space E_{2n+2}.

* Sasaki [402] suggested that the tensor N^i_{jk} be called the torsion tensor of the almost-contact manifold.

Tanno [435] has shown that every three-dimensional K-contact metric manifold is normal and that every three-dimensional simply-connected complete K-contact metric manifold with a constant scalar curvature $R > -2$ is diffeomorphic to a sphere.

4. A very close interrelation exists between manifolds with an almost-contact structure and with an almost-complex structure. Boothby and Wang [175] have already shown that a compact simply connected homogeneous contact manifold is a principal T^1-fiber space over a homogeneous Kahlerian manifold (here T^1 is the group of a one-dimensional torus). One-dimensional fibrations over manifolds with a complex structure were studied also by Morimoto [336, 337]. He showed that if X_{2n+1} is a principal fiber space with a circle fibration over a complex manifold M and if on X_{2n+1} there exists a connection form η possessing the property $d\eta = p^*\Omega$, where $p: X_{2n+1} \rightarrow M$ is the projection of X_{2n+1} onto M and Ω is a 2-form on M, invariant relative to the almost-complex structure operator, then a normal almost-contact structure (φ, ξ, η) exists on X_{2n+1}. The converse proposition is also valid. Morimoto called a principal fiber space X_{2n+1} with such an almost-contact structure a contact bundle over the complex manifold M and studied the properties of contact bundles. A more general treatment was given by Hatakeyama [253] and Ogiue [362]. Thus, Ogiue studied regular almost-contact (φ, ξ, η)-structures on a manifold X_{2n+1}, in which all the integral curves (orbits) of the vector field ξ are homeomorphic to each other. Furthermore, φ and η should be invariant under the action of the one-parameter group G generated by field ξ. In such a case X_{2n+1} turns out to be a principal fiber space $X_{2n+1}(M, p, G)$ with structure group G over a differentiable manifold M of the orbits of field ξ and, moreover, the form η yields a connection on X_{2n+1}. The manifold M turns out to be equipped with an almost-complex structure I defined by the equality

$$I_m(X) = p_*[\varphi_x(X_x^*)],$$

where X is a vector field on M, X^* is its lift in the connection η, $m = p(x)$, p_* is the differential of the projection $p: X_{2n+1} \rightarrow M$. The normality of a (φ, ξ, η)-structure on X_{2n+1} leads to the integrability of the almost-complex structure I and to the condition of invariance of the form Ω noted by Morimoto (see above). However, the integrability of a (φ, ξ, η)-structure is equivalent to the integrability of the structure I and to the total integrability of the equation $\eta = 0$ on the manifold X_{2n+1}. Ogiue also examined an almost-contact metric $(\varphi, \xi, \eta, \tilde{g})$-structure on X_{2n+1} with a G-invariant metric $\tilde{g}$: it induces an almost-Hermitian structure on M, and the fundamental 2-form Ω of an almost-Hermitian structure is related to the fundamental 2-form Φ of an almost-contact metric structure by the relation $\Phi = p^*\Omega$. Here an almost-Sasakian $(\varphi, \xi, \eta, \tilde{g})$-structure induces an almost-Kählerian structure on M, while the fundamental 2-form Φ of an almost-Sasakian structure is the curvature form of the connection η on X_{2n+1}. Tanno [425] carried out precisely the same analysis for K-contact manifolds X_{2n+1}.

In turn almost-complex manifolds may be fibered by one-dimensional layers over almost-contact manifolds. One-dimensional fibrations over manifolds with almost-contact structures, in particular over homogeneous contact manifolds, were studied by Ogawa [357, 358]. He showed, by generalizing several results of Sasaki and Hatakeyama, that if X_{2n+1} is an almost-contact manifold with a (φ, ξ, η)-structure, while A is a one-dimensional Lie group, then an almost-complex structure I, invariant relative to right shifts, arises on the principal fiber space $P(X_{2n+1}, p, A)$ equipped with an infinitesimal connection η. An almost-contact metric structure on X_{2n+1} here generates an almost-Hermitian structure on P. Ogawa studied the conditions for an almost-complex (I) and an almost-Hermitian structure on P to be integrable and Kahlerian, and also studied the properties of the lifts of vector fields on X_{2n+1}. Precisely similar analysis for one-dimensional fibrations over almost-complex and almost-contact manifolds was carried out by Tashiro [440] and Yokote [493]. Tanno studied almost-complex structures in principal fiber spaces with odd-dimensional layers over almost-contact manifolds [429] (also see §6).

5. Many authors have studied various transformations of contact and almost-contact manifolds. If X_{2n+1} is a contact manifold with a contact form η, then the diffeomorphism $f: X_{2n+1} \to X_{2n+1}$ is called a contact transformation of manifold X_{2n+1} if

$$f^*\eta = \sigma \cdot \eta,$$

where σ is a nonzero function on X_{2n+1} (when $\sigma = 1$ we call it a strictly contact transformation). A diffeomorphism f of a contact manifold will be a contact transformation if and only if it transforms every highest-dimensional integral submanifold of a contact distribution into another exactly identical submanifold [401]. An infinitesimal transformation X of a contact manifold is called a contact (respectively, exactly or strictly contact) transformation if $L_X\eta = \sigma \cdot \eta$ (respectively, $L_X\eta = 0$), where σ is a differentiable function called the associated function of transformation X. An infinitesimal transformation X is called an automorphism of a (φ, ξ, η, g)-structure if $L_X\varphi^i_j = 0$, $L_X\xi^i = 0$, $L_X g_{ij} = 0$; the first two of the conditions stated define an infinitesimal automorphism of a (φ, ξ, η)-structure. Finite automorphisms of these structures are defined as diffeomorphisms preserving the corresponding tensors.

We note a number of results obtained for transformations of contact and almost-contact manifolds. Sasaki and Hatakeyama [404] elucidated that the Lie algebra of the infinitesimal contact transformations of a contact metric manifold has infinite dimension. For a manifold X_{2n+1} with a contact metric structure we denote by Ψ the group of transformations which leaves the field φ invariant, by A the group of all automorphisms of a contact metric structure, and by I the isometry group of manifold X_{2n+1}. Tanno proved [424, 426] that Ψ is a Lie group and that the relations $\Psi \subset I$, and $\Psi = A$ hold for a K-contact manifold X_{2n+1} whose scalar curvature either is constant and different from $-n/2$ or is nonconstant and bounded. He also pointed out other types of contact Riemann manifolds for which the equality $\Psi = A$ is fulfilled (also see [427]). He has also shown [423] that if an infinitesimal contact transformation X of a contact metric manifold leaves the tensor φ_{ij} invariant, then X is an exact infinitesimal contact transformation; in a compact contact manifold an infinitesimal transformation X is a φ-transformation ($L_X\varphi^i_j = 0$) if and only if it is an isometry and at the same time an exact infinitesimal contact transformation. In connection with the last result we note Mizusawa's theorem [335] according to which an infinitesimal φ-transformation X satisfies in a contact metric space the equations $L_X\eta_i = \sigma\eta_i$, $L_X g_{ij} = \sigma(g_{ij} + \eta_i\eta_j)$, where σ = const, and conversely; Mizusawa proved also that in a K-contact metric space an infinitesimal affine transformation is an automorphism of the structure. A number of other properties of infinitesimal contact transformations of contact metric spaces have been established in [367, 423, 335]. Infinitesimal and finite automorphisms of contact and other structures have been treated also by Libermann [319], Chu and Kobayashi [195], and Hatakeyama [255].

We now present several results pertaining to normal contact metric structures. Okumura [367, 372] and Tanno [434] showed that under certain natural assumptions the presence of nonisometric conformal transformations (infinitesimal or finite) on such manifolds X_{2n+1} $(n > 1)$ stipulates the isometry of these manifolds to a sphere. In a normal contact metric manifold every infinitesimal transformation preserving the curvature tensor is an infinitesimal isometry (Okumura [372]). Okumura [367, 372], Mizusawa [335], Ogiue [361], and Sakai [392] have also pointed out other properties of infinitesimal isometric, conformal, and projective transformations of normal contact metric spaces and of K-contact spaces. Also see Tanno [438].

6. Passing on to almost-contact manifolds we note the results of Tanno [424], showing that on an almost-contact metric manifold a conformal transformation which leaves the characteristic 2-form Φ invariant is an automorphism of a (φ, ξ, η, g)-structure. Morimoto [336] proved that on a compact normal almost-contact manifold the group A(M) of its automorphisms is a Lie group of the transformations of M; he also proved that the groups A(M) and $A(\overline{M})$ of the automorphisms of two almost-contact manifolds M and $\overline{M}$ induce a subgroup of the automorphism group of the almost-complex manifold $M \times \overline{M}$. The properties of infinitesimal conformal and projective transformations in manifolds with a (φ, ξ, η, g)-structure were derived by Nakagawa [345]. Infinitesimal automorphisms of an almost-contact manifold and the properties of their liftings in a one-dimensional fibration with connection over such manifolds were considered by Ogawa [357]. Vector fields on contact and almost-contact manifolds were studied by Tanno [430, 431] in connection with their global properties. An analog of a geodesic mapping for Sasakian manifolds (the so-called CL-mapping) was studied by Koto and Nagao [296].

The significance of almost-contact structures for geometry is apparent also in that they arise naturally on hypersurfaces in complex and almost-complex spaces, and the properties of such hypersurfaces are of interest both in their own right as well as in applications (for example, see [421]). The differential-geometric study of hypersurfaces in almost-complex spaces was initiated by Tashiro (see [439]). In a $(2n + 2)$-dimensional differentiable manifold M with an almost-complex structure F^{α}_{β} let there be given a hypersurface X_{2n+1} by the equations, in local coordinates,

$$\xi^\alpha = \xi^\alpha(x^i)\,(\alpha, \beta, \ldots = 1, \ldots, 2n+2;\ i, j, \ldots = 1, \ldots, 2n+1).$$

We assume that the hypersurface X_{2n+1} has been rigged with a vector C^α possessing the property that the vector $F^\alpha_\sigma C^\sigma$ lies in the tangent hyperplane to the hypersurface. Denoting $B^\alpha_i = \frac{\partial \xi^\alpha}{\partial x^i}$, we have

$$F^\alpha_\sigma B^\sigma_i = \varphi^s_i B^\alpha_s + \eta_i C^\alpha, \quad F^\alpha_\sigma C^\sigma = -\xi^s B^\alpha_s.$$

In such a case, taking the equality $F^\sigma_\alpha F^\beta_\sigma = -\delta^\beta_\alpha$, we find the relations

$$\varphi^s_j \varphi^l_s = -\delta^l_j + \eta_j \xi^l, \ \varphi^l_s \xi^s = 0, \quad \varphi^s_i \eta_s = 0, \ \eta_s \xi^s = 1,$$

whence it follows that an almost-contact (φ, ξ, η)-structure arises on a specially rigged hypersurface. The collection $\{X_{2n+1}, C^\alpha\}$ is called an almost-contact hypersurface in an almost-complex manifold M. If the manifold M is almost-Hermitian and if the rigging vector C^α is the basis vector of the normal, then the metric tensor g_{ij} induced on X_{2n+1} is associated with a (φ, ξ, η)-structure and converts it to a (φ, ξ, η, g)-structure so that an almost-Gray structure is induced on the hypersurface in the almost-Hermitian manifold. The papers by Tashiro [439], Kurita [298], Okumura [368, 369, 371], Nagai and Kôjyô [340], Yano and Ishihara [480], and Sasagawa [394] contain many results on the properties of almost-contact structures and on the intrinsic geometry of rigged hypersurfaces in various complex and almost-complex spaces. We note several theorems relating to hypersurfaces in Kählerian spaces. In a Kählerian manifold an almost-Gray structure with a closed fundamental 2-form Φ is induced on a hypersurface. This structure will be Sasakian if and only if the second fundamental tensor h_{ij} of the hypersurface has the form

$$h_{ij} = g_{ij} + \mu \eta_i \eta_j,$$

where μ is a scalar field on the hypersurface. Every Sasakian manifold can be immersed in a Kahlerian manifold in the form of a completely umbilical hypersurface (Tashiro [439]). An almost-contact metric hypersurface of a Kahlerian space satisfies the condition $\triangle_k \varphi^i_j = 0$ if and only if its second fundamental tensor has the form $h_{ij} = \alpha \eta_i \eta_j$ (Ichijyo [271]). An almost-contact structure induced on a hypersurface of a Kahlerian space will be normal if and only if η_i is a Killing vector. A normal almost-contact structure is induced on a completely geodesic hypersurface of a Kahlerian manifold. A normal almost-contact hypersurface in the Euclidean space R^{2n+2}, equipped with a natural Kahlerian structure, is locally isometric to one of the following: S^{2n+1}, R^{2n+1}, $S^r \times R^s$ $(r + s = 2n + 1)$ (Okumura [368]). In particular, a hypersurface of the Euclidean space R^{2n+2}, on which a Sasakian structure is induced, is the hypersphere S^{2n+1} (Kurita [298], Okumura [368]). Okumura has devoted a lot of attention to normal almost-contact hypersurfaces in a Kählerian space of constant holomorphic curvature [369]. He showed that such a hypersurface has not more than three different principal normal curvatures which are all constant, and, moreover, the vector η_i corresponds to a simple root of the characteristic equation $\operatorname{Det}(h_{ij} - \lambda g_{ij}) = 0$. If only two different principal normal curvatures exist, then the hypersurface possesses a Sasakian structure and locally is a so-called C-Fubinian manifold (see [441]). Okumura [371] obtained analogous results for general contact hypersurfaces in a Kählerian space of constant holomorphic curvature.

7. The so-called f-structures, first considered by Yano [474, 476], are an immediate generalization of almost-contact structures. An f-structure is defined on a differentiable manifold X_n of arbitrary dimension by giving the field of an affinor f satisfying the equation $f^3 + f = 0$. If we construct two new affinors

$$l = -f^2, \quad m = f^2 + 1,$$

then they satisfy the relations

$$l + m = 1, \quad l^2 = l, \quad m^2 = m, \quad lm = ml = 0,$$

where 1 denotes the identity operator. Thus, at each point $x \in X_n$ the operators l and m effect the projection of the tangent space T_x onto two complementary area elements L_x and M_x so that two mutually complementary distributions L and M arise in the manifold X_n. The dimension r of the area elements L_x is always even, and in these area elements the affinor f acts as an operator of an almost-complex structure; the number r is called the rank of the f-structure. In the area elements M_x the affinor f acts as a null operator. If $n = 2p + 1$ and $r = 2p$, then the affinor f can be taken as the affinor φ of the almost-contact structure (φ, ξ, η), where the equation $\eta = 0$ yields the distribution L, while the vector field ξ yields the distribution M.

Let N denote the Nijenhuis tensor of a structure affinor f. In this case we can state the following mutually equivalent necessary and sufficient conditions for the integrability of distribution L:

$$m[lX, lY] = 0,$$
$$mN(X, Y) = 0,$$
$$mN(lX, lY) = 0,$$
$$mN(fX, fY) = 0$$

for any two vector fields X and Y.

Analogously we can write the mutually equivalent conditions for the integrability of distribution M:

$$l(mX, mY) = 0,$$
$$N(mX, mY) = 0,$$
$$lN(mX, mY) = 0.$$

The necessary and sufficient integrability conditions for both distributions L and M can be represented as

$$N(X, Y) = lN(lX, lY) + N(lX, mY) + N(mX, lY)$$

for any two vector fields X and Y.

If distribution L is integrable and if on each of its integral manifolds of maximal dimension r the induced affinor f of an almost-complex structure also is integrable, then the f-structure is called partially integrable. The necessary and sufficient condition for the partial integrability of an f-structure can be represented in one of two equivalent forms:

$$N(lX, lY) = 0, \quad N(fX, fY) = 0$$

for any two vector fields X and Y.

When the tensor N vanishes identically the f-structure is called integrable; in this case, in the neighborhood of every point of manifold X_n there exist local coordinates relative to which the matrix of the structure affinor takes the form

$$f = \begin{pmatrix} 0 & -1_m & 0 \\ 1_m & 0 & 0 \\ 0 & 0 & 0 \end{pmatrix},$$

where 1_m is an m-th order unit block and r = 2m (see Yano and Ishihara [479, 481]).

Yano showed [476] that on a manifold with an f-structure we can define a positive-definite Riemann metric g_{ij} possessing the property

$$f_i^s f_j^t g_{st} = g_{ij} - m_{ij}, \text{ where}$$

$$m_{ij} = m_i^s g_{sj}.$$

The distributions L and M are mutually orthogonal relative to this metric. An f-structure with a metric added on in this fashion is called an (f, g)-structure; it serves as a natural generalization of an almost-contact metric (φ, ξ, η, g)-structure. On a manifold with such a structure there arises a field of a skew-symmetric tensor

$$f_{ij} = f_i^s g_{sj}.$$

An (f, g)-structure is called normal if the relations

$$\nabla_{[i} f_{jk]} = 0,$$

$$\nabla_{(i} m_{jk)} = 0$$

are fulfilled, from which it follows that the distribution M is integrable and flat relative to the metric g_{ij}. A manifold with a normal (f, g)-structure is a natural generalization of a Sasakian manifold (Koto [295]).

Ishihara and Yano [276, 481] introduced the notion of normal f-structures, generalizing normal (φ, ξ, η)-structures and defined by a method close to the method of Sasaki and Hatakeyama (see [403] and **3** of the present section): they construct an almost-complex structure in the vector bundle belonging to distribution M and require the integrability of this structure.

In recent times there have come to be studied also the so-called framed f-structures (see [344, 481]); by a framed f-structure on a differentiable manifold X_n is meant the collection of f-

structures of rank r, of the (n − r) independent vector fields f_y spanning distribution M, and of the (n − r) covector fields f^x, (x, y = 1, . . . , n − r), possessing the properties $f^x(f_y) = \delta^x_y$, $m = f_x \otimes f^x$.

Just as almost-contact structures arise on the hypersurfaces of almost-complex spaces, so also f-structures are induced on surfaces of other dimensions in these spaces. If X_n is a surface in a convolute manifold M with an almost-complex structure I, and if $T(X_n)$ is a tangent fiber of this surface, then under the condition $\dim(T(X_n)_x \cap I(T(X_n)_x)) = n - r > 0$ there arises, after a suitable rigging, an f-structure of rank r at every point $x \in X_n$ on the surface X_n (see [482, 484, 342, 395]).

Infinitesimal transformations of manifolds with a normal (f, g)-structure and framed f-manifolds were considered by Koto [295] and Nakagawa [343].

8. Among other papers referring to almost-contact structures or to f-structures not touched upon above, we note the investigations by Tachibana [414, 415], Nakagawa [346], Lin [327], Takizawa [418], Fujitani [224], Ichijyo [270], Hsu [265], Boothby [174], and Tanno [436, 437].

In the present section we have completely ignored the papers on structures in the tangent fibers of differentiable manifolds because we shall be speaking of them in §6.

In conclusion, we particularly note the survey articles by Sasaki [402] and by Yano and Ishihara [481], devoted to almost-contact structures and to f-structures.

§6. Geometry of Tangent Fibers

Tangent fibers (tangent bundles) of differentiable manifolds, i.e., the fiber bundles of all their possible tangent vectors, have already been studied for a long time in topology (see [128]). As far as the differential-geometric study of these fibers is concerned, it started comparatively recently. In the first place here we should note the researches of Sasaki on the differential geometry of the tangent fibers of Riemann manifolds [396, 399]. Let X_n be a differentiable manifold and let x^i be its local coordinates; if v^i are the coordinates of some vector v at the point $x(x^i)$, then the collection (x, v) of the point x and the vector v comprises an element of the tangent fiber $T(X_n)$, so that in this fiber there arise the local coordinates (x^i, v^i). By assuming X_n to be a Riemann manifold V_n with the line element $ds^2 = g_{ij}dx^i dx^j$, Sasaki introduced in $T(V_n)$ the Riemann metric with the line element

$$d\sigma^2 = ds^2 + g_{ij}Dv^i Dv^j,$$

where Dv^i are the coordinates of the absolute differential of vector v. Sasaki's main attention was directed at the study of the liftings (lifts) of the geodesic lines of the Riemann manifold V_n, i.e., the families of curves in $T(V_n)$, described in local coordinates by the system of differential equations

$$\frac{dx^i}{dt} = v^i, \ \frac{dv^i}{dt} = -\left\{ {i \atop jk} \right\} v^j v^k.$$

He also studied the geometry of the tangent sphere fiber of manifold V_n, the $(2n-1)$-dimensional hypersurface in $T(V_n)$, representing the set of unit vectors of manifold V_n; a contact structure arises naturally in this fiber [399].

Simultaneously with this it was clarified that in tangent fibers $T(X_n)$ we can study different structures generated by structures or connections and given on the base manifold X_n. Nagano [341] has shown that an almost-complex structure can be defined in the tangent fiber of a Riemann manifold*; next, Tachibana and Okumura [417] established that relative to this structure the above-mentioned metric of Sasaki turns out to be almost-Kählerian; it will be Kählerian if and only if the original base manifold is locally Euclidean.

Dombrowski [203] obtained more general results, showing that the giving of a general linear connection in the base manifold X_n permits the construction of an almost-complex structure in the tangent fiber $T(X_n)$. If the connection on X_n has been defined by its own components Γ^i_{jk} relative to the local coordinates x^i, then by using the local coordinates (x^i, v^i) we can define on $T(X_n)$ the linear mapping $K: TT(X_n) \to T(X_n)$ which associates the vector

$$A = a^s \frac{\partial}{\partial x^s} + a^{n+s} \frac{\partial}{\partial v^s} \in TT(X_n)$$

* Nagano's results refer even to Finsler manifolds.

with the vector

$$KA = (a^{n+s} + \Gamma^s_{pq} a^p v^q) \frac{\partial}{\partial x^s}\,.$$

The kernel of this mapping at each point (x, v) $\in T(X_n)$ makes up the so-called horizontal vectors whose coordinates (a^i, a^{n+i}) are related by the relations

$$a^{n+i} + \Gamma^i_{pq} a^p v^q = 0.$$

Besides these horizontal vectors, forming an n-dimensional area element at the point (x, v), there exist vertical vectors B = $a^{n+s}\frac{\partial}{\partial v^s}$, characterized by the equalities $a^i = 0$ and forming a complementary n-dimensional area element at the point (x, v) of the tangent bundle $T(X_n)$. Taking an arbitrary vector $W = w^s \frac{\partial}{\partial x^s}$, tangent to X_n at point x, we can define at each point (x, v) of the layer over x a horizontal vector W^H and a vertical vector W^V called, respectively, the horizontal and the vertical liftings of the vector W at the point (x, v):

$$W^H = w^s \frac{\partial}{\partial x^s} - \Gamma^s_{pq} w^p v^q \frac{\partial}{\partial v^s}\,,$$

$$W^V = w^s \frac{\partial}{\partial v^s}\,.$$

We have: $KW^V = W$. Further, taking an arbitrary vector $A = a^s \frac{\partial}{\partial x^s} + a^{n+s}\frac{\partial}{\partial v^s} \in TT(X_n)$, tangent to $T(X_n)$ at the point (x, v), we construct its horizontal and vertical components A' and A":

$$A' = a^s \frac{\partial}{\partial x^s} - \Gamma^s_{pq} a^p v^q \frac{\partial}{\partial v^s}\,,$$

$$A'' = (a^{n+s} + \Gamma^s_{pq} a^p v^q) \frac{\partial}{\partial v^s}\,.$$

The horizontal component A' is the horizontal lifting of the vector $P = a^s \frac{\partial}{\partial x^s}$ tangent to X_n: $A' = P^H$. The vertical component A" is the vertical lifting of the vector Q = KA also tangent to X_n: $A'' = Q^V$. Next we construct the vectors P^V and Q^H and to the vector $A = P^H + Q^V$ we correspond the vector $IA = -Q^H + P^V$. It is clear that the operator I constructed possesses the property $I^2 = -1$, so that it yields an almost-complex structure in $T(X_n)$. We introduce the Nijenhuis tensor N of the affinor of an almost-complex structure I and by π we denote the natural projection $T(X_n) \to X_n$ which maps the point $(x, v) \in T(X_n)$ into the point $x \in X_n$. By assigning three arbitrary vectors X, Y, Z at the point x, for brevity we denote the point $(x, Z) \in T(X_n)$ by the symbol z. Then the relations

$$\pi_*(N(X^V, Y^V)_z) = S(X,Y),$$
$$K(N(X^V,Y^V)_z) = R(X, Y)Z$$

are valid, where S and R are the torsion and curvature tensors of the connection Γ. Hence it follows (see [203]) that the structure I will be integrable if and only if the connection Γ has no curvature and torsion. The presence of a Riemann metric g on X_n permits the construction in $T(X_n)$ of a metric G, almost Hermitian relative to I, defined by the equality

$$G(A, B) = g(\pi_* A, \pi_* B) + g(KA, KB)$$

for any vectors A and B tangent to $T(X_n)$ at one point. In particular, if the mapping K has been defined by means of the Levi — Civita connection for the metric g, then, as was noted above, the metric G becomes almost-Kählerian, while for a locally Euclidean metric g it becomes Kählerian.

The generalization of these constructions has been made by Kandatu [281] and by Yano and Ishihara [483], where an almost-complex structure of the tangent fiber $T(X_n)$ is constructed with the aid of a nonlinear connection on the base manifold X_n. A still more general analysis was carried out by Ako [159].

Other methods also exist for introducing almost-complex and other structures in tangent fibers. Thus, if an almost-complex structure f and a linear connection Γ are given on a manifold X_n, then on $T(X_n)$ we can construct an almost-complex structure J by defining it by the following equalities (Tanno [429]):

$$\pi_*(JW) = f(\pi_* W),$$
$$K(JW) = f(KW), \tag{1}$$

where W is an arbitrary tangent vector to $T(X_n)$ at the point z, while on the right-hand sides are considered the action of the operator f on the vectors tangent to X_n at the point $\pi(z)$; the operator K has been constructed with the aid of the connection Γ. For this structure to be integrable the connection Γ now should not be locally Euclidean.

We can act analogously in the case of an almost-contact manifold X_{2n+1} with a (φ, ξ, η)-structure, on which there is additionally given a linear connection Γ (Tanno [428, 429]). By a special choice of connection Γ we can construct an almost-complex structure J in the fiber $T(X_{2n+1})$, whose integrability is equivalent to the normality of an almost-contact structure on X_{2n+1}. The presence of an almost-contact metric (φ, ξ, η, g)-structure on X_{2n+1} allows us to introduce an almost-Hermitian (relative to J) structure on $T(X_{2n+1})$. Conditions for this structure to be almost-Kählerian have been found [429].

Another approach to structures on tangent fibers arises in the theory of complete, vertical, and horizontal liftings (lifts) of tensor fields and connections, worked out in recent years by a number of authors (see Yano and Ledger [489, 299], Yano and Kobayashi [487, 488], and Yano and Ishihara [485, 486]).* Thus, in [487], it was shown that the complete liftings of an almost-complex structure or of an f-structure on a base manifold X_n are, respectively, an almost-complex structure or an f-structure on the tangent fiber $T(X_n)$. Analogous results were obtained for horizontal liftings defined after the giving of a linear connection on X_n (see [486]). Also note [494].

Yano and Patterson constructed vertical, complete, and horizontal liftings of tensor fields and connections in cotangent fibers and examined the properties of the complete liftings of an almost-complex structure and of an f-structure (see [477, 491, 492]); almost-complex structures in cotangent fibers were studied also by Tondeur [446].

Nagano [341], Akbar-Zadeh and Bonan [158], Yano and Davies [478], Matsumoto [332], and Ichijyo [273] have considered the applications of tangent fibers and of the almost-complex structures in them to Finsler geometry.

* Also see Kagan [54].

Rakhula [115], in connection with the theory of derivations, studied a sequence of manifolds of which each succeeding one is the tangent fiber of the one preceding.

§ 7. Almost-Tangent Structures

On a tangent fiber $T(X_n)$ of a differentiable manifold X_n there arises in a natural way the field of an affinor I under whose action a vector W tangent to $T(X_n)$ at the point z goes into the vertical lifting of the projection $\pi_* W$,

$$IW=(\pi_* W)^V.$$

If x^i are local coordinates on X_n and (x^i, v^i) are local coordinates in the fiber $T(X_n)$, then the vector $W=w^s\frac{\partial}{\partial x^s}+w^{n+s}\frac{\partial}{\partial v^s}$ is taken into the vector $IW=w^s\frac{\partial}{\partial v^s}$ by the operator I. Obviously, the operator I satisfies the equation $I^2 = 0$, and in local coordinates its matrix reduces to the form

$$\begin{pmatrix} 0 & 0 \\ 1_n & 0 \end{pmatrix}, \tag{1}$$

where 1_n is the n-th order unit block.

By generalizing this situation, Clark and Bruckheimer [197] introduced the notion of an almost-tangent structure on manifolds of even dimension. An almost-tangent structure on a differentiable manifold X_{2n} can be defined as a G-structure on X_{2n}, whose structure group G is represented by matrices of the form

$$\begin{pmatrix} A & 0 \\ B & A \end{pmatrix},$$

where A is a nonsingular block and B is an arbitrary n-th order block. Such a structure is completely defined by giving on X_{2n} the field of the structure affinor I whose matrix has the form (1) at each point relative to the adapted frames. Thus, the tangent fiber of any differentiable manifold possesses an integrable almost-tangent structure.

A linear connection on X_{2n}, in which the tensor I is covariantly constant, is called a connection of almost-tangent structure or an almost-tangent connection; such a torsion-free connection exists if and only if the Nijenhuis tensor of the structure affinor I equals zero.

The greatest number of papers in the area of almost-tangent structures are by Eliopoulos, who considered complex almost-tangent structures and introduced the characteristic forms of almost-tangent connections [213]; he studied the properties of the holonomy group of an almost-tangent connection and showed that the Bernard structure tensor of the almost-tangent structure can be identified (to within sign) with the Nijenhuis tensor of the structure affinor I [215]; he considered singular Riemann metrics compatible with almost-tangent structures [214]. Klein [287], on the basis of the investigations of Frölicher and Nijenhuis [217], studied differential operators on almost-tangent manifolds, mentioning their applications to analytical dynamics.

Finally, Eliopoulos constructed a generalization of almost-tangent structures, the so-called r-tangent structures (see [216]).

§ 8. Almost-Quaternian Structures

The survey article by Beklemishev [4] considered in great detail the fundamental papers on the theory of manifolds with almost-quaternion structures written up to 1963. These were the investigations of Libermann [315], Obata [353, 354, 355], Ishihara [274], and Wakakuwa [452, 453]. Therefore, we shall not repeat the discussions of these but will remark on the other investigations in this area.

Bonan has written a number of papers studying almost-Hermitian quaternion structures [171] and almost-quaternion connections [170], proposing a new definition of almost-quaternion structures [169, 173], and finding the structure tensor (in the sense of Bernard) of an almost-quaternion structure [172]. If I, J, K are structure affinors of an almost-quaternion structure (denoted in [4] by the symbols $\varphi, \psi, \varkappa$), satisfying the conditions $I^2 = J^2 = K^2 = -1$, $K = IJ = -JI$, then the structure tensor of an almost-quaternion manifold has the form

$$T = \frac{2}{3}([II] + [JJ] + [KK]).$$

There exists a unique almost-quaternion connection with torsion tensor T.

Affine connections on manifolds with an almost-quaternion structure have been analyzed also by Fukami [225], Wakakuwa [451, 452, 454], and Pinl [382].

Martinelli [328-331] studied generalized quaternion structures and indicated a real metric model of a projective quaternion space. Martinelli's generalization consists in that he considers on a 4n-dimensional differentiable manifold X_{4n} a G-structure with a structure group $\widetilde{L}_n^Q$ or $\widetilde{U}_n^Q$, where $\widetilde{L}_n^Q$ is the general linear group over quaternion field and $\widetilde{U}_r^Q$ is the general unitary quaternion group. These groups are defined in the following manner. If a vector of an n-dimensional right vector space Q^n over the quaternion field is represented by a one-column quaternion matrix Ξ, then the transformations from $\widetilde{L}_n^Q$ and $\widetilde{U}_n^Q$ are given by the formula

$$\Xi' = A\Xi q,$$

where A is, respectively, a general nonsingular or a unitary quaternion matrix and q is a quaternion with unit norm.

Tallini [419, 420] has considered the differential geometry of a projective quaternion space with an elliptic Hermitian metric; Hirzebruch [259] has studied the structure of a projective quaternion space.

Uniform quaternion spaces have been discussed by Tits [444]. The structure of a vector quaternion space has been studied by Bruni [178].

§ 9. Symplectic Structures; Other Structures

If on an even-dimensional differential manifold X_{2n} we are given the field of a nonsingular skew-symmetric tensor a_{ij} or, equivalently, we are given a 2-form Ω of complete rank 2n, then we say that the manifold X_{2n} has an almost-symplectic structure (see Libermann [316]). This structure is called symplectic if the form Ω is closed: $d\Omega = 0$. In the above-mentioned survey article by Libermann there is a detailed discussion of almost-symplectic structures and, therefore, here we shall only list briefly the papers of the last decade relating to almost-symplectic structures and their analogs.

Along with symplectic structures it turns out to be very appropriate (in particular, from the point of view of applications to analytical dynamics) to examine the so-called almost-cosymplectic structures given on an odd-dimensional manifold X_{2n+1} by a system of two forms $\{\omega, \Omega\}$, where ω is a Pfaffian form and Ω is a 2-form of rank 2n where, moreover, there is satisfied the condition

$$\omega \wedge \Omega^n \neq 0.$$

Such structures arise, for example, on hypersurfaces in almost-symplectic manifolds (also see [370]). If the forms ω and Ω are closed, then we speak of a cosymplectic structure on X_{2n+1}. If we set $\Omega = d\omega$ a contact structure turns out to be a special case of an almost-symplectic structure.

Infinitesimal automorphisms of symplectic and cosymplectic structures were studied by Libermann [317, 319, 321] and by Lichnerowicz [325, 326]; Akbar-Zadeh [157] analyzed the infinitesimal automorphisms of almost-cosymplectic structures and Lefebvre [300] the conformal transformations and homotheties of almost-symplectic structures. The transitivity of the automorphism group of a compact symplectic manifold was established by Hatakeyama [255].

Affine connections in almost-symplectic manifolds were studied by Lemlein [70, 71], Tondeur [445], Gaukhman [21], and Miron [334]. Among the papers devoted to the study of affine symplectic spaces we note that by Yaglom [152]. Libermann [322], Bruschi [179], and Zwart [495] also have considered cosymplectic (in particular, uniform) spaces. See Lichnerowicz [326], Tondeur [445], and Akbar-Zadeh [157] regarding the applications of symplectic and cosymplectic structures to analytical dynamics.

Barras [163, 164] has constructed a generalization of almost-symplectic and cosymplectic structures.

Passing on to other structures we stop to consider in detail the T-structures of Legrand [304, 305, 307]. A T-structure on a differentiable manifold X_m is given by a distribution S of complex tangent area elements of dimension n (n < m). At each point of the manifold X_m there arise a set of adapted frames (in which the first n vectors lie in an area element of distribution S) and a set of so-called S-frames (a system of n linearly independent vectors comprising a basis in the area element being considered). The collections of these frames form, respectively, the principal fiber spaces $E_S(X_m)$ and $E_S(X_m)$.

Every infinitesimal connection on $\hat{E}_S(X_m)$ is naturally identified with a complex linear connection on X_m, which is characterized by the fact that its forms ω in the adapted frames possess the property $\omega_\alpha^A = 0$ ($\alpha = 1, \ldots, n$; $A = n+1, \ldots, m$). Such a connection induces an S-connection, an infinitesimal connection on $E_S(X_m)$ defined by a system of forms (ω_α^β). Using the curvature forms of the S-connection $\Omega_\beta^\alpha = d\omega_\beta^\alpha + \omega_\sigma^\alpha \wedge \omega_\beta^\sigma$ we can construct the so-called characteristic form $\psi = \Omega_\alpha^\alpha = d\omega_\alpha^\alpha$, belonging (independently of the choice of the S-connection) to one and the same cohomology class, namely, the characteristic class of T-structures. The characteristic classes for real T-structures equal zero. The vanishing of the characteristic form is reflected by the properties of the holonomy group of the S-connection [304]. Legrand also introduced the concept of the torsion form of the T-structure [306], whose vanishing is a necessary, and for an analytical T-structure also a sufficient, condition for the integrability of distribution S.

A special case of an almost-product structure, namely, a paracomplex structure, was analyzed by Libermann [316] (also see Legrand [303]). Libermann also considered an almost-quaternion structure of second kind [316] which could be called an almost-antiquaternion structure.

We further note Gaukhman's paper [22] in which he constructs a generalization of the notion of affinor structure to the case of an arbitrary principal fiber space, and Tsalenko's paper [136] on flag structures.

Teleman [443] and Hangan [241, 242] investigated Grassmann and almost-Grassmann manifolds.

Among other structures studied by various authors we can name Sasaki's (φ, ψ)-structure [398, 268], generalizing almost-product structures and (φ, ξ, η)-structures; the affinor structures φ_j^i, possessing the property $\varphi^p = \pm 1$ (Ishihara [275]); manifolds with several independent almost-complex structures (Wakakuwa [456]). Okubo [365] discovered the existence of an affinor structure H, satisfying the equation $H^3 - H = 0$, in the linear frame bundle of a differentiable manifold. Furthermore, structures of a special form have been studied in [243-250, 262, 263, 291, 308, 455]. We must also note the interesting investigations of Sasaki on the homogeneous contact transformations of differentiable manifolds

[400], and also of Yano and Muto [490] on homogeneous contact structures on even-dimensional manifolds.

§ 10. Structure Defined by Second-Order Algebras and by General Associative Algebras. Biaxial and Biaffine Geometries. Their Generalizations

It is well known that there exist three types of second-order algebras with unity: the algebras of complex, binary, and dual numbers [118]. The bases of these algebras can be represented in the form $\{e_1 = 1, e_2\}$ and, moreover, $e^2 = \omega$, where $\omega = -1, +1$, or 0, depending on whether we have the algebra of complex, binary, or dual numbers. Thus, in the formulas $e_i e_j = \gamma^s_{ij} e_s$ we obtain the following structure constants: $\gamma^i_{j1} = \gamma^i_{1j} = \delta^i_j$, $\gamma^1_{22} = \omega$, $\gamma^2_{22} = 0$.

If in a 2m-dimensional real vector space B_{2m} we are given, relative to some frame, two affinors $\underset{1}{\gamma}$, and $\underset{2}{\gamma}$ with the matrices

$$(\gamma^{(a-1)2+i}_{(b-1)2+j}) = (\delta^a_b \gamma^i_{jk}), \tag{1}$$

where (a, b = 1, . . . , m; i, j, k = 1, 2), then the first one of them (k = 1) will represent an identity transformation, while the second (k = 2) will define in the space B_{2m} an involution of elliptic, hyperbolic, or parabolic type depending on the type of algebra. In the space B_{2m} this involution yields a structure defined by the algebra considered. The adapted frames will be those for which the matrix of the structure affinor $f = \underset{2}{\gamma}$ retains its form.

When m = 2 we obtain a four-dimensional vector space with structure, and the set of its pseudovectors defines a three-dimensional projective space with structure, namely, the so-called biaxial space of elliptic, hyperbolic, or parabolic type. In the hyperbolic case (ω = +1) the investigation of such a space was started by Mayer [333], while in the elliptic case by Salzert [393] and, independently of him, by Norden [83, 84]*; moreover, Norden carried out

* The beginnings of Norden's research on biaxial geometry go back to 1937-1938, as is apparent from the little we know of the activities of the Seminar on Vector and Tensor Analysis during the period from January 15, 1937 to January 15, 1942 (see Proc. Seminar Vector Tensor Analysis, Issue 6, Moscow (1948)).

the most systematic development of the foundations of biaxial geometry and its applications. When $m > 2$ the set of pseudovectors of the space B_{2m} defines a projective space B^{J}_{2m-1} with structure, namely, the so-called biplanar space of elliptic, hyperbolic, or parabolic type.

An affine space of an even number 2m of dimensions, whose fundamental transformation group consists of all affine transformations which take the above-mentioned involution f into itself, is called a biaffine space B^{J}_{2m}. Such a space is the real model of an m-dimensional affine space A_m over the algebra of complex, binary, or dual numbers. The geometry of biaffine spaces also was studied by Norden and his students. A detailed exposition of these papers (up to 1963) has been given in three survey articles by Norden [91, 96, 98]. We do not repeat these considerations and we do not devote ourselves to any detailed analysis of the later analogous papers, we merely list them systematically, giving preference to the last decade.

The differential geometry of the curves and surfaces of a biaxial space of elliptic type was studied by Norden [83, 84], Tret'yakov [135], and Zuev [39, 40, 42, 48]. The corresponding analysis for a space of parabolic type was made by Talantova [130]. Petkanchin [101, 102], Ivanov [127-277-279], Stanilov [125-127], Havlicek [257], and Papuc [374, 375] studied the differential geometry of a biaxial space of hyperbolic type. Zuev applied the concepts of biaxial geometry of elliptic type to the theory of algebraic curves and surfaces and to certain types of ordinary differential equations and of flat vector fields [41, 43-47, 49, 50]. Among other papers we note further [3, 131, 144]. Also see [133].

The numerous investigations on the line geometry of biaxial spaces by Bukharaev, Kovantsov, Stanilov, and Petkanchin were covered in Shcherbakov's survey [151] and are not reproduced here in the Literature Cited section.

The differential geometry of a four-dimensional biaffine space of elliptic type was investigated by Norden who developed a theory of curves and of two-dimensional surfaces in such a space [85, 86, 90, 95] and also by Bukharaev [6, 7]. Podkovyrin [108-113] studied analytical hypersurfaces (of codimension two) in multidimensional biaffine spaces. Pinl's paper [381] also is pertinent here. Kriščiūnaitė [64] analyzed the three-dimensional hypersurfaces of four-dimensional biaffine spaces of elliptic and hyperbolic type from the point of view of the contact structures arising on them. Also see [63].

Norden studied the question of the complex representation of the tensors of a biaffine space [87, 89]; see [4] for analogous investigations by other authors. Complex representation of tensors was used by Izotov to classify second-order hypersurfaces of a biplanar space because such a classification is equivalent to the problem of the simultaneous reduction of a quadratic and a Hermitian form to a canonical form [52, 53]; the problem, in a more general form, was afterwards solved by Ermolaev [34-36]. Norden himself used the complex representation of tensors to study the bitensors of a Lorentz space [89, 93], to represent conveniently the differential invariants of Riemann spaces ([18]; the paper was written jointly with Vishnevskii), and to give a geometric interpretation of spinors [92]. The article by Droz-Vincent [204] also refers to these same aspects. An analogous question on the representation of bivalent tensors in the case of a biaffine space of parabolic type was analyzed by Talantova [129, 132].

Very interesting applications of structures defined by second-order algebras were established by Norden in the line geometry of spaces of constant curvature [94, 96]; Vishnevskii's papers [11, 15] also are devoted to this question. Here we should mention the work of Zhelezina [37, 38].* Distinctive applications of the concepts of biaxial geometry to the theory of Riemann spaces were given by Vishnevskii [13] and Prvanovich [114].

Generalizations of biaxial and biplanar spaces, not using algebras, were constructed by Papuc (for example, see [376, 377]).

Structures defined by associative algebras are of interest in connection with the study of spaces over algebras being intensively carried out in recent years. Rozenfel'd, Yaglom, Dzhavadov, and their students have written many papers in which they study affine, projective, and non-Euclidean spaces over algebras and in which they indicate the applications of algebras to various aspects of geometry (with no pretense to completeness we mention the inves-

* Structures defined by second-order algebras are treated also in [8, 9, 81].

tigations in [25-31, 55, 56-60, 103, 104, 118, 119, 120, 153]).* Among the work of foreign authors in this area during recent years we have that of Teleman [442], Tits [444], Depunt [202], Crumeyrolle [199], Pernet [378, 380], and Hubaut [269]. There also exist many investigations on spaces over nonassociative algebras; however, we do not refer to these papers here.†

Let A_n be an associative algebra with basis $\{e_1, \ldots, e_n\}$ and with structure constants $\gamma^k_{ij}: e_i e_j = \gamma^s_{ij} e_s$. Then we can consider a space with a linear connection which has the constant coefficients $\Gamma^k_{ij} = \gamma^k_{ij}$ in some local coordinate system. If in addition the algebra A_n is commutative, then such a space will be locally Euclidean. These spaces have been studied by Vrănceanu [449, 450], Badea [161, 162], Văduva [448], and Gorciu [228]. Here Vrănceanu and Gorciu have also used nonassociative algebras.

Norden's students, continuing the researches he started on the geometry of biaffine spaces and their generalizations, came to examining spaces with structures defined by associative algebras. Here, algebras possessing a principal unity (the so-called unital algebras) were used. Thus, Neifel'd [79, 80] studied the geometry of a three-dimensional equiaffine space with a structure defined by an affinor g^i_j satisfying the equation $g^3 = 1$. Calling this space an Appell space, Neifel'd then went on to study the so-called conformal Appell spaces, i.e., three-dimensional torsion-free affinely-connected spaces possessing a field of the covariantly constant affinor g with the property $g^3 = 1$. In all these considerations an essential role is played by a third-order commutative associative algebra with basis $\{1, \theta, \theta^2\}$, $\theta^3 = 1$. This algebra is isomorphic with the direct sum of the complex and real number fields: $A_3 = C + R$; it permits us to introduce into consideration a space with a cubic metric. The hyperbolic analog of an Appell space is obtained in just a similar manner if as a base we set an algebra which is the direct sum of three real number fields. Neifel'd's investigations were generalized and extended by Vyzhgina [19, 20] by constructing metrics in general commutative algebras and, in particular, in

*An idea of the nature of the very latest researches in this area can be obtained from the abstracts of the reports presented at the Third Interuniversity Scientific Conference on Problems in Geometry held in Kazan (see [1, 5, 61, 76, 82, 124, 140, 154]).

†Octave geometries have been studied, for example, by Persits [100].

cyclic number algebras with basis $\{1, \theta, \theta^2, \ldots, \theta^{n-1}\}$, $\theta_n = 1$. Shirokov examined spaces of dimension mn, where n is the order of algebra A_n, and in them gave a structure defined by algebra A_n [142, 143, 145-150]. For example, if A_n is a commutative associative initial algebra and B_{mn} is a vector space, then the correspond-structure in B_{mn} is given by a system of n structure affinors $\underset{k}{f}{}^{\alpha}_{\beta}$, whose coordinates in adapted frames once again have the form (1), where now the Latin indices range over values from 1 to n. Using the n fields of such affinors we can give a structure defined by an algebra over a differentiable manifold X_{mn}. Connections were explicitly derived in which the structure affinors are covariantly constant. In special cases this manifold X_{mn} becomes a real model of an m-dimensional space over the algebra A_n. Vishnevskii undertook to make a detailed study of Riemann spaces with a structure defined by the algebra of plural numbers [10, 12, 14, 16] (also see [17]). This algebra has the basis $\{1, \varepsilon, \varepsilon^2, \ldots, \varepsilon^{-1}\}$, where the unit ε has the property $\varepsilon^n = 0$. Starting from the results of Kruchkovich and Solodovnikov [65] he studied irreducible B-spaces of proper structure, i.e., Riemann spaces with a metric tensor g_{ij} and a covariantly constant symmetric tensor γ_{ij}, in which the affinor γ^i_j has, in the Jordan canonical form, a single eigenvalue and all the cells are of one and the same order m. If the dimension of the space equals n = pm, then it turns out to be a real model of a p-dimensional Riemann space over an m-th order algebra of plural numbers. Furthermore, Vishnevskii has considered analogs of Kählerian spaces over algebras of plural numbers and real models of Riemann spaces over direct sums of algebras.* Structures on differentiable manifolds, defined by a plural number algebra, are encountered in Eliopoulos [216].

LITERATURE CITED

1. N. T. Abbasov, "Motion groups and symmetry images in elliptical spaces over alternion algebras," in: Third Interuniv. Scient. Conf. Prob. Geom. (report abstracts), Kazan (1967), pp. 3-4.
2. M. A. Akivis, "The structure of multidimensional surfaces carrying a net of curvature lines," Dokl. Akad. Nauk SSSR, 149(6): 1247-1249 (1963).
3. G. S. Barkhin, "Motion subgroups of a biaxial space," Uch. Zap. Kazansk. Univ., 111(8): 105-124 (1951).

* The geometric applications of algebras of plural and cyclic numbers have recently been pointed out by Eremina [33], Kuznetsova [66], Lobanova [75], and Pevzner [99].

4. D. V. Beklemishev, "Differential geometry of spaces with an almost-complex structure," in: Geometry. 1963, Progress in Science, VINITI, Akad. Nauk SSSR, Moscow (1965), pp. 165-212.
5. T. M. Boguslovskaya, "Symmetry images of complex and quaternion semielliptic spaces," in: Third Interuniv. Scient. Conf. Prob. Geom. (report abstracts), Kazan (1967), p. 21.
6. R. G. Bukharaev, "Theory of surfaces of a biaffine space and theory of congruences of a biaxial space," Uch. Zap. Kazansk. Univ., 116(1): 7-9 (1956).
7. R. G. Bukharaev, "On the theory of surfaces in a biaffine space. 1," Izv. Vysshikh. Uchebn. Zavedenii, Matematika, No. 3, pp. 81-92 (1960).
8. V. V. Vishnevskii, "Certain questions in the geometry of B-spaces," in: Collection of Aspirants' Papers at Kazan Univ., Exact Sciences, Kazan (1962), pp. 137-147.
9. V. V. Vishnevskii, "A class of Kahlerian spaces," in: Summaries of the Scient. Conf. at Kazan Univ. in 1961, Math. Sci. Sect., Kazan (1962).
10. V. V. Vishnevskii, "Complex structures in B-spaces," Uch. Zap. Kazansk. Univ., 123(1): 24-48 (1963).
11. V. V. Vishnevskii, "The Klein-Norden mapping and its application to the line geometry of non-Euclidean spaces," in: Summaries of the Scient. Conf. at Kazan Univ. in 1962, Math. Sci. Sect., Kazan (1963), pp. 27-30.
12. V. V. Vishnevskii, "Complex structures of a class of Kahler-Rashevskii spaces," Dokl. Akad. Nauk SSSR, 149(2): 233-236 (1963).
13. V. V. Vishnevskii, "A biplanar space connected with the Riemann space V_4," in: Collection of Aspirants' Papers at Kazan Univ., Math., Mech., Phys., Kazan (1964), pp. 41-45.
14. V. V. Vishnevskii, "A generalization of Shirokov-Rashevskii spaces," Uch. Zap. Kazansk. Univ., 125(1): 60-73 (1965).
15. V. V. Vishnevskii, "More on complex structures in line geometry," in: Proc. Seminar Vector Tensor Analysis and Their Applications to Geom., Mech., Phys., Issue 13, Moscow University (1966), pp. 467-492.
16. V. V. Vishnevskii, "A property of analytical functions over algebras and its application to the study of complex structures in Riemann spaces," Uch. Zap. Kazansk. Univ., 126(1): 5-12 (1966).
17. V. V. Vishnevskii, "On the theory of Riemann spaces over tensor products of algebras," in: Third. Interuniv. Scient. Conf. Prob. Geom. (report abstracts), Kazan (1967).
18. V. V. Vishnevskii and A. P. Norden, "Complex representation of the invariants of a four-dimensional Riemann space," Izv. Vysshikh. Uchebn. Zavedenii, Matematika, No. 2, pp. 176-182 (1959).
19. L. B. Vyzhgina, "The geometry of commutative algebras," Uch. Zap. Mosk. Obl. Ped. Inst., 149(1): 67-78 (1964).
20. L. B. Vyzhgina, "Projective spaces over algebras of cyclic, anticyclic, and plural numbers," Uch. Zap. Mosk. Obl. Ped. Inst., 173(1): 17-25 (1967).
21. V. A. Gaukhman, "Symmetric almost-symplectic affine connections on Hermitian manifolds," Dokl. Akad. Nauk SSSR, 146(2): 277-280 (1962).
22. V. A. Gaukhman, "Γ-structures on principal fiber spaces," Tr. Tomskogo Univ., 176: 20-27 (1964).

23. V. A. Gaukhman, "Diagonalizable almost-tensor structures," Usp. Mat. Nauk, 19(6): 226-228 (1964).
24. V. W. Guillemin and S. Sternberg, "Algebraic model of a transitive differential geometry," Bull. Am. Math. Soc., 70(1): 16-47 (1964).
25. M. A. Dzhavadov, "Non-Euclidean geometries over algebras of alternions," Uch. Zap. Kazansk. Gos. Univ., 115(10): 8-9 (1955).
26. M. A. Dzhavadov, "Projective spaces over algebras," Uch. Zap. Azerb. Univ., No. 2, pp. 3-18 (1957).
27. M. A. Dzhavadov, "Non-Euclidean geometries over algebras," Uch. Zap. Azerb. Univ., No. 4, pp. 3-16 (1957).
28. M. A. Dzhavadov, "Affine spaces over algebras," Uch. Zap. Azerb. Univ., No. 7, pp. 3-21 (1957).
29. M. A. Dzhavadov, "Linear congruences in projective spaces over algebras of alternions," Izv. Akad. Nauk Azerbad.SSR, No. 7, pp. 3-15 (1957).
30. M. A. Dzhavadov, "An application of a non-Euclidean geometry over the algebra of binary matrices to the projective geometry over the algebra of real matrices," Uch. Zap. Azerb. Univ., Fiz.-Mat. i Khim. Ser., No. 3, pp. 11-16 (1960).
31. M. A. Dzhavadov and N. T. Abbasov, "Bicomplex and biquaternion hyperbolic spaces," Uch. Zap. Azerb. Univ., Ser. Fiz.-Mat. Nauk, No. 1, pp. 9-15 (1964).
32. I. P. Egorov, "Motions in generalized differential-geometric spaces," in: R. V. Gamkrelidze (ed.), Progress in Mathematics, Vol. 6: Topology and Geometry, Plenum Press, New York (1970), pp. 00.
33. L. V. Eremina, "The geometry of cosymmetry images," in: Third Interuniv. Scient. Conf. Prob. Geom. (report abstracts), Kazan (1967), p. 53.
34. Yu. B. Ermolaev, "Simultaneous reduction of a pair of bilinear forms to canonical form," Dokl. Akad. Nauk SSSR, 132(2): 257-259 (1960).
35. Yu. B. Ermolaev, "Simultaneous reduction of a symmetric and a Hermitian form," Izv. Vysshikh. Uchebn. Zavedenii, Matematika, No. 2, pp. 10-23 (1961).
36. Yu. B. Ermolaev, "Simultaneous reduction of a pair of bilinear forms to canonical form over an arbitrary perfect field of characteristic $\neq 2$," in: Summaries of the Scient. Conf. at Kazan Univ. in 1962, Math. Sci. Sect., Kazan (1963), pp. 25-27.
37. I. I. Zhelezina, "Quasi-Euclidean line geometry as a complex Euclidean geometry," Uch. Zap. Mosk. Gos. Ped. Inst., No. 243, pp. 233-240 (1965).
38. I. I. Zhelezina, "Galillean line geometry as a dual Euclidean geometry," Uch. Zap. Mosk. Gos. Ped. Inst., No. 243, pp. 241-247 (1965).
39. I. V. Zuev, "The real interpretation of the curves on a complex centroaffine plane," Uch. Zap. Kazansk. Gos. Univ., 115(10): 13-15 (1955).
40. I. V. Zuev, "The real interpretation of the curves on a complex centroaffine plane," Uch. Zap. Elabuzhsk. Gos. Ped. Inst., 3: 17-34 (1958).
41. I. V. Zuev, "Straight lines on an algebraic surface with a double Königs net," Izv. Vysshikh. Uchebn. Zavedenii, Matematika, No. 5, pp. 52-61 (1960).
42. I. V. Zuev, "Application of analytic functions to the problem of the straight lines of a surface containing an imaginary double Königs net," Tr. Kazansk. Aviats. Inst., No. 71, pp. 97-105 (1962).

43. I. V. Zuev, "Cyclotomic integral curves of the equation $f(x, y, y') = 0$," Izv. Vysshikh. Uchebn. Zavedenii, Matematika, No. 6, pp. 64-71 (1963).
44. I. V. Zuev, "A class of vector fields," in: Summaries of the Scient. Conf. at Kazan Univ. in 1963, Math., Cybernet. and Probability Theory, Mech. Sect., Kazan (1964), pp. 34-35.
45. I. V. Zuev, "A class of solenoidal fields," in: Summaries of the Scient. Conf. at Kazan Univ. in 1963, Math., Cybernet. and Probability Theory, Mech. Sect., Kazan (1964), pp. 35-36.
46. I. V. Zuev, "A class of differential equations," in: Summaries of the Scient. Conf. at Kazan Univ. in 1963, Math., Cybernet. and Probability Theory, Mech. Sect., Kazan (1964), p. 37.
47. I. V. Zuev, "On the geometry of the zeros and poles of a rational function," in: Summaries of the Scient. Conf. at Kazan Univ. in 1963, Math., Cybernet. and Probability Theory, Mech. Sect., Kazan (1964).
48. I. V. Zuev, "A biaxial invariant of a ruled surface," Uch. Zap. Kazansk. Univ., 125(1): 74-78 (1965).
49. I. V. Zuev, "A class of flat vector fields," Uch. Zap. Kazansk. Univ., 125(1): 79-87 (1965).
50. I. V. Zuev, "On the Lagrange equation," Uch. Zap. Kazansk. Univ., 125(1): 88-99 (1965).
51. M. A. Ivanova, "Fields of parallel area elements in a pseudo-Riemann space," Uch. Zap. Arkhang. Gos. Ped. Inst., No. 4, pp. 85-109 (1959).
52. G. E. Izotov, "The joint reduction of a quadratic and a Hermitian form," Izv. Vysshikh. Uchebn. Zavedenii, Matematika, No. 1, pp. 143-159 (1957).
53. G. E. Izotov, "Second-order surfaces of a biplanar space," Izv. Vysshikh. Uchebn. Zavedenii, Matematika, No. 1, pp. 89-102 (1958).
54. F. I. Kagan, "Tensors in tangent bundles," in: Third Interuniv. Scient. Conf. Prob. Geom. (report abstracts), Kazan (1967), pp. 71-72.
55. L. M. Karpova, "Semi-Riemann symmetric spaces over algebras," Izv. Akad. Nauk Azerb. SSR, Ser. Fiz.-Tekh. i Mat. Nauk, No. 4, pp. 45-49 (1964).
56. S. S. Kasimova, "Quadri-quaternion hyperbolic spaces," Uch. Zap. Azerb. Univ., Ser. Fiz.-Mat. Nauk, No. 4, pp. 19-23 (1963).
57. S. S. Kasimova, "Quadri-quaternion symplectic spaces," Uch. Zap. Azerb. Univ., Ser. Fiz.-Mat. Nauk, No. 2, pp. 48-55 (1964).
58. S. S. Kasimova, "Euclidean spaces over the algebra of real matrices," Uch. Zap. Azerb. Univ., Ser. Fiz.-Mat. Nauk, No. 2, pp. 68-73 (1966).
59. S. S. Kasimova, "Quasi-Euclidean spaces over the algebra of real matrices," Izv. Akad. Azerbad.SSR, Ser. Fiz.-Tekh. i Mat. Nauk, No. 4, pp. 16-22 (1966).
60. S. S. Kasimova, L. M. Ezhovo-Guseva, and N. D. Zablotskikh, "Dimensions of manifolds of tangent planes of Hermitian quadrics and of planes intersecting with the corresponding planes in Hermitian null-systems, in spaces over algebras," Uch. Zap. Azerb. Univ., Ser. Fiz.-Mat. Nauk, No. 5, pp. 27-32 (1966).
61. S. S. Kasimova, "Limit non-Euclidean and symplectic spaces over the algebra of real matrices," in: Third Interuniv. Scient. Conf. Prob. Geom. (report abstracts), Kazan (1967), pp. 75-76.
62. V. V. Kolyan, "On the theory of curves of a composition space," Izv. Vysshikh. Uchebn. Zavedenii, Matematika, No. 3, pp. 43-55 (1968).

63. A. L. Kriščiūnaitė, "Almost-contact structures and their analogs," in: Third Interuniv. Scient. Conf. Prob. Geom. (report abstracts), Kazan (1967), p. 83.

64. A. L. Kriščiūnaitė, "Almost-contact structures on the hypersurfaces of a four-dimensional centrobiaffine space of elliptic and hyperbolic type," Litovsk. Mat. Sb., 7(3): 423-438 (1967).

65. G. I. Kruchkovich and A. S. Solodovnikov, "Constant symmetric tensors in Riemann spaces," Izv. Vysshikh. Uchebn. Zavedenii, Matematika, No. 3, pp. 147-158 (1959).

66. L. N. Kuznetsova, "Automorphisms of pairs of null-systems in a complex projective space," in: Third Interuniv. Scient. Conf. Prob. Geom. (report abstracts), Kazan (1967), p. 88.

67. B. L. Laptev, "Petr Alekseevich Shirokov. Survey of published papers," in: P. A. Shirokov, Selected Papers on Geometry, Kazan University, Kazan (1966), 432 pp.

68. G. F. Laptev, "Manifolds immersed in generalized spaces," in: Proc. Fourth All-Union Math. Congr., Vol. 2, "Nauka," Leningrad (1964), pp. 226-233.

69. G. F. Laptev, "Basic higher-order infinitesimal structures on a smooth manifold," in: Proc. Geom. Seminar, Inst. Nauchn. Inform. Akad. Nauk SSSR, 1: 139-189 (1966).

70. V. G. Lemlein, "Spaces of symmetric almost-symplectic connection," Dokl. Akad. Nauk SSSR, 115(4): 655-658 (1957).

71. V. G. Lemlein, "The curvature tensor and certain types of spaces of symmetric almost-symplectic connection," Dokl. Akad. Nauk SSSR, 117(5): 755-758 (1957).

72. E. K. Leont'ev, "On Chebyshev compositions," Uch. Zap. Kazansk. Univ., 126(1): 23-40 (1966).

73. E. K. Leont'ev, "On the general theory of compositions in spaces of affine-connection," in: Summaries Scient. Aspirants Conf. at Kazan Univ. in 1965, Math., Mech. (reports abstracts), Kazan (1967), pp. 26-28.

74. E. K. Leont'ev, "The classification of special sheaves and compositions of multidimensional spaces," Izv. Vysshikh. Uchebn. Zavedenii, Matematika, No. 5, pp. 40-51 (1967).

75. L. B. Lobanova, "Non-Euclidean spaces over algebras of cyclic, anticyclic, and plural numbers," in: Third Interuniv. Scient. Conf. Prob. Geom. (report abstracts), Kazan (1967), pp. 98-99.

76. L. M. Markina, "Dual non-Euclidean spaces," in: Third Interuniv. Scient. Cong. Prob. Geom. (report abstracts), Kazan (1967), p. 110.

77. L. A. Matevosyan, "Surfaces in reducible spaces," Aikakan SSR Gitutyunneri Akademian Tegekagir, Matematika, 1(6): 374-391 (1966).

78. L. A. Matevosyan, "A class of surfaces in reducible spaces," Aikakan SSR Gitutyunneri Akademian Tegekagir, Matematika, 2(2): 83-89 (1967).

79. E. G. Neifel'd, "The geometry of Appell space," in: Collections of Aspirants' Papers at Kazan Univ., Exact Sciences, Kazan (1962), pp. 148-157.

80. E. G. Neifel'd, "The geometry of conformal-Appell space," Uch. Zap. Kazansk. Univ., 123(1): 128-140 (1963).

81. E. G. Neifel'd, "Normalized space over the complex number field and its generalizations," in: Third Interuniv. Scient. Conf. Prob. Geom. (report abstracts), Kazan (1967), p. 122.

82. L. S. Nikitina, "Semi-antiquaternion projective geometry," in: Third Interuniv. Scient. Conf. Prob. Geom. (reports abstracts), Kazan (1967), p. 122.
83. A. P. Norden, "Zero-curvature surfaces of a biaxial space," Dokl. Akad. Nauk SSSR, 58:1597-1600 (1947).
84. A. P. Norden, "Space of linear congruence," Mat. Sb., 24:429-455 (1949).
85. A. P. Norden, "A biaffine space and its mapping onto itself," Uch. Zap. Kazansk. Univ., 112(10):3-11 (1952).
86. A. P. Norden, "Theory of curves of a biaffine space," Uch. Zap. Kazansk. Univ., 112(10):13-26 (1952).
87. A. P. Norden, "Self-conjugate images of a biaxial space," Uch. Zap. Kazansk. Univ., 114(2):3-12 (1954).
88. A. P. Norden, "Theory of surfaces of a biaffine space," Uch. Zap. Kazansk. Univ., 114(2):13-28 (1954).
89. A. P. Norden, "Complex representation of the tensors of a biplanar space," Uch. Zap. Kazansk. Univ., 114(8):45-53 (1954).
90. A. P. Norden, "The geometry of biaxial and biaffine spaces," Uch. Zap. Kazansk. Gos. Univ., 115(10):12 (1955).
91. A. P. Norden, "The method of normalization and its applications to the geometry of affinely connected spaces," in: Proc. Third All-Union Math. Congr., 1956, Vol. 3, Akad. Nauk SSSR, Moscow (1958), pp. 418-423.
92. A. P. Norden, "Geometric interpretation of certain concepts of spinor analysis," in: Proc. Third All-Union Math. Congr., Vol. 1, Akad. Nauk SSSR, Moscow (1956), p. 160.
93. A. P. Norden, "Complex representation of the tensors of a Lorentz space," Izv. Vysshikh. Uchebn. Zavedenii, Matematika, No. 1, pp. 156-163 (1959).
94. A. P. Norden, "A class of four-dimensional A-spaces," Izv. Vysshikh. Uchebn. Zavedenii, Matematika, No. 4, pp. 145-157 (1960).
95. A. P. Norden, "A class of minimal surfaces of a four-dimensional space," Izv. Vysshikh. Uchebn. Zavedenii, Matematika, No. 3, pp. 91-97 (1962).
96. A. P. Norden, "Certain possible directions of development of line geometry," Uch. Zap. Kazansk. Univ., 123(1):141-151 (1963).
97. A. P. Norden, "Spaces of Cartesian Composition," Izv. Vysshikh. Uchebn. Zavedenii, Matematika, No. 4, pp. 117-128 (1963).
98. A. P. Norden, "Biaxial geometry and its generalizations," in: Proc. Fourth All-Union Math. Congr., 1961, Vol. 2, "Nauka," Leningrad (1964), pp. 236-243.
99. S. L. Levzner, "The geometry of a pair of quadrics in a projective space," in: Third Interuniv. Scient. Conf. Prob. Geom. (reports abstracts), Kazan (1967), pp. 127-128.
100. D. B. Persits, "Octave geometries," in: Third Interuniv. Scient. Conf. Prob. Geom. (report abstracts), Kazan (1967), p. 129.
101. B. Petkanchin, "An elliptic system of lines in a biaxial geometry," Izv. Mat. Inst., Bulgar. Akad. Nauk, 2(2):136-161 (1957).
102. B. Petkanchin, "A hyperbolic system of lines in a parabolic biaxial geometry," Godishnik Sofiisk. Univ. Mat. Fak., 58:171-194 [1963-1964 (1965)].
103. N. D. Petsko, "Projective metrizations and complex numbers," Uch. Zap. Kolomensk. Ped. Inst., 8:127-143 [1964 (1965)].

104. N. D. Petsko, "Biquaternion elliptic spaces and their application to real geometries. Projective metrics," Uch. Zap. Kolomensk. Ped. Inst., 8: 144-164 [1964 (1965)].

105. N. M. Pisareva, "Symmetric, almost-reducible, and reducible Weyl spaces," Izv. Vysshikh. Uchebn. Zavedenii, Matematika, No. 6, pp. 137-142 (1964).

106. N. M. Pisareva, "Almost-reducible and symmetric almost-reducible spaces of affine connection," Mat. Sb., 66(1): 119-126 (1965).

107. N. M. Pisareva, "Affinely connected spaces admitting of a transitive motion group whose stationary linear subgroup is completely reducible," Mat. Sb., 68(1): 75-80 (1965).

108. A. S. Podkovyrin, "A class of four-dimensional minimal surfaces of a six-dimensional biaffine space," in: Materials Scient. Conf. Kazan State Pedagogical Inst., 1962, Kazan (1963), pp. 427-429.

109. A. S. Podkovyrin, "A class of four-dimensional minimal surfaces in a six-dimensional biaffine space," Volzhsk. Mat. Sb., No. 2, pp. 245-250 (1964).

110. A. S. Podkovyrin, "Minimal X_4 of the space B_6, admitting of a mapping onto developable X_2 of a complex A_3," Uch. Zap. Kazansk. Univ., 125(1): 121-129 (1965).

111. A. S. Podkovyrin, "A class of minimal surfaces of a six-dimensional space," KVKIU, Trudy Ychilishcha, Issue 1, Kazan (1967).

112. A. S. Podkovyrin, "Hypersurfaces of a unitary space. I," Izv. Vysshikh. Uchebn. Zavedenii, Matematika, No. 8, pp. 41-52 (1967).

113. A. S. Podkovyrin, "Hypersurfaces of a unitary space. II," Izv. Vysshikh. Uchebn. Zavedenii, Matematika, No. 9, pp. 75-85 (1967).

114. M. Prvanovich, "Riemann extensions as generalized biplanar spaces," Publ. Inst. Math., 6: 9-16 (1966).

115. M. Rakhula, "On a higher-order differential geometry," Tartu Ulikooli Toimetised, Issue 150, pp. 122-131 (1964).

116. P. K. Rashevskii, Geometric Theory of Differential Equations, Moscow (1947).

117. P. K. Rashevskii, "Scalar field in fiber space," in: Proc. Seminar Vector Tensor Analysis, Issue 6, pp. 225-248 (1948).

118. B. A. Rozenfel'd, Non-Euclidean Geometries, Gostekhizdat, Moscow (1955), 744 pp.

119. L. V. Rumyantseva, "Quaternion symplectic geometry," in: Proc. Seminar Vector Tensor Analysis with Appl. to Geom., Mech., Phys., Issue 12, Moscow University (1963), pp. 287-314.

120. L. V. Rumyantseva, "Quadriquaternion elliptic spaces," Uch. Zap. Azerb. Univ., Ser. Fiz.-Mat. Nauk, No. 3, pp. 35-38 (1963).

121. L. V. Sabinin, "The geometry of subsymmetric spaces," Nauchn. Dokl. Vysshikh. Shkoly, Fiz.-Mat. Nauk, No. 3, pp. 46-49 (1958).

122. L. V. Sabinin, "The structure of the motion groups of uniform Riemann spaces with axial symmetry," Nauchn. Dokl. Vysshikh. Shkoly, Fiz.-Mat. Nauk, No. 6, pp. 127-138 (1958).

123. L. V. Sabinin, "Geometry of trisymmetric Riemann spaces," Sibirsk. Mat. Zh., 2(2): 266-278 (1961).

124. I. N. Semenova, "Symmetry images in binary Hermitian quasielliptic spaces," in: Third Interuniv. Scient. Conf. Prob. Geom. (report abstracts), Kazan (1967), pp. 149-150.
125. G. Stanilov, "Geometric interpretation of certain differential invariants of one-parameter systems of lines in a biaxial geometry," Izv. Mat. Inst., Bulgar. Akad. Nauk, 6: 95-99 (1962).
126. G. Stanilov, "Minimal lines in a biaxial geometry," Fiz.-Mat. Spisanie, 7(1): 51-53 (1964).
127. G. Stanilov and Yu. B. Ivanov, "Geometric interpretations of certain differential forms and invariants in the biaxial theory of surfaces," Dokl. Bulgar. Akad. Nauk, 18(7): 601-602 (1965).
128. N. E. Steenrod, The Topology of Fibre Bundles, Princeton University Press, Princeton, N. J. (1951).
129. N. V. Talantova, "Biaxial space of parabolic type," Uch. Zap. Kazansk. Univ., 117(9): 30-34 (1957).
130. N. V. Talantova, "Biaxial space of parabolic type," Izv. Vysshikh. Uchebn. Zavedenii, Matematika, No. 3, pp. 214-228 (1959).
131. N. V. Talantova, "One-term motion subgroups of a biaxial space of parabolic type," Tr. Kazansk. Aviats. Inst., No. 68, pp. 21-33 (1962).
132. N. V. Talantova, "Null-conjugate images of a biaxial space of parabolic type," Uch. Zap. Kazansk. Univ., 123(1): 152-171 (1963).
133. N. V. Talantova, "Classification of the subgroups of the motion group of a biaxial space of parabolic type," in: Third Interuniv. Scient. Conf. Prob. Geom. (report abstracts), Kazan (1967), p. 165.
134. T. Temirov, "Developable direction field and motions in affinely connected spaces," Tr. Samarkandsk. Univ., No. 151, pp. 127-138 (1964).
135. V. D. Tret'yakov, "On the theory of curves of a biaxial space," Uch. Zap. Kazansk. Univ., 115(10): 15-16 (1955).
136. M. M. Tsalenko, "Flag structures," Uch. Zap. Mosk. Gos. Ped. Inst., No. 243, pp. 186-191 (1965).
137. M. M. Tsalenko, "Differentiable manifolds with a tensor structure," Usp. Mat. Nauk, 22(1): 173-176 (1967).
138. Ya. L. Shapiro, "Several path systems imbedded in a Riemann space," Izv. Vysshikh. Uchebn. Zavedenii, Matematika, No. 3, pp. 166-172 (1963).
139. B. N. Shapukov, "Invariant manifold of a bilinear metric space," Uch. Zap. Kazansk. Univ., 125(1): 141-160 (1965).
140. L. V. Shestyreva, "Symmetry images in a three-dimensional quaternion symplectic space," in: Third Interuniv. Scient. Conf. Prob. Geom. (report abstracts), Kazan (1967), pp. 192-193.
141. A. P. Shirokov, "A property of covariantly constant affinors," Dokl. Akad. Nauk SSSR, 102(3): 461-464 (1955).
142. A. P. Shirokov, "A class of spaces over algebras," Izv. Vysshikh. Uchebn. Zavedenii, Matematika, No. 1, pp. 163-170 (1961).
143. A. P. Shirokov, "Certain real realizations of spaces over algebras," Izv. Vysshikh. Uchebn. Zavedenii, Matematika, No. 5, pp. 117-127 (1961).
144. A. P. Shirokov, "Classification of the motion groups of a biaxial space of elliptic type," Uch. Zap. Kazansk. Univ., 123(1): 208-221 (1963).

145. A. P. Shirokov, "Spaces over associative unital algebras," Uch. Zap. Kazansk. Univ., 123(1): 222-247 (1963).
146. A. P. Shirokov, "Symmetric spaces defined by algebras," Izv. Vysshikh. Uchebn. Zavedenii, Matematika, No. 6, pp. 159-171 (1963).
147. A. P. Shirokov, "On the theory of spaces defined by commutative algebras," Uch. Zap. Kazansk. Univ., 125(1): 165-182 (1965).
148. A. P. Shirokov, "A type of G-structure defined by algebras," in: Proc. Geom. Seminar, Inst. Sci. Inform. Akad. Nauk SSSR, 1: 425-456 (1966).
149. A. P. Shirokov, "Symmetric spaces defined by fourth-order commutative algebras," Uch. Zap. Kazansk. Univ., 126(1): 60-80 (1966).
150. A. P. Shirokov, "On pure tensors and invariant subspaces in manifolds with almost-algebraic structure," Uch. Zap. Kazansk. Univ., 126(1): 81-89 (1966).
151. R. N. Shcherbakov, "Differential line geometry of a three-dimensional space," in: R. V. Gamkrelidze (ed.), Progress in Mathematics, Vol. 6: Topology and Geometry, Plenum Press, New York (1970), pp.
152. I. M. Yaglom, "Some algebraic peculiarities of a real symplectic space," in: Proc. Third All-Union Math. Congr., Vol. 1, Akad. Nauk SSSR, Moscow (1956), pp. 40-41.
153. I. M. Yaglom, Complex Numbers and Their Application in Geometry, Fizmatgiz, Moscow (1963), p. 192.
154. E. U. Yasinskaya, "Metric invariants of linear complexes of real non-Euclidean and seminon-Euclidean spaces and Hermitian quadrics and linear complexes of Hermitian non-Euclidean and seminon-Euclidean spaces," in: Third Interuniv. Scient. Conf. Prob. Geom. (report abstracts), Kazan (1967), p. 199.
155. T. Adati, "On a Riemannian space admitting a field of planes," Tensor, 14: 60-67 (1963).
156. T. Adati and A. Kandatu, "On a Riemannian space with a field of null planes," Math. Japon., Vol. 7 (May): 95-102 (1962).
157. H. Akbar-Zadeh, "Sur les automorphismes de certaines structures presque cosymplectiques," Can. Math. Bull., 8(1): 39-57 (1965).
158. H. Akbar-Zadeh and E. Bonan, "Structure presque kahlérienne naturelle sur le fibré tangent à une variété finslérienne," Compt. Rend. Acad. Sci., 258(23): 5581-5582 (1964).
159. M. Ako, "Nonlinear connection in vector bundles," Kodai Math. Sem. Repts., 18(4): 307-316 (1966).
160. A. Aragnol, "Connexions euclidiennes canoniquement associées à certaines structures presque - produit," Compt. Rend. Acad. Sci., 242(3): 339-341 (1956).
161. M. Badea, "Spatii A_3 cu connexiune affină constată, local euclidiene si algebrele asociate," Studii Si Cercetări Mat. Acad. RPR, 9(2): 559-590 (1958).
162. M. Badea, "Spatii A_3 cu conexiune afină constantă local euclidiene si algebrele asociate," Lucrările Consf. Geometrie Si Topol. 1958. Bucuresti, Acad. RPR, pp. 153-158 (1962).
163. C. de Barras, "Variétés hor - symplectiques," Compt. Rend. Acad. Sci., 259(6): 1291-1294 (1964).
164. C. de Barras, "Variétés presque hor-complexes," Compt. Rend. Acad. Sci., 260(6): 1543-1546 (1965).

165. D. Bernard, "Sur la structure des pseudogroupes de Lie," Compt. Rend. Acad. Sci., 239(20): 1263-1265 (1954).
166. D. Bernard, "Sur les G-structures complexes," Compt. Rend. Acad. Sci., 243(23): 1821-1824 (1956).
167. D. Bernard, "Définition globale du tenseur de structure d'une G-structure," Compt. Rend. Acad. Sci., 247(19): 1546-1549 (1958).
168. D. Bernard, "Sur la géométrie différentielle des G-structures," Ann. Inst. Fourier, 10: 151-270 (1960).
169. E. Bonan, "Structures presque quaternioniennes," Compt. Rend. Acad. Sci., 258: 792-795 (1964).
170. E. Bonan, "Connexions presque quaternioniennes," Compt. Rend. Acad. Sci., 258(6): 1696-1699 (1964).
171. E. Bonan, "Structures presque hermitiennes quaternioniennes," Compt. Rend. Acad. Sci., 258(7): 1988-1991 (1964).
172. E. Bonan, "Tenseur de structure d'une variété presque quaternionienne," Compt. Rend. Acad. Sci., 259(1): 45-48 (1964).
173. E. Bonan, "Structure presque quaternale sur une variété différentiable," Compt. Rend. Acad. Sci., 260(21): 5445-5448 (1965).
174. W. Boothby, "Homogeneous complex contact manifolds," Different Creow Providence, R. I., Am. Math. Soc. (1961), pp. 144-54.
175. W. Boothby and H. C. Wang, "On contact manifolds," Ann. Math., 68(3): 721-734 (1958).
176. F. Brickell and R. S. Clark, "Conformally Riemannian structures. I," Bull. Soc. Math. France, 90(1): 15-26 (1962).
177. F. Brickell and R. S. Clark, "Conformally Riemannian structures. II," Bull. Soc. Math. France, 90(1): 27-38 (1962).
178. M. Bruni, "Relazioni tra metrica euclidea ed hermitiana in uno spazio vettoriale quaternionale," Atti Accad. Naz. Lincei. Rend. Cl. Sci. Fis., Mat. Natur., 38(4): 488-491 (1965).
179. M. L. Bruschi, "Un théorème de décomposition pour les variétés cosymplectiques homogenes," Compt. Rend. Acad. Sci., 257(26): 4120-4121 (1963).
180. E. Cartan, "Sur la structure des groupes infinis de transformations," Ann. Ecole Normale, 21: 153-206 (1904); 21: 219-308 (1905).
181. E. Cartan, "La structure des groupes infinis," in: Oeuvres Complètes, Paris (1953), Vol. II, pp. 1335-1358.
182. C. Cattaneo, "Proiezioni naturali e derivazione transversa in una varietà riemanniana a metrica iperbolica normale," Ann. Mat. Pura ed Appl., 48: 361-386 (1959).
183. C. Cattaneo, "Dèrivation transverse et grandeurs relatives en relativité générale," Compt. Rend. Acad. Sci., 248(2): 197-199 (1959).
184. C. Cattaneo, "Principi di conservazione e teoremi di Gauss in relatività generale," Rend. Mat. e Applic., 21(3-4): 373-405 (1962).
185. I. Cattaneo-Gasparini, "Connessioni adattate a una struttura quasi prodotto," Ann. Mat. Pura ed Appl., 64: 133-150 (1963).
186. I. Cattaneo-Gasparini, "Dérivée covariante 'liée' dans une V_{n+1} riemannienne à structure presque produit," Compt. Rend. Acad. Sci., 256(10): 2089-2091 (1963).

187. I. Cattaneo-Gasparini, "Proiezioni dei tensori du curvatura di una varietà riemanniana a metrica iperbolica normale," Rend. Mat. e Applic., 22(1): 127-146 (1963).

188. I. Cattaneo-Gasparini, "Struttura metrica adattata a una struttura quasi prodotto," Atti Accad. Naz. Lincei. Rend. Cl. Sci. Fis., Mat. Natur., 36(5): 623-628 (1964).

189. I. Cattaneo-Gasparini, "Sulle G-strutture di una V_n definite da una 11-formo complessa a valori vettoriali," Ann. Mat. Pura ed Appl., 65: 81-96 (1964).

190. I. Cattaneo-Gasparini, "Su una condizione necessaria per l'esistenza globale di un campo di r-piani su una varietà differenziabile," Atti Accad. Naz. Lincei. Rend. Cl. Sci. Fis., Mat. Natur, 42(5): 634-639 (1967).

191. B. Cenkl, "On the G-structure of higher order," Časop. Pěstov. Mat., 90(1): 26-32 (1965).

192. Shiing-shen Chern, "Pseudo-groupes continus infinis," in: Colloq. Internat. Centre Nat. Rech. Scient. 52, Strasbourg, 1953, Paris (1953), pp. 119-136.

193. Shiing-shen Chern, "On a generalization of Kähler geometry," in: Algebra, Geometry, and Topology, Princeton University Press, Princeton, N. J. (1957), pp. 103-121.

194. Shiing-shen Chern, "Sur les métriques riemanniennes compatibles avec une réduction du groupe structural," Semin. Topol. et Géom. Différent. Ch. Ehresmann. Fac. Sci. Paris, 1958-1960 (1961).

195. H. Chu and S. Kobayashi, "The automorphism group of a geometric structure," Trans. Am. Math. Soc., 113(1): 141-150 (1964).

196. R. S. Clark, "Special geometric object fields on a differentiable manifold," Tensor, 13: 60-70 (1963).

197. R. S. Clark and M. R. Bruckheimer, "Sur les structures presque tangentes," Compt. Rend. Acad. Sci., 251(5): 627-629 (1960).

198. R. S. Clark and M. R. Bruckheimer, "Tensor structures on a differentiable manifold," Ann. Math. Pura ed Appl., 54: 123-141 (1961).

199. A. Crumeyrolle, "Variétés différentiables à coordonnées hypercomplexes. Application à une géométrisation et à une généralisation de la théoriè d'Einstein-Schrodinger," Ann. Fac. Sci. Univ. Toulouse Sci. Math. et Sci. Phys., 26: 105-137 [1962 (1964)].

200. P. Dazord, "Tenseur de structure d'une G-structure dérivée," Compt. Rend. Acad. Sci., 258(10): 2730-2733 (1964).

201. R. Debever, "Connexions métriques et champs d'éléments plans paralléles dans les variétés quatre dimensions," Bull. Cl. Sci. Acad. Roy. Belg., 44(1): 56-61 (1958).

202. J. Depunt, "Sur la géométrie ternionienne dans le plan," Bull. Soc. Math. Belg., 11(2): 123-133 (1959).

203. P. Dombrowski, "On the geometry of the tangent bundle," J. Reine und Angew. Math., 210(1-2): 73-88 (1962).

204. Vincent P. Droz, "Sur certaines structures résultant de la notion d'échangeabilité," Compt. Rend. Acad. Sci., 262(26): A1484-A1487 (1966).

205. C. Ehresmann, "Espaces fibrés de structures comparatables," Académie des Sciences, Paris, Compt. Rend. Acad. Sci., 214: 141-147 (1942).

206. C. Ehresmann, "Les prolongements d'une variété différentiable. I. Calcul des jets, prolongement principal," Compt. Rend. Acad. Sci., 233: 598-600 (1951).
207. C. Ehresmann, "Les prolongements d'une variété différentiable. II. L'espace des jets d'ordre r de V_n dans V_m," Compt. Rend. Acad. Sci., 233: 777-779 (1951).
208. C. Ehresmann, "Les prolongements d'une variété différentiable. III. Transitivité des prolongements," Compt. Rend. Acad. Sci., 233: 1081-1083 (1951).
209. C. Ehresmann, "Les prolongements d'une variété différentiable. IV. Élements de contact et éléments d'enveloppe," Compt. Rend. Acad. Sci., 234: 1028-1030 (1952).
210. C. Ehresmann, "Les prolongements d'une variété différentiable. V. Covariants différentiels et prolongements d'une structure infinitésimale," Compt. Rend. Acad. Sci., 234: 1124-1425 (1952).
211. C. Ehresmann, "Structures locales et structures infinitésimales," Compt. Rend. Acad. Sci., 234: 587-589 (1952).
212. C. Ehresmann, "Sur les structures infinitésimales réguliéres," in: Proc. Internat. Congr. Mathematicians, 1954, Amsterdam, Vol. 1, Croningen, Amsterdam (1957).
213. H. A. Eliopoulos, "Structures presque tangentes sur les variétés différentiables," Compt. Rend. Acad. Sci., 255(14): 1563-1565 (1962).
214. H. A. Eliopoulos, "Euclidean structures compatible with almost-tangent structures," Bull. Cl. Sci. Acad. Roy. Belg., 50(10): 1174-1182 (1964).
215. H. A. Eliopoulos, "On the general theory of differentiable manifolds with almost-tangent structure," Can. Math. Bull., 8(6): 721-748 (1965).
216. H. A. Eliopoulos, "Structures r-tangentes sur les variétés différentiables," Compt. Rend. Acad. Sci., 263(13): A413-A416 (1966).
217. A. Frölicher and A. Nijenhuis, "Theory of vector-valued differential forms. Part I. Derivations in the graded ring of differential forms," Proc. Koninkl. Ned. Akad. Wetenschap., A59(3): 338-359 (1956).
218. A. Fröhlicher and A. Nijenhuis, "Theory of vector-valued differential forms. Part II. Almost-complex structures," Proc. Koninkl. Ned. Akad. Wetenschap., A61(4): 414-429 (1958).
219. A. Fröhlicher and A. Nijenhuis, " Invariance of vector form operations under mappings," Comment. Math. Helv., 34(3): 227-248 (1960).
220. A. Fujimoto, "On the structure tensor of G-structure," Mem. Coll. Sci. Univ. Kyoto, A33(1): 157-169 (1960).
221. A. Fujimoto, "On automorphisms of G-structures," J. Math. Kyoto Univ. (formerly Mem. Coll. Sci. Univ. Kyoto, Ser. A, Math.), 1(1): 1-20 (1961).
222. A. Fujimoto, "Conformal geometry of G-structures. I," Tensor, 15(3): 181-190 (1964).
223. A. Fujimoto, "Conformal geometry of G-structures. II," Tensor, 15(3): 191-199 (1964).
224. T. Fujitani, "Complex-valued differential forms on normal contact riemannian manifolds," Tohoku Math. J., 18(4): 349-361 (1966).
225. T. Fukami, "Invariant tensors under the real representation of symplectic group and their applications," Tohoku Math. J., 10(1): 81-90 (1958).
226. T. Fukami, "Affine connections in almost-product manifolds with some structures," Tohoku Math. J., 11(3): 430-446 (1959).

227. S. Goldberg, "Rigidité de variétés de contact à courbure positive," Compt. Rend. Acad. Sci., 261(9): 1936-1939 (1965).

228. V. G. Gorciu, "Asupra unor algebre neasociative A_3," Studii Si Cercetări Mat. Acad. RPR, 16(6): 747-751 (1964).

229. J. W. Gray, "A theory of pseudogroups with applications to contact structures," Thesis, Stanford University (1957). Technical Report ONR.

230. J. W. Gray, "Contact structures," Abstr. Short Communs. Internat. Congress Math. in Edinburgh, University of Edinburgh (1958), p. 113.

231. J. W. Gray, "Some global properties of contact structures," Ann. Math., 69(2): 421-450 (1959).

232. P. Griffiths, "Deformations of G-structures. Part A. General theory of deformations," Math. Ann., 155(4): 292-315 (1964).

233. P. Griffiths, "Deformations of G-structures. Part B. Deformations of geometric G-structures," Math. Ann., 158(5): 328-351 (1965).

234. P. Griffiths, "On the existence of a locally complete germ of deformation of certain G-structures," Math. Ann., 159(3): 151-171 (1965).

235. V. Guillemin, "The integrability problem for G-structures," Trans. Am. Math. Soc., 116(4): 544-560 (1965).

236. V. Guillemin and I. Singer, "Differential equations and G-structures," in: Proc. US—Japan Seminar Different. Geometry, Kyoto, 1965, Nippon Hyoronsha Co., Ltd., Tokyo (1966), pp. 34-36.

237. V. Guillemin and S. Sternberg, "Sur les systèmes de formes différentielles," Ann. Inst. Fourier, 13(2): 61-74 (1963).

238. T. Hangan, "Asupra diferentierilor formelor tensoriale asociate unui tensor mixt de ordinul al doilea," An. Univ. Bucuresti. Ser. Stint. Natur., 10(29): 209-224 (1961).

239. T. Hangan, "Sur la forme canonique d'un tenseur mixte du second ordre," Compt. Rend. Acad. Sci., 255(3): 452-453 (1962).

240. T. Hangan, "On the canonical form of a second-order mixed tensor," Rev. Math. Pures. et Appl. (RPR), 7(4): 617-626 (1962).

241. T. Hangan, "Géométrie différentielle grassmannienne," Rev. Roumaine Math. Pures et Appl., 11(5): 519-531 (1966).

242. T. Hangan, "On the Riemannian structure of the real and complex Grassmann manifolds," Tensor, 18(1): 26-31 (1967).

243. S. Hashimoto, "Operators on the almost-complex manifolds which admit quaternion structure," Sci. Rept. Fac. Literat. and Sci. Hirosaki Univ., 7(2): 37-41 (1960).

244. S. Hashimoto, "On differentiable manifolds which admit $(\Phi_{(1)}\Phi_{(2)}\Phi_{(3)}\cdots\Phi_{(r)})$ structures," Sci. Rept. Fac. Literat. and Sci. Hirosaki Univ., 8(2): 67-73 (1961).

245. S. Hashimoto, "On the differentiable manifolds with certain almost-contact structures which are similar to the quaternion structures," Sci. Rept. Fac. Literat. and Sci. Hirosaki Univ., 8(2): 74-78 (1961).

246. S. Hashimoto, "On differentiable manifold with almost-quaternion contact structure. I," Sci. Rept. Fac. Literat. and Sci. Hirosaki Univ., 8(1): 1-7 (1961).

247. S. Hashimoto, "On differentiable manifold with almost-quaternion contact structure. II," Sci. Rept. Fac. Literat. and Sci. Hirosaki Univ., 8(2): 79-87 (1961).

248. S. Hashimoto, "On differentiable manifold with almost-quaternion contact structure. III," Sci. Rept. Fac. Literat. and Sci. Hirosaki Univ., 10(1): 5-9 (1961).
249. S. Hashimoto, "On differentiable manifold with almost-quaternion contact structure," Tensor, 15(3): 258-268 (1964).
250. S. Hashimoto, "On the differentiable manifold M^n admitting tensor fields (F, G) of type (1, 1) satisfying $F^3 + F = 0$, $G^3 + G = 0$, $FG = -GF$, and $F^2 = G^2$," Tensor, 15(3): 269-274 (1964).
251. S. Hashimoto and T. Ichinohe, "On a manifold induced from the differentiable manifold with the almost-contact metric structure," Sci. Rept. Fac. Literat. and Sci. Hirosaki Univ., 9(2): 62-71 (1962).
252. Y. Hatakeyama, "On the existence of Riemann metrics associated with a 2-form of rank 2r," Tohoku Math. J., 14(2): 162-166 (1962).
253. Y. Hatakeyama, "Some notes on differentiable manifolds with almost-contact structures," Tohoku Math. J., 15(2): 176-181 (1963).
254. Y. Hatakeyama, "On the integrability of a structure defined by two semisimple 0-deformable vector 1-forms which commute with each other," Tohoku Math. J., 17(2): 171-177 (1965).
255. Y. Hatakeyama, "Some notes on the group of automorphisms of contact and symplectic structures," Tohoku Math. J., 18(3): 338-347 (1966).
256. Y. Hatakeyama, Y. Ogawa, and S. Tanno, "Some properties of manifolds with contact metric structure," Tohoku Math. J., 15(1): 42-48 (1963).
257. K. Havlicek, "O diferenciálni geometrii jednoparametrické soustavy přimek v dvojosé geometrii," Časop. Pěstov. Mat., 87(1): 103 (1962).
258. R. Hermann, "Sur les automorphismes infinitésimaux d'une G-structure," Compt. Rend. Acad. Sci., 239(25): 1760-1761 (1954).
259. F. Hirzebruch, "Über die quaternionalen projektiven Räume," Sitzber. Math.-Nat. Kl. Bayer. Akad. Wissenschap., pp. 301-312 [1953 (1954)].
260. Chorng-Shi Houh, "Affine connections in an almost-product manifold," Tensor, 12(2): 167-184 (1962).
261. Chorng-Shi. Houh, "The integrability of a structure on a differentiable manifold," Tohoku Math. J., 17(1): 72-75 (1965).
262. Chen-Jung Hsu, "Note on the integrability of a certain structure on differentiable manifold, Tohoku Math. J., 12(2): 349-360 (1960).
263. Chen-Jung Hsu, "On some structures which are similar to the quaternion structure," Tohoku Math. J., 12(3): 403-428 (1960).
264. Chen-Jung Hsu, "On some properties of π-structures on differentiable manifold," Tohoku Math. J., 12(3): 429-454 (1960).
265. Chen-Jung Hsu, "Note on (φ, ξ, η)-structure," Tohoku Math. J., 13(3):434-442 (1961).
266. Chen-Jung Hsu, "On Walker differentiation in differentiable manifolds with $r-\pi$-structure," Tohoku Math. J., 14(4): 332-343 (1962).
267. Chen-Jung Hsu, "Remarks on certain almost-product spaces," Pacif. J. Math., 14(1): 163-176 (1964).
268. Chen-Jung Hsu and Chorng-Shi Houh, "Note on the integrability conditions of (φ, ψ)-structures," Tohoku Math. J., 18(4): 368-377 (1966).
269. X. Hubaut, "Construction d'une droite projective sur une algèbre associative," Bull. Cl. Sci. Acad. Roy. Belg., 50(6): 618-623 (1964).

270. Y. Ichijyo, "Differentiable manifolds admitting complex distributions," J. Math. Kyoto Univ., 5(1): 67-85 (1965).
271. Y. Ichijyo, "On almost-contact metric manifolds admitting parallel fields of null planes," Tohoku Math. J., 16(2): 123-129 (1964).
272. Y. Ichijyo, "Analytic manifolds admitting parallel fields of complex planes," J. Math. Kyoto Univ., 4(2): 369-380 (1965).
273. Y. Ichijyo, "Almost-complex structures of tangent bundles and Finsler metrics," J. Math. Kyoto Univ., 6(3): 419-452 (1967).
274. S. Ishihara, "Holomorphically projective changes and their groups in an almost-complex manifold," Tohoku Math. J., 9(3): 273-297 (1957).
275. S. Ishihara, "On a tensor field φ_i^h satisfying φ^p w ± 1," Tohoku Math. J., 13(3): 443-454 (1961).
276. S. Ishihara, "Normal structure f satisfying $f^3 + f = 0$," Kodai Math. Semin. Rept., 18(1): 36-47 (1966).
277. Ju. B. Ivanov, "Flächen in der zweiachsigen Geometrie," Compt. Rend. Bulgare Acad. Sci., 14(1): 11-13 (1961).
278. Ju. B. Ivanov, "Die geodätischen Linien erster Art in der zweiachsigen Geometrie als Minimallinien," Compt. Rend. Bulgare Acad. Sci., 15(7): 697-698 (1962).
279. Ju. B. Ivanov, "Über die Kongruenz der Flächen in der hyperbolischen zweiachsigen Geometrie," Compt. Rend. Bulgare Acad. Sci., 17(8): 685-688 (1964).
280. A. Kandatu, "Integrability of a field of partially null planes in a Riemannian space," Tensor, 14: 68-70 (1963).
281. A. Kandatu, "Tangent bundle of a manifold with a nonlinear connection," Kodai Math. Semin. Rept., 18(4): 259-270 (1966).
282. S. Kashiwabara, "The structure of a Riemannian manifold admitting a parallel field of one-dimensional tangent vector subspaces," Tohoku Math. J., 11(3): 327-350 (1959).
283. S. Kashiwabara, "The structure of a Riemannian manifold admitting a parallel field of tangent vector subspaces," Tohoku Math. J., 12(1): 102-119 (1960).
284. S. Kashiwabara, "On the parallelisability under Riemannian metrics of direction fields over 3-dimensional manifolds," Tohoku Math. J., 14(1): 24-47 (1962).
285. S. Kashiwabara, "On the parallelisability under Riemannian metrics of direction fields over 3-dimensional manifolds. II," Tohoku Math. J., 15(4): 332-334 (1963).
286. Y. Kerbrat, "Variétés à structure presque-produit tangente et connexions spéciales," Compt. Rend. Acad. Sci., AB262(4): A237-A239 (1966).
287. J. Klein, "Opérateurs différentiels sur les variétés presque tangentes," Compt. Rend. Acad. Sci., 257(17): 2392-2394 (1963).
288. E. Kobayashi, "A remark on the Nijenhuis tensor," Pacif. J. Math., 12(3): 963-977 (1962).
289. E. Kobayashi, "Errata to the paper 'A remark on the Nijenhuis tensor'," Pacif. J. Math., 12(4): 1466 (1962).
290. E. Kobayashi, "A remark on the existence of a G-structure," Proc. Am. Math. Soc., 16(6): 1329-1331 (1965).
291. S. Kobayashi, "Remarks on complex contact manifolds," Proc. Am. Math. Soc., 10(1): 164-167 (1959).

292. S. Kobayashi, "Canonical forms on frame bundles of higher-order contact," in: Proc. Symposia in Pure Math., Vol. 3, Differential Geometry. A. M. S. (1961), pp. 186-193.

293. S. Kobayashi and T. Nagano, "On a fundamental theorem of Weyl-Cartan on G-structures," J. Math. Soc. Japan, 17(1): 84-101 (1965).

294. K. Kodaira, "On deformations of some complex pseudo-group structures," Ann. Math., 71(2): 224-302 (1960).

295. S. Koto, "Infinitesimal transformations of a manifold with f-structure," Kodai Math. Semin. Rept., 16(2): 116-126 (1964).

296. S. Koto and M. Nagao, "On an invariant tensor under a CL-transformation," Kodai Math. Semin. Rept., 18(2): 87-95 (1966).

297. M. Kurita, "Tensor fields and their parallelism," Nagoya Math. J., 18: 133-151 (1961).

298. M. Kurita, "On normal contact metric manifolds," J. Math. Soc. Japan, 15(3): 304-318 (1963).

299. A. J. Ledger and K. Yano, "The tangent bundle of a locally symmetric space," J. London Math. Soc., 40(3): 487-492 (1965).

300. J. Lefebvre, "Transformations conformes et automorphismes de certaines structures presque symplectique," Compt. Rend. Acad. Sci., AB262(13): A752-A754 (1966).

301. G. Legrand, "Sur les variétés à structure de presque-produit complexe," Compt. Rend. Acad. Sci., 242(3): 335-337 (1956).

302. G. Legrand, "Structures presque hermitiennes au sens large," Compt. Rend. Acad. Sci., 243(19): 1392-1395 (1956).

303. G. Legrand, "Etude d'une généralisation des structures presque complexes sur les variétés différentiables," Rend. Circolo Mat. Palermo, 7(3): 323-354 (1958). 8(1): 5-48 (1959).

304. G. Legrand, "T-structures sur les variétés différentiables," Compt. Rend. Acad. Sci., 250(20): 3266-3268 (1960).

305. G. Legrand, "Une interprétation de la forme de courbure d'une connexion infinitésimale," Compt. Rend. Acad. Sci., 250(21): 3441-3442 (1960).

306. G. Legrand, "Notions diverses de formes de torsion," Compt. Rend. Acad. Sci., 256(10): 2087-2088 (1963).

307. G. Legrand, "T-structures homogènes," Compt. Rend. Acad. Sci., 258(19): 4648-4650 (1964).

308. D. Lehmann, "Remarques sur la connexion canonique d'une variété de Stiefel," Compt. Rend. Acad. Sci., 259(17): 2754-2757 (1964).

309. J. Lehmann-Lejeune, "Sur l'intégrabilite de certaines G-structures," Compt. Rend. Acad. Sci., 258(22): 5326-5329 (1964).

310. J. Lehmann-Lejeune, "Intégrabilité des G-structures défines par une 1-forme à valeurs dans le module des champs de vecteurs, O-déformable," Compt. Rend. Acad. Sci., 259(23): 4216-4219 (1964).

311. J. Lehmann-Lejeune, "Intégrabilité des G-structures définites par une 1-forme a valeurs dans le module des champs de vecteurs, O-déformable et nilpotente," Compt. Rend. Acad. Sci., 260(3): 772-775 (1965).

312. J. Lehmann-Lejeune, "Etude des formes différentialles liée à certaines G-structures," Compt. Rend. Acad. Sci., 260(7): 1838-1841 (1965).

313. J. Lehmann-Lejeune, "Intégrabilité des G-structures défines par une 1-forme 0-deformable à valéurs dans le fibre tangent," Ann. Inst. Fourier, 16(2):329-387 [1966 (1967)].

314. J. Lewis, "Invariants of higher-order G-structures," Doctoral Dissertation, Northwestern University, 1964, 42 pp.; Dissertation Abstr., 25(11):6658 (1965).

315. P. Libermann, "Sur le problème d'équivalence de certaines structures infinitésimales," Ann. Mat. Pura ed Appl., 36:27-120 (1954).

316. P. Libermann, "Sur les structures presque complexes et autres structures infinitésimales réguliéres," Bull. Soc. Math. France, 83(3):195-224 (1955).

317. P. Libermann, "Automorphismes infinitésimaux d'une structure symplectique," Compt. Rend. Acad. Sci., 242:1114-1117 (1956).

318. P. Libermann, Pseudogroupes infinitésimaux. Applications aux G-structures," Compt. Rend. Acad. Sci., 246(9):1365-1368 (1958).

319. P. Libermann, "Sur les automorphismes infinitésimaux des structures symplectiques et des structures de contact," in: Colloq. Géometrie Différent. Globale, Brussels, 1958, Paris (1959), pp. 37-59.

320. P. Libermann, "Pseudogroupes infinitésimaux attachés aux pseudogroupes de Lie," Bull. Soc. Math. France, 87(4):409-425 (1959).

321. P. Libermann, "Sur certaines résolutions de faisceaux," Compt. Rend. Acad. Sci., 254(10):1735-1737 (1962).

322. P. Libermann, "Sur quelques exemples de structures pfaffiennes et presque cosymplectiques," Ann. Mat. Pura ed Appl., 60:153-172 (1962).

323. P. Libermann, "On sprays and higher-order connections," Proc. Nat. Acad. Sci. USA, 49(4):459-462 (1963).

324. P. Libermann, "Sur la géometrie des prolongements des espaces fibrés vectoriels," Ann. Inst. Fourier, 14(1):145-172 (1964).

325. A. Lichnerowicz, "Sur la réductivité de certaines algébres d'automorphismes," Compt. Rend. Acad. Sci., 253(13):1302-1304 (1961).

326. A. Lichnerowicz, "Théorémes de réductivité sur des algébres d'automorphismes," Rend. Mat. e Applic., 22(1-2):197-244 (1963).

327. T. Lin, "On complex manifolds with certain structures which are related to complex contact structures," Tohoku Math. J., 15(3):187-202 (1963).

328. E. Martinelli, "Varietà a struttura quaternionale generalizzata," Atti Accad. Naz. Lincei. Rend. Cl. Sci. Fis., Mat. Natur., 26(3):353-362 (1959).

329. E. Martinelli, "Modello metrico reale dello spazio projettivo quaternionale," Ann. Mat. Pura ed Appl., 49: 73-89 (1960).

330. E. Martinelli, "Metriche hermitiane sulle varietà a struttura quasi quaternionale generalizzata," Atti Accad. Naz. Lincei. Rend. Cl. Sci. Fis., Mat. Natur, 39(6):400-407 (1965).

331. E. Martinelli, "Variétés à structure quaternionienne généralisée," Rev. Roumaine Math. Pures et Appl., 10(7):915-922 (1965).

332. M. Matsumoto, "Connections, metrics, and almost-complex structures of tangent bundles," J. Math. Kyoto Univ., 5(3):251-278 (1966).

333. O. Mayer, "Geometrie biaxiale différentielle des courbes," Bull. Math. Soc. Roum. des Sci. (1938), p. 40.

334. R. Miron, "Connexions compatibles aux structures conformes-presque symplectiques," Compt. Rend. Acad. Sci., 265(21):A685-A687 (1967).

335. H. Mizusawa, "On infinitesimal transformations of K-contact and normal contact metric spaces," Sci. Rept. Niigata Univ., A(1): 5-18 (1964).
336. A. Morimoto, "On normal almost-contact structures," J. Math. Soc. Japan, 15(4): 420-436 (1963).
337. A. Morimoto, "On normal almost-contact structures with a regularity," Tohoku Math. J., 16(1): 90-104 (1964).
338. Y. Muto, "On some (1, 1) tensor field and connections," Sci. Rept. Yokohama Nat. Univ., Sec. 1, No. 10, pp. 1-11 (1963).
339. Y. Muto, "On integrability conditions of equations of contravariant almost-analytic vectors and some related ones," Sci. Rept. Yokohama Nat. Univ., Sec. 1, No. 11, pp. 1-9 (1964).
340. T. Nagai and H. Kôjyô, "On some considerations of hypersurfaces in certain almost-complex spaces," J. Fac. Sci. Hokkaido Univ., Ser. 1, 18(3-4): 114-123 (1965).
341. T. Nagano, "Isometries on complex-product spaces," Tensor, 9(1): 47-61 (1959).
342. H. Nakagawa, "f-Structures induced on submanifolds in spaces almost hermitian or kaehlerian," Kodai Math. Semin. Rept., 18(2): 161-183 (1966).
343. H. Nakagawa, "On the automorphism groups of f-manifolds," Kodai Math. Sem. Rept., 18(3): 251-257 (1966).
344. H. Nakagawa, "On framed f-manifolds," Kodai Math. Sem. Rept., 18(4): 293-306 (1966).
345. H. Nakagawa, "On differentiable manifolds with certain almost-contact structures," Sci. Rept. Tokyo Kyoiku Daigaku, A8: 144-163 (1964).
346. H. Nakagawa, "On the set of (φ, ξ, η)-structures over a differentiable manifold," Sci. Rept. Tokyo Kyoiku Daigaku, A7(26): 146-153 (1962).
347. T. Nasu, "A remark on a certain G-structure," J. Sci. Hiroshima Univ., Ser. A, Div. 1, 29(2): 253-270 (1965).
348. Ngo van-Que, "G-structure unvariante," Compt. Rend. Acad. Sci., 253(22): 2454-2456 (1951).
349. H. Nickerson, "On differential operators and connections," Trans. Am. Math. Soc., 99(3): 509-539 (1961).
350. A. Nijenhuis, "X_{n-1}-forming sets of eigenvectors," Proc. Koninkl. Ned. Akad. Wetenschap., A54(2): 200-212 (1951); Indagationes Math., 13(2): 200-212 (1951).
351. A. Nijenhuis, "Jacobi-type identities for bilinear differential concomitants of certain tensor fields. I," Proc. Koninkl. Ned. Akad. Wetenschap., A58(3): 390-397 (1955); Indagationes Math., 17(3): 390-397 (1955).
352. A. Nijenhuis, "Jacobi-type identities for bilinear differential concomitants of certain tensor fields. II," Proc. Koninkl. Ned. Akad. Wetenschap., A58(3): 398-403 (1955); Indagationes Math., 17(3): 398-403 (1955).
353. M. Obata, "Affine transformations in an almost-complex manifold with a natural affine connection," J. Math. Soc. Japan, 8(4): 345-362 (1956).
354. M. Obata, "Affine connexions in a quaternion manifold and transformations preserving the structure," J. Math. Soc. Japan, 9(4): 406-416 (1957).
355. M. Obata, "Hermitian manifolds with quaternion structure," Tohoku Math. J., 10(1): 11-18 (1958).
356. T. Ochiai, "On the automorphism group of a G-structure," J. Math. Soc. Japan, 18(2): 189-193 (1966).

357. Y. Ogawa, "Some properties on manifolds with almost-contact structures," Tohoku Math. J., 15(2): 148-161 (1963).
358. Y. Ogawa, "On principal torus bundles over a homogeneous contact manifold," Tohoku Math. J., 16(3): 297-308 (1964).
359. Y. Ogawa, "On 5-dimensional Sasaki-Einstein space with sectional curvature $\geq 1/3$," Tohoku Math. J., 19(2): 103-109 (1967).
360. Y. Ogawa, "On a (2m + 1)-dimensional sasakian space with sectional curvature $> (4m - 3)/4m(2m - 1)$," Natur. Sci. Rept. Ochanomizu Univ., 18(1): 7-13 (1967).
361. K. Ogiue, "On almost-contact manifolds admitting axiom of planes of axiom of free mobility," Kodai Math. Semin. Rept., 16(4): 223-232 (1964).
362. K. Ogiue, "On fiberings of almost-contact manifolds," Kodai Math. Semin. Rept., 17(1): 53-62 (1965).
363. K. Ogiue, "Theory of conformal connections," Kodai Math. Semin. Rept., 19(2): 193-224 (1967).
364. K. Ogiue, "On the geometry of G-structures of higher order," Proc. Japan Acad., 43(4): 255-257 (1967).
365. T. Okubo, "On the differential geometry of frame bundles," Ann. Math. Pure ed Appl., 72: 29-44 (1966).
366. M. Okumura, "Some remarks on space with a certain contact structure," Tohoku Math. J., 14(2): 135-145 (1962).
367. M. Okumura, "On infinitesimal conformal and projective transformations of normal contact spaces," Tohoku Math. J., 14(4): 398-412 (1962).
368. M. Okumura, "Certain almost-contact hypersurfaces in Euclidean spaces," Kodai Math. Semin. Rept., 16(1): 44-54 (1964).
369. M. Okumura, "Certain almost-contact hypersurfaces in kaehlerian manifolds of constant holomorphic sectional curvatures," Tohoku Math. J., 16(3): 270-284 (1964).
370. M. Okumura, "Cosymplectic hypersurfaces in kaehlerian manifolds of constant holomorphic sectional curvature," Kodai Math. Semin. Rept., 17(2): 63-73 (1965).
371. M. Okumura, "Contact hypersurfaces in certain Kaehlerian manifolds," Tohoku Math. J., 18(1): 74-102 (1966).
372. M. Okumura, "Certain infinitesimal transformation of normalcontact metric manifold," Kodai Math. Semin. Rept., 18(2): 116-119 (1966).
373. M. Okumura, "Totally umbilical hypersurfaces of a locally product Riemannian manifold," Kodai Math. Semin. Rept., 19(1): 35-42 (1967).
374. D. I. Papuc, "Sur la théorie des surfaces d'un espace axial à 3 dimensions," An. Stint. Univ. Iasi, Sec. 1, 4(2): 49-80 (1958).
375. D. I. Papuc, "Classes de surfaces spéciales de l'espace axial à 3 dimensions," Ann. Stint. Univ. Iasi, Sec. 1, 7(2): 307-320 (1961).
376. D. I. Papuc, "Sur les variétés des espaces kelinéens à groupe linéaire complétement reductible," Compt. Rend. Acad. Sci., 256(1): 62-64 (1963).
377. D. I. Papuc, "Sur la théorie des variétés des espaces kelinéens à groupe linéaire complètement réductible," Compt. Rend. Acad. Sci., 257(3): 589-591 (1963).
378. R. Pernet, "Une géométrie conforme quaternionienne et son extension," Bull. Cl. Sci. Acad. Roy. Belg., 49(10): 1116-1129 (1963).

379. R. Pernet, "Une géométrie conforme quaternionienne et son extension," Bull. Cl. Sci. Acad. Roy. Belg., 49(11): 1214-1224 (1963).
380. R. Pernet, "Une géométrie conforme quaternionienne et son extension," Bull. Cl. Sci. Acad. Roy. Belg., 49(11): 1271-1280 (1963).
381. M. Pinl, "Zur Differentialgeometrie im totalisotropen R_2 und R_3 eines komplexen euklidischen R_4 und R_6," Monatsh. Math., 63(3): 256-264 (1959).
382. M. Pinl, "Über eine konstante lineare Übertragung, welche dem Schiefkörper der Hamiltonschen Quaternionen entspricht," Ann. Mat. Pura ed Appl., 55:13-21 (1961).
383. H. Reichardt, "Eine Aufspaltung von Windung und Krümmung in affin zusammenhängenden Räumen," Acta Sci. Math., 21(3-4): 300-308 (1960).
384. H. Reichardt, "Über die Teilräume affin zusammengängender Räume," Abhandl. Math. Seminar Univ. Hamburg, 25(3-4): 146-154 (1962).
385. B. L. Reinhart, "Harmonic integrals on almost-product manifolds," Trans. Am. Math. Soc., 88(1): 243-276 (1958).
386. B. L. Reinhart, "Foliated manifold with bundle-like metrics," Ann. Math., 69(1): 119-132 (1959).
387. B. L. Reinhart, "Closed metric foliations," Michigan Math. J., 8(1): 7-9 (1961).
388. E. Ruh, "On the automorphism group of a G-structure," Comment. Math. Helv., 39(3): 189-204 (1964).
389. H. S. Ruse, "On parallel fields of planes in a Riemannian space," Quart. J. Math., 20(80): 218-234 (1949).
390. H. S. Ruse, "Parallel planes in a Riemannian V_n," Proc. Roy. Soc. Edinburgh, Sec. A, 62(1): 78-92 (1949/50).
391. T. Saeki, "Affine transformations in a differentiable manifold with π-structure," J. Math. Soc. Japan, 14(3): 341-349 (1962).
392. T. Sakai, "Some transformation on K-contact and normal contact Riemannian manifolds," Tohoku Math. J., 18(2): 216-224 (1966).
393. M. Salzert, "Die Eigenschaften derjenigen Kollineationen, die zwei konjugiert imaginären windschiefen Geraden im Raume festlassen," Schr. der Math. Inst. und des Inst. für Angew. Math. der Univ. Berlin, 5(2): 133-177 (1940).
394. H. Sasagawa, "Certain almost-contact hypersurfaces in the almost-tachibanan space," Res. Rept. Nagaoka Techn. Coll., 2(2): 101-113 (1966).
395. H. Sasagawa, "The f-structure induced on the submanifold of an *O-space," Res. Rept. Nagaoka Techn. Coll., 2(3): 183-201 (1966).
396. S. Sasaki, "On the differential geometry of tangent bundles of Riemannian manifolds," Tohoku Math. J., 10(3): 338-354 (1958).
397. S. Sasaki, "On differentiable manifolds with certain structures which are closely related to almost-contact structure. I," Tohoku Math. J., 12(3): 459-476 (1960).
398. S. Sasaki, "On differentiable manifolds with (φ, ψ)-structures," Tohoku Math. J., 13(1): 132-153 (1961).
399. S. Sasaki, "On the differential geometry of tangent bundles of Riemannian manifolds. II," Tohoku Math. J., 14(2): 146-155 (1962).
400. S. Sasaki, "Homogeneous contact transformations," Tohoku Math. J., 14(4): 369-397 (1962).
401. S. Sasaki, "A characterization of contact transformations," Tohoku Math. J., 16(3): 285-290 (1964).

402. S. Sasaki, "On almost-contact manifolds," in: Proc. US-Japan Seminar Different. Geometry, Kyoto, 1965, Nippon Hyoronsha Co., Ltd., Tokyo (1966), pp. 128-136.

403. S. Sasaki and Y. Hatakeyama, "On differentiable manifolds with certain structures which are closely related to almost-contact structure. II," Tohoku Math. J., 13(2): 281-294 (1961).

404. S. Sasaki and Y. Hatakeyama, "On differentiable manifolds with contact metric structures," J. Math. Soc. Japan, 14(3): 249-271 (1962).

405. S. Sasaki and Chen-Jung Hsu, "On the integrability of almost-contact structure," Tohoku Math. J., 14(2): 167-176 (1962).

406. I. M. Singer and S. Sternberg, "The infinite groups of Lie and Cartan. Part I. The transitive groups," J. Analyse Math., 15: 1-114 (1965).

407. W. Slebodzinski, "Contribution à la géométrie différentielle d'un tenseur mixte de valence deux," Colloq. Math., 13(1): 49-54 (1964).

408. D. C. Spencer, "Deformation of structures on manifolds defined by transitive, continuous pseudogroups. Part 1. Infinitesimal deformations of structure," Ann. Math., 76(2): 306-398 (1962).

409. D. C. Spencer, "Deformation of structures on manifolds defined by transitive, continuous pseudogroups. Part II. Deformations of structure," Ann. Math., 76(3): 399-445 (1962).

410. D. C. Spencer, "Deformation of structures on manifolds defined by transitive, continuous pseudogroups. Part III. Structures defined by elliptic pseudogroups," Ann. Math., 81(3): 389-450 (1965).

411. K. Srinivasacharyulu, "Sur certaines familles Différentiables de G-structures," Compt. Rend. Acad. Sci., 250(7): 1171-1173 (1960).

412. S. Tachibana, "Some theorems on locally product Riemannian spaces," Tohoku Math. J., 12(2): 281-292 (1960).

413. S. Tachibana, "Analytic tensor and its generalization," Tohoku Math. J., 12(2): 208-221 (1960).

414. S. Tachibana, "On harmonic tensors in compact Sasakian spaces," Tohoku Math. J., 17(3): 271-284 (1965).

415. S. Tachibana, "On a decomposition of C-harmonic forms in a compact Sasakian space," Tohoku Math. J., 19(2): 198-212 (1967).

416. S. Tachibana and S. Koto, "On almost-analytic functions, tensors, and invariant subspaces," Tohoku Math. J., 14(2): 177-186 (1962).

417. S. Tachibana and M. Okumura, "On the almost-complex structure of tangent bundles of Riemannian spaces," Tohoku Math. J., 14(2): 156-161 (1962).

418. S. Takizawa, "On contact structures of real and complex manifolds," Tohoku Math. J., 15(3): 227-252 (1963).

419. S. M. Tallini, "Metrica hermitiana ellittica in uno spazio proiettivo quaternionale," Ann. Mat. Pura ed Appl., 60: 203-233 (1962).

420. S. M. Tallini, "Métreque hermitienne elliptique dans un espace projectif quaternionien," Ann. Univ. Sci. Budapest. Sec. Math., 7: 109-112 (1964).

421. N. Tanaka, "On the pseudo-conformal geometry of hypersurfaces of the space of n-complex variables," J. Math. Soc. Japan, 14(4): 397-429 (1962).

422. N. Tanaka, "On the equivalence problems associated with a certain class of homogeneous spaces," J. Math. Soc. Japan, 17(2): 103-139 (1965).
423. S. Tanno, "Note on infinitesimal transformations over contact manifolds," Tohoku Math. J., 14(4): 416-430 (1962).
424. S. Tanno, "Some transformations on manifolds with almost-contact and contact metric structures," Tohoku Math. J., 15(2): 140-147 (1963).
425. S. Tanno, "On fiberings of some noncompact contact manifolds," Tohoku Math. J., 15(3): 289-297 (1963).
426. S. Tanno, "Some transformations on manifolds with almost-contact and contact metric structures. II," Tohoku Math. J., 15(4): 322-331 (1963).
427. S. Tanno,"A remark on transformations of a K-contact manifold," Tohoku Math. J., 16(2): 173-175 (1964).
428. S. Tanno, "An almost-complex structure of the tangent bundle of an almost-contact manifold," Tohoku Math. J., 17(1): 7-15 (1965).
429. S. Tanno, "Almost-complex structures in bundle spaces over almost-contact manifolds," J. Math. Soc. Japan, 17(2): 167-186 (1965).
430. S. Tanno, "A theorem on regular vector fields and its application to almost-contact structures," Tohoku Math. J., 17(3): 235-238 (1965).
431. S. Tanno, "Sur une variété munie d'une structure de contact admettant certains transformations," Tohoku Math. J., 17(3): 239-243 (1965).
432. S. Tanno, "Partially conformal transformations with respect to (m − 1)-dimensional distributions of m-dimensional Riemannian manifolds," Tohoku Math. J., 17(4): 358-409 (1965).
433. S. Tanno, "Partially conformal transformations with respect to (m − 1)-dimensional distributions of m-dimensional Riemannian manifolds. II," Tohoku Math. J., 18(4): 378-392 (1966).
434. S. Tanno, "A conformal transformation of certain contact Riemannian manifolds," Tohoku Math. J., 18(3): 270-273 (1966).
435. S. Tanno, "Sur une variété de K-contact métrique de dimension 3," Compt. Rend. Acad. Sci., 263(9): A317-A319 (1966).
436. S. Tanno, "Harmonic forms and Betti numbers of certain contact Riemannian manifolds," J. Math. Soc. Japan, 19(3): 308-316 (1967).
437. S. Tanno, "Topology of compact Sasakian manifolds," Proc. Japan. Acad. 43(6): 423-425 (1967).
438. S. Tanno, "Curvature preserving transformations of K-contact Riemannian manifolds," Kodai Math. Sem. Rept., 19(2): 156-158 (1967).
439. Y. Tashiro, "On contact structure of hypersurfaces in complex manifolds. I," Tohoku Math. J., 15(1): 62-78 (1963).
440. Y. Tashiro, "On contact structure of hypersurfaces in complex manifolds. II," Tohoku Math. J., 15(2): 167-175 (1963).
441. Y. Tashiro and S. Tachibana, "On Fubinian and C-Fubinian manifolds," Kodai Math. Semin. Rept., 15: 176-183 (1963).
442. C. Teleman, "Asupra spatiilor projective cuaternionice," in: Lucrarile Const. Geom. Different., 1955, Timisoara Acad. RPR (1956), pp. 249-252.
443. C. Teleman, "Sur les variétés de Grassmann," Bull. Math. Soc. Sci. et Phys. RPR, 2(2): 203-224 (1958).

444. J. Tits, "Isotropie des espaces de Klein," in: Colloq. Géométrie Différent. Globale, Brussels, 1958, Paris-Louvain (1959), pp. 153-161.
445. P. Tondeur, "Affine Zusammenhänge auf Mannigfaltigkeiten mit fast-symplektischer Structur," Comment. Math. Helv., 36(3): 234-244 (1961).
446. P. Tondeur, "Structure presque kählérienne naturelle sur la fibré des vecteurs covariants d'une variété riemannienne," Compt. Rend. Acad. Sci., 254(3): 407-408 (1962).
447. P. Tondeur, "Champs invariants de p-plans sur un espace homogéne," Compt. Rend. Acad. Sci., 259(25): 4473-4475 (1964).
448. I. Văduva, "Asupra spatiilor cu conexiune afină constantă atasate unor algebre reale," Studii Si Cercetari Mat. Acad. RPR, 12(2): 535-541 (1961).
449. G. Vrănceanu, Lecons de Géométrie Différentielle, Vol. I, Vol. II, Bucharest, Acad. RPR (1957).
450. G. Vrănceanu, "Spazi a connessione affine e le algebre di numeri ipercomplessi," Ann. Scuola Normale Super. Pisa, 12(1-2): 5-20 (1958).
451. H. Wakakuwa, "On Riemannian manifolds with homogeneous holonomy group Sp (n)," Tohoku Math. J., 10(3): 274-303 (1958).
452. H. Wakakuwa, "On affinely connected manifolds with homogeneous holonomy group CL (n, Q) $\otimes$ T^1," Tohoku Math. J., 11(3): 364-375 (1959).
453. H. Wakakuwa, "On almost complex symplectic manifolds and affine connections with restricted homogeneous holonomy group Sp (n, C)," Tohoku Math. J., 12(2): 175-202 (1960).
454. H. Wakakuwa, "On some affine connections in manifolds with almost quaternion structure and homogeneous holonomy groups," Tensor, 11(1): 57-73 (1961).
455. H. Wakakuwa, "On affine connections in an almost-complex symplectic manifold," Tensor, 11(2): 110-126 (1961).
456. H. Wakakuwa, "On linearly independent almost-complex structures in a differentiable manifold," Tohoku Math. J., 13(3): 393-422 (1961).
457. A. G. Walker, "On parallel fields of partially null vector spaces," Quart. J. Math., 20(16): 135-145 (1949).
458. A. G. Walker, "Canonical form for a Riemannian space with a parallel field of null planes," Quart. J. Math., Ser. 2, 1(1): 69-79 (1950).
459. A. G. Walker, "Canonical forms (II): Parallel partially null planes," Quart. J. Math., Ser. 2, 1(2): 147-152 (1950).
460. A. G. Walker, "The fibering of Riemannian manifolds," Proc. London Math. Soc., 3(9): 1-19 (1953).
461. A. G. Walker, "Connexions for the parallel distributions in the large," Quart. J. Math., 6(24): 301-308 (1955).
462. A. G. Walker, "Connexions for parallel distributions in the large, II," Quart. J. Math., 9(35): 221-231 (1958).
463. A. G. Walker, "Distributions and global connexions," in: Colloq. Géométrie Différent. Globale, Brussels, 1958, Paris-Louvain (1959), No. 52, p. 285-298.
464. A. G. Walker, "Almost-product structures. Differential geometry," in: Proc. Sympos. Pure Math., Am. Math. Soc., Vol. 3, pp. 94-100 (1961).
465. A. West, "The projection of infinitesimal connections," J. London Math. Soc., 40(3): 551-564 (1965).

466. T. I. Willmore, "Connexions for systems of parallel distributions," Quart. J Math., 7(28): 269-276 (1956).
467. T. I. Willmore, "Parallel distributions on manifolds," Proc. London Math. Soc., 6(22): 191-204 (1956).
468. T. I. Willmore, "Systems of parallel distributions," J. London Math. Soc., 32(2): 153-156 (1957).
469. T. I. Willmore, "Global theorems on manifolds which admit distributions," Sémin. Topol. et Géom. Différent.Ch. Ehresmann, Fac. Sci. Paris, 1957-1958, Issue 1, Paris (1959), pp. 18/1-18/7.
470. T. I. Willmore, "Connexions associated with foliated structures," Ann. Inst. Fourier, 14(1): 43-47 (1964).
471. Y. G. Wong, "Fields of parallel planes in affinely connected spaces," Quart. J. Math., 4(16): 241-253 (1953).
472. K. Yano, "On Walker differentiation in almost-product or almost-complex spaces," Proc. Koninkl. Ned. Akad. Wetenschap., A61(5): 573-580 (1958); Indagationes Math., 20(5): 573-580 (1958).
473. K. Yano, "Affine connexions in an almost-product space," Kodai Math. Semin. Rept., 11(1): 1-24 (1959).
474. K. Yano, "On a structure f satisfying $f^3 + f = 0$," Technical Report, University of Washington, No. 2 (1961).
475. K. Yano, "Eckmann–Frölicher connexions on almost-analytic submanifolds," Kodai Math. Semin. Rept., 14(1): 53-58 (1962).
476. K. Yano, "On a structure defined by a tensor field f of type (1.1) satisfying $f^3 + f = 0$," Tensor, 14: 99-109 (1963).
477. K. Yano, "Tensor fields and connections on cross sections in the cotangent bundle," Tohoku Math. J., 19(1): 32-48 (1967).
478. K. Yano and E. T. Davies, "On the tangent bundles of Finsler and Riemannian manifolds," Rend. Circolo Mat. Palermo, 12(2): 211-228 (1963).
479. K. Yano and S. Ishihara, "On integrability conditions of a structure f satisfying $f^3 + f = 0$," Quart. J. Math., Oxford, 15: 217-222 (1964).
480. K. Yano and S. Ishihara, "Almost-contact structures induced on hypersurfaces in complex and almost-complex spaces," Kodai Math. Semin. Rept., 17(3): 222-249 (1965).
481. K. Yano and S. Ishihara, "Structure defined by F satisfying $F^3 + F = 0$," in: Proc. US-Japan Seminar Different. Geometry, Kyoto, 1965, Nippon Hyoronsha Co., Ltd., Tokyo (1966), pp. 153-166.
482. K. Yano and S. Ishihara, "The f-structure induced on submanifolds of complex and almost-complex spaces," Kodai Math. Semin. Rept., 18(2): 120-160 (1966).
483. K. Yano and S. Ishihara, "Differential geometry in tangent bundle," Kodai Math. Semin. Rept., 18(4): 271-292 (1966).
484. K. Yano and S. Ishihara, "Correction to the paper 'The f-structure induced on submanifolds of complex and almost-complex spaces'," Kodai Math. Semin. Rept., 18(4): 386 (1966).
485. K. Yano and S. Ishihara, "Almost-complex structures induced in tangent bundles," Kodai Math. Semin. Rept., 19(1): 1-27 (1967).
486. K. Yano and S. Ishihara, "Horizontal lifts of tensor fields and connections to tangent bundles," J. Math. Mech., 16(9): 1015-1029 (1967).

487. K. Yano and S. Kobayashi, "Prolongations of tensor fields and connections to tangent bundles. I. General theory," J. Math. Soc. Japan, 18(2): 194-210 (1966).

488. K. Yano and S. Kobayashi, "Prolongations of tensor fields and connections to tangent bundles. II. Infinitesimal automorphisms," J. Math. Soc. Japan, 18(3): 236-246 (1966).

489. K. Yano and A. J. Ledger, "Linear connections on tangent bundles," J. London Math. Soc., 39(3): 495-500 (1964).

490. K. Yano and Y. Muto, "Homogeneous contact structures," Math. Ann., 167(3): 195-213 (1966).

491. K. Yano and E. M. Patterson, "Vertical and complete lifts from a manifold to its cotangent bundle," J. Math. Soc. Japan, 19(1): 91-113 (1967).

492. K. Yano and E. M. Patterson, "Horizontal lifts from a manifold to its cotangent bundle," J. Math. Soc. Japan, 19(2): 185-198 (1967).

493. I. Yokote, "On a foliated manifold with 1-dimensional distribution" Sci. Rept. Tokyo Kyoiku Daigaku, A9(202-208): 1-23 (1965).

494. I. Yokote, "On some properties of the tangent bundles of Riemannian manifolds which admit a skew-symmetric covariant tensor of order 2," Sci. Rept. Tokyo Sect. A, 9: 143-147 (1967).

495. P. B. Zwart, "Compact homogeneous spaces possessing invariant contact, symplectic or cosymplectic structures," Doctoral Dissertation, Washington University, 1965; Dissertation Abstr., 26(11): 6756 (1966).

Packings, Coverings, Partitionings, and Certain Other Distributions in Spaces of Constant Curvature

E. P. Baranovskii

Introduction

A set $\{K_i\}$ of bodies K_i forms a packing in a domain D if $\bigcup_i K_i \subset D$ and no two bodies of set $\{K_i\}$ have common interior points. Sometimes a packing is called a filling and in the Russian translation of Cassels' book [26] the term "stacking" is used. If no two bodies K_i and K_j have common interior points and $\bigcup_i K_i = D$, then we speak of the partitioning of domain D into bodies K_i. The set $\{K_i\}$ forms a covering of domain D, or covers domain D, if $\bigcup_i K_i \supset D$. In the definitions listed above it is understood that the domain D can coincide with the whole space in which it is located.

The subject of the present survey is the results concerning the above-named distributions — packings, coverings, and partitionings, as well as some others, in Euclidean spaces E^n, on surfaces of constant curvature in these spaces, and in hyperbolic and elliptic planes and spaces. The survey is devoted to the papers published in the past five to six years; however, for a more complete and clear description of the situation we have also cited certain very essential results obtained earlier. In this connection a list of the literature cited has been compiled: the author of the survey has strived to list all the publications on the survey topic since 1962; earlier papers are cited only in the event that their results have been referred to.

The branch of geometry considered in the survey — we shall briefly call it the theory of distributions, another name is discrete geometry — is a comparatively new one, its existence as an independent entity having started only several decades ago. Therefore, in spite of its rather intensive development — the number of journal articles devoted to it grows each year — very few books have been written wholly on the theory of distributions. The first such book is that by the Hungarian mathematician Fejes-Tóth [82], published in 1953. A Russian translation of it appeared in 1958. The book was devoted to the theory of distributions in a Euclidean plane and in a three-dimensional space. The Russian translation [82] has an appendix written by Yaglom which adds both new results not appearing in the original book on distributions in E^2 and E^3 as well as a survey of the information on the theory of packings and coverings in E^n for arbitrary n.

The decade following the appearance of Fejes-Tóth's book was characterized by very substantial achievements in the theory of packings and coverings in spaces E^n for arbitrary n. This period culminated in the remarkable book by Rogers [180], who is also the author of many of the most important results in the solution of some of the main problems in the theory of distributions, namely, the finding of the densest packings and of the sparsest coverings by equal spheres in spaces E^n. Taken together, the books by Fejes-Tóth and by Rogers form a textbook which leaves nothing better to be desired and which introduces the reader to the circle of ideas and methods of the theory of distributions. No special background is needed to read these books.

A considerable part of another book by Fejes-Tóth [90] is devoted to the theory of distributions, mainly, of partitionings. In particular, questions of packings and coverings in non-Euclidean spaces are discussed there.

A popular exposition of general topics of discrete geometry is contained also in survey articles and in articles of a systematic nature [67, 85, 87, 88, 135, 137, 153].

In a certain sense a singular position in the theory of distributions is taken by the problems of the densest lattice packing by congruent spheres in the space E^n, of the least dense lattice covering of space E^n by such spheres, and of the partitionings of spaces E^n, generated by given point lattices in them. These problems, couched in the language of quadratic forms, simultaneously comprise an important part of the geometry of numbers also. It is noteworthy that many achievements in recent years in the solution of the problems on lattice distributions are based on the researches in the theory of quadratic forms by Voronoi [11], published in 1908-1909, and on the papers of Delone [69-71] in the Twenties and early Thirties. Delone's article [15] is a manual for familiarizing oneself with the concepts and methods of the theory of lattice distributions.

When regular partitionings are concerned, then here discrete geometry intersects with the theory of discrete groups and with the geometry of crystals. Therefore, although in the survey we have listed the results obtained in the last five to six years in the theory of regular partitionings, very little space is devoted to discussing them, in spite of their significance, in comparison with packings, coverings, and other distributions which are specific for discrete geometry.

§1. Packings in Spaces E^n and in Their Subsets

1. Densest Packing of Spheres of Equal Radii in E^n. The problem of the most dense packing of equal spheres whose centers form a lattice is equivalent to the problem of finding the largest value of the minima of positive-definite quadratic forms with a specified determinant. This is the oldest problem in the theory of distributions and in the geometry of numbers. Many of the most important achievements in its solution are due to Korkin and Zolotarev in the 1870's, although the very first results go back to Lagrange and Gauss. A survey of the results known up to 1962-1963 on lattice packings, and their bibliography, can be found in Rogers' book [180] (pp. 1-5).

Considerably newer is the problem of the densest packing of equal spheres in the general case when no preliminary constraints are imposed on the locations of the sphere centers. If we discount the almost trivial result on the packing of equal circles on a plane (Thue, 1892), then all the achievements in its solution have taken place in the last twenty years.

Let R be the radius of a sphere C in E^n with a center at some point $\overline{a}$ and let $\{K + \overline{a}_i\}$ be the set of open spheres, forming the packing, of unit radius with centers at the points $\overline{a}_i$. The upper limit, independent of the choice of the point $\overline{a}$,

$$\rho_{+} = \limsup_{R\to\infty} \frac{N(R)}{R^n}, \tag{1.1}$$

where N(R) is the number of spheres $K + \overline{a}_i$ such that $(K+\overline{a}_i)\cap C \neq \varnothing$ is called the density of the packing. Since the density ρ_+, as well as the other densities to be encountered below, do not change under similarity transformations of space E^n, the choice of a unit radius for the spheres does not diminish the generality of the definitions.

The quantity

$$\delta(n) = \sup \rho_{+}, \tag{1.2}$$

where the upper bound is taken over all packings in E^n, is called the density of the densest packing of equal spheres. If we restrict ourselves only to lattice packings, the analogous quantity $\delta_L(n)$ is called the density of the densest lattice packing. On the other hand,

$$\delta_L(n) = I_n \cdot \sup r^n, \tag{1.3}$$

where r is the minimum distance between the points of a lattice with a unit volume of the fundamental domain, the sup is taken over all such lattices, and I_n is the volume of an n-dimensional sphere of unit radius.

The values of $\delta_L(n)$ have been found for $n \leq 8$ (Lagrange: n = 2; Gauss: n = 3; Korkin and Zolotarev: n = 4, 5; Blichfeldt: n = 6, 7, 8; see the Introduction in the book [180]). The value of the quantity $\delta(n)$ is known only for n = 2. For a three-dimensional space it is most probable that $\delta(3) = \delta_L(3)$; however, even here the problem has not been solved, to say nothing for a general n.

In the main the papers in recent years on the most dense packing of equal spheres have been devoted to estimating the quantities $\delta(n)$ and $\delta_L(n)$ and to the possibility of constructing more dense packings for certain values of n.

The best upper bound thus far for $\delta(n)$ was obtained more than ten years ago (1958) by Rogers [177] (also see Chapter 7 of his book [180] and the article by Baranovskii [3]). The following proposition is the basis for Rogers' estimate: the density of any packing of equal spheres in E^n does not exceed the density σ_n of the packing of a regular n-dimensional simplex of side 2 by the parts falling inside this simplex of unit spheres with centers at the vertices of the simplex. From it ensues

$$\delta(n) \leqslant \sigma_n = \frac{n}{e}\left(\frac{1}{2}\right)^{n/2}\left[1 + O\left(\frac{1}{n}\right)\right]. \tag{1.4}$$

The exact value of σ_n takes the form

$$\sigma_n = \frac{(n+1)!}{\sqrt{n}\,\Gamma\left(1+\frac{n}{2}\right)}\left(\frac{e}{2}\right)^{n/2}\int\limits_{-\infty}^{\infty} e^{-u^2}\left[\int\limits_0^{\infty} e^{-y^2-\sqrt{2n}y+2iuy-iu\sqrt{\frac{2}{n}}}\times dy\right]^n du. \tag{1.5}$$

Estimate (1.4) also refers to the case of lattice packings; obviously, $\delta_L(n) \leq \delta(n)$. If we compare estimate (1.4), taking it as referring to lattice packings, with the upper bound

$$\delta_L(n) \leqslant \frac{n+2}{2}\left(\frac{1}{2}\right)^{n/2} \tag{1.6}$$

obtained by Blichfeldt [50] in 1914 and if we keep the lower bound (1.7) in mind, then we see that the progress here is trifling.

A better lower bound for $\delta(n)$ for arbitrary n is the estimate obtained by Rogers [176] in 1947 for lattice packings:

$$\delta(n) \geqslant \delta_L(n) \geqslant \frac{n\zeta(n)}{e(1-e^{-n})\,2^{n-1}}, \quad \zeta(n) = \sum_{k=1}^{\infty} k^{-n}. \tag{1.7}$$

An estimate for $\delta(n)$, somewhat worse than (1.7), is obtained as a simple corollary of the work by Coxeter, Few, and Rogers [65] on the upper bounds of the density of the least dense coverings of space E^n by equal spheres. We shall dwell on the contents of [65] below (§ 2.1), but we here state only the magnitude of the estimate:

$$\delta(n) \geqslant \left[\frac{n}{2(n+1)}\right]^{n/2}\sigma_n = \left(\frac{1}{2}\right)^n \frac{n}{e\sqrt{e}}\left[1 + O\left(\frac{1}{n}\right)\right]. \tag{1.8}$$

The lower bounds (1.7) and (1.8) relate to n sufficiently large; for spaces E^n of not very high dimension sufficiently dense packings have been obtained in which the values of ρ_+ exceed both the estimate (1.8) as well as (1.7).

Ignat'ev, occupying himself with the problem of densest packings as applied to the theory of optimal coding [23], proposed a very simple method for the practical construction of lattice packings [24, 25]. By taking as the initial elements two lattices on a straight line and three on a plane and by treating the space E^n as a direct product of lower-dimensional spaces, Ignat'ev constructed, by means of a relatively small number of trials, sufficiently dense lattice packings from the given elementary lattices. In [25] these lattice packings have been described up to $n = 32$, while density values have been derived up to $n = 64$. The packings found for $n \leq 8$ coincide with the densest packings. In the spaces E^9 and E^{10}, Ignat'ev's packings have precisely the same densities as the most dense packings of Chaundy [57]. However, when $n = 11$ and $n = 12$, they turn out to be less dense than the ones found earlier by Barnes [42] in E^{11} and by Coxeter and Todd [64] in E^{12} [the corresponding density values are: $I_{11}(1/32)$ and $I_{11}(1/(18\sqrt{3})$; $I_{12}(1/32)$ and $I_{12}(1/27)$].

Leech's two articles [150, 151] are devoted both to the finding of the most dense packings of equal spheres as well as to the problem of the number of spheres of a packing, which are tangent to a given sphere. Leech's results on the second problem will be considered in §4.1. Here we note his achievements in the construction of dense packings.

In the space E^{24} Leech found a lattice packing Λ_{24} with density $\rho(\Lambda_{24}) = I_{24}$; this is 16 times denser than in Ignat'ev [25]. If we compute the value of σ_n [see (1.5)] for $n = 24$, we obtain [151]:

$$\sigma_{24} = I_{24} \cdot 1.2724. \tag{1.9}$$

Thus, the upper bound (1.4) for the density $\delta(24)$ is sufficiently close to $\rho(\Lambda_{24})$ and Λ_{24} is possibly the densest lattice packing in E^{24}. On the other hand, the closeness of $\rho(\Lambda_{24})$ and σ_{24} directs us to the conjecture that the values of $\delta(n)$ are considerably closer to the upper bound (1.4) than to the lower bound (1.7) or (1.8).

On the basis of the packing Λ_{24} Leech constructed lattice packings in the spaces E^n with $n < 24$. Their densities coincide when $13 \leq n \leq 16$, but are greater than the densities of the corresponding Ignat'ev packings [25] when $17 \leq n \leq 23$.

Leech also constructed packings in the spaces E^{2^m} for all positive integral values of m. However, when $m \to \infty$, the densities of these packings are less than the lower bound (1.7).

2. Densest Packings of Equal Spheres in Subsets of Spaces E^n. The problem of the densest packing of k spheres in a cube (or in a sphere) is equivalent to the problem of the distribution of k points in a unit cube (sphere) in such a way that the minimum of the distances between any two of them would be largest. Schaer and Meir [185] and Schaer [181] solved this problem in the case of a square on a plane for $2 \leq k \leq 9$. For these same values of k, Schaer [182-184] indicated the densest packings of spheres in a cube in three-dimensional space.

Kravitz [144] considered the densest packing of circles in a circle in connection with the packing of cylinders of equal diameter in a cylindrical container. He indicated dense packings for $2 \leq k \leq 19$, which he says were found by empirical means. It is apparent that all of them are also the densest; although no proof of this is presented in the article, it is easy to obtain for the majority of the k's.

Let M(r) be the largest number of spheres of radius r which can be put into a sphere of radius r + 1. Blachman [44] found an explicit expression for M(r) in spaces E^n for $r \geq 1/\sqrt{2}$, and gave estimates for M(r) from above and from below for $r < 1/\sqrt{2}$.

Molnar [160] investigated the question of the density of equal sphere packings in a polytope of a special form (the Segre-Maler polytope) in the space E^3.

3. Multiple Packings of Equal Spheres. A set of spheres of equal radii form a k-fold packing in a domain D if each point of the domain falls inside no more than k spheres. The density $\rho_+(k)$ of the given packing is determined from formula (1.1); the density $\delta_k(n)$ of the densest k-fold packing in space E^n is the upper bound of the density $\rho_+(k)$ on the set of all possible k-fold packings in E^n.

Few [101] found the following estimate for the quantity $\delta_k(n)$ for arbitrary n:

$$\delta_1(n)\left(\frac{2k}{k+1}\right)^{n/2} \leqslant \delta_k(n) \leqslant \left(1+\frac{1}{n}\right)\frac{(n+1)^k-1}{\left(1+\frac{1}{k}\right)^{n/2}}. \qquad (1.10)$$

The upper bound in (1.10) when $k \geq \sqrt{n}$ is no improvement over the trivial inequality $\delta_k(n) < k$.

Continuing the investigations of Heppes [127] on densest multiple lattice packings of circles on a plane, Blundon [51, 52] found the densest 5- and 6-fold packings of this kind; their densities equal:

$$\delta_5^L(2) = \frac{8}{7}\sqrt{21}\cdot\delta(2),\ \delta_6^L(2) = \frac{35}{8}\sqrt{2}\cdot\delta(2)\left(\delta(2) = \frac{\pi}{2\sqrt{3}}\right). \tag{1.11}$$

Heppes [127] obtained the values of $\delta_2^L(2),\ \delta_3^L(2),\ \delta_4^L(2)$.

Woods [195] proved the theorem: if in a lattice L of space E^4 there is not even one pair of points, the distance between which is less than unity, and if d(L) = 1, then any sphere of unit radius contains at least one point of L. From this theorem it follows that the densest lattice packing of unit spheres in E^4 is at the same time also the densest double lattice packing of such spheres.

Few's article [102] contains a survey of results on multiple packing; also included were his results, then at the stage of being in press, on the double packing of spheres.

4. Packings of Unequal Spheres. One of the characteristics of the packing of spheres of radii r_i is the quantity

$$p = \inf_{i,j}\frac{r_i}{r_j}. \tag{1.12}$$

For the case of a plane Fejes-Tóth [82] proved for values of p sufficiently close to unity ($0 < p_0 \le p < 1$) the packing of unequal spheres cannot be more than the densest packing of equal spheres.

Florian [105] found that $p_0 \le 0.906$.

Gilbert [108] considered a random collection of points in E^{n+1} (n variables are the Cartesian coordinates, one variable is t, time) satisfying a Poisson distribution with a density of α points per unit volume in E^{n+1}; he also considered various methods for constructing the packings of spheres with centers at the points of the space $E^n \subset E^{n+1}$, whose radii are determined by the values of t. For such packings he obtained formulas for the distribution of the radii on a unit volume of space E^n and estimates of the density, as well as a number of other results of value in applications.

Random packings in E^3 of spheres (with two to five different radii) were analyzed also by Hogendijk [138]. His packings satisfied the condition that the partitioning of the space into tetrahedra with vertices at the centers of spheres pairwise tangent to one another is permissible.

5. Sphere Packings with Unit Density. Let D be some bounded domain in the space E^n. In this paragraph we agree to give the name dense packing to the packing with unit density of spheres of different radii in D. The Apollonian packing of circles on a plane is dense; the arcs of three circles tangent to each other externally form a curvilinear triangle D in which a circle can be inscribed; in the three curvilinear triangles thus obtained once again circles can be inscribed; and so on to infinity.

Let U be a circle of unit radius and let $C=\{\theta_m\}_{m=1}^{\infty}$ be a dense packing of the circles θ_m in the circle U. Then

$$\sum_{m=1}^{\infty} r^2(\theta_m) = 1, \tag{1.13}$$

where $r(\theta_m)$ is the radius of circle θ_m. Mergelyan [29] proved that

$$\sum_{m=1}^{\infty} r(\theta_m) = \infty. \tag{1.14}$$

Melzak [157] examined, for an arbitrary dense packing C, the function

$$F_\alpha(C) = \sum_{m=1}^{\infty} r^\alpha(\theta_m), \quad 1 \leqslant \alpha \leqslant 2. \tag{1.15}$$

The number e(C) such that series (1.15) converges when $\alpha > e(C)$ and diverges when $\alpha < e(C)$ is called the exponent of the packing. The existence of a packing for which $e(C) = 2$ was proven in [157]; a class of packings for which $1.035 < e(C) < 1.999971$ was found.

A number of the problems on dense packings is concerned with the Besicovitch and Hausdorff dimensions of set S. A function depending on a real variable α is constructed for S:

$$\Lambda^\alpha(S) = \varliminf_{\delta\to 0} \left\{ \sum_{i=1}^{\infty} (d(G_i))^\alpha : d(G_i) < \delta, \ \bigcup_{i=1}^{\infty} G_i \supset S \right\}, \tag{1.16}$$

where $d(G_i)$ is the diameter of set G_i and the G_i are arbitrary open sets. The Hausdorff dimension is defined as the upper bound of those numbers α for which $\Lambda^\alpha(S) = +\infty$, and the Besicovitch dimension as the lower bound of those numbers α for which $\Lambda^{\alpha}(S) = 0$.

Hirst [136] estimated the Hausdorff dimension χ of the set S of points not falling within the circles of the Apollonian packing:

$$1.001 < \chi \leqslant 1.43113\ldots = \frac{\ln 3}{\ln\left(1+\frac{2}{\sqrt{3}}\right)} \tag{1.17}$$

Larman [145-148] considered all possible dense packings of n-dimensional spheres θ_i ($i = 1, 2, \ldots$) in the unit cube $I_n \subset E^n$. Let s_n be the lower bound of the Besicovitch dimension of the set $I_n - \bigcup_{i=1}^{\infty} \theta_i$ over all dense packings. In [145, 148] Larmon proved that

$$s_2 \geqslant 1.03, \quad s_n \geqslant s_{n-1} + 1 \tag{1.18}$$

The inequality $s_2 > 1$ signifies that the set not covered by circles consists not just of the junction points of the boundaries of the circles of the packing.

The number τ = sup t, where t are all those numbers for which the series $\sum_{i=1}^{\infty} d^t(\theta_i)$ converges, is called the index of convergence of the dense packing $\{\theta_i\}_{i=1}^{\infty}$ of spheres in the cube I_n. Let τ_n = inf τ over all packings. In [146] Larman proved that $\tau_n \geq s_n$.

If in the sequence $\{\theta_i\}_{i=1}^{\infty}$ the circles are distributed in the order of decreasing radius, then, as Florian has shown, the area Σ_m of the set $I_2 - \bigcup_{i=1}^{m} \theta_i$ is not less than $O(r(\theta_m))$ for any packing. Larman found the estimate:

$$\Sigma_m \geqslant 2\cdot 10^{-16} (r(\theta_m))^{0.97}. \tag{1.19}$$

6. Other Equal Sphere Packings: with Free Gaps, without Blocking, k-Layered Clouds. In [161, 162, 165] Molnar considered equal circle packings on a plane (and on other surfaces of constant curvature) with free gaps or, in other words, with the requirement of space between the circles. On the

packings there is imposed the requirement that to every circle of the packing there abut a circle-free domain of the packing (a free gap) into which one or several circles of specified radii could be placed. Free gaps of different circles may overlap partially or completely. A number of inequalities were obtained relative to the greatest density of such packings under various conditions on the free gaps.

The first paper in which the concept of blocking is connected with an equal circle packing on a plane is that of Fejes-Tóth [81]. A circumference C is drawn around a packing of n circles on a plane. In order to take a certain circle outside of the boundary C with a continuous motion not intersecting other circles, it is first necessary to take out (not more than) x other circles of the packing.

The largest value of x is called the blocking number B(n) of the packing. Fejes-Tóth [81] gave a lower bound for B(n) for $n < 30$. Heppes [132, 133] treated the question of the densest packings of circles which do not block each other [$B(n) = 0$] and obtained some estimates of the density.

A packing of convex figures in a strip between two parallel straight lines is called a layer. Fejes-Tóth [94] called the ratio $\omega/\inf l$ its permeability, where ω is the width of the strip and l is the length of the curve connecting the boundary lines of the strip without intersecting the interior points of the figures in the packing. He obtained a lower bound for the permeability of a layer of like circles.

Let there be given a sphere A of radius r with center at a point S. The packing of a finite number of unit spheres of the same dimension as A, not having any points in common with A, is called a k-layered spherical cloud of the sphere A if any ray Sx intersects at least k spheres of the packing and if at least one ray intersects precisely k spheres. Let B be the smallest sphere with center at point S containing a spherical cloud. The quantity $d(k; r) = R - r$, where R is the radius of sphere B, is called the thickness of the cloud. The sphere A can have a radius $r = \infty$. Then, the smallest distance between two planes between which the cloud is contained is called the thickness of the cloud. The first spherical clouds were considered by Fejes-Tóth [84]. For circular clouds Jucovič [142] found an estimate for $\inf d(k; 1)$ with $k \geq 2$:

$$\sqrt{3}(k-2)+2\sqrt{2+\sqrt{3}}\leqslant \inf d(k;1) \leqslant\sqrt{k^2(2+\sqrt{3})+m(2-\sqrt{3})}, \qquad (1.20)$$

where m equals 0 or 1 and $k \equiv m \pmod 2$.

The problem of the number of spheres in a single-layered cloud has been considered for a three-dimensional space. Heppes [131] found that $N \geq 24$, improving the result $N \geq 19$ obtained by Fejes-Tóth [84]. The upper bound: $N \leq 42$ (see [128]). For the case of a flat cloud, when $r = \infty$, Heppes [128] showed that its thickness $d(k; \infty) \geq (k-1)\sqrt{3} + 2$. Another proof of this inequality was given by Hajós [124].

7. Packings of Convex and Star Figures and Bodies. A nonempty set of points $\bar{x}$ is called a star domain S if for each $\bar{x} \in S$, $\lambda\bar{x} \in S$ is true for all $0 \leq \lambda \leq 1$. Unlike convex flat domains K for which the densities $\delta(K)$ and $\delta_L(K)$ of the densest arbitrary and densest lattice packings coincide ([82, 180]), von Wolff [194] presented an example of a flat star domain S whose densest packing P has a density $\delta(S)$ greater than the density $\delta_L(S)$ of its densest lattice packing.

Groemer [121] proved more: for any $\varepsilon > 0$ we can find a star domain S_ε and a packing P such that

$$\varepsilon \cdot \rho_+(P, S_\varepsilon) \geqslant \delta_L(S_\varepsilon). \qquad (1.21)$$

Minkowski, because of an error in the proof, took it that the greatest density $\delta_L(T)$ of a lattice packing of tetrahedra in E^3 was equal to $9/38$. Groemer [114] derived an example showing that $\delta_L(T) \geq 18/49$.

Courant's article [60] contains an elegant proof of Fáry's theorem (for example, see [82]) on the fact that the least density of the densest lattice packing of convex figures is achieved in the case of triangles and equals $2/3$.

Fejos-Tóth [98] solved the problem of the densest packing of flat figures each of which is a set of points distant from a given rectangle by a distance not exceeding some specified value.

In a space E^3 a finite set P is considered of closed pairwise-nonintersecting parallelepipeds whose edges are parallel to the coordinate axes, and the projection of this set on each of the coordinate planes is a rectangle. If we exclude the trivial case of one parallelepiped, then what is the smallest number N of elements of the set P? Rademacher, Dickson, and Plotkin [173] proved that $N = 12$. For E^2 the analogous number is $N = 4$ and for E^4, $N \geq 30$.

A set U is called a minimal (maximal) packing in the class of sets τ, $U \in \tau$, if $T \in \tau$, $T \subset U$, $T \neq U$, implies $\Delta(T) < \Delta(U)$ ($\Delta(T) > \Delta(U)$), where $\Delta(M)$ is the critical determinant of set M.

Hans [125, 126] obtained a number of results concerning extremal packings in certain classes of convex and star bodies.

Oler's article [170] contains a proof of the theorem on the nonnegativeness of the slackness function, introduced by him in [169], of the packing of convex discs in a domain of a special form.

Lindecrantz [154] described a method for the densest packing of equal rectangles in a rectangle in connection with the optimum cutting of rectangular sheets.

Moon and Moser [168] proved a number of theorems on the smallest parallelepiped of space E^n, into which convex bodies may be packed, whose diameters are bounded by a number D and the total volume by a number V.

Inequalities concerning the sums of radii and of their squares of a finite number k ($k \geq 3$) of circles with radii $r_1, r_2, \ldots, r_k$, packed in a unit circle, have been obtained by László [149].

Lekkerkerker's article [152] is devoted in part to the packings of flat convex figures.

8. Stable (Firm) Packings. The packing $\{K + \bar{a}_i\}$ of bodies is said to be stable, or firm, if not one of the bodies $K + \bar{a}_i$ can be displaced from its position without destroying the packing, i.e., without the interior points of the displaced body overlapping the interior points of some other body of the packing. On this basis we can define packings which are stable or unstable in the direction of a vector $\bar{l}$.

Böröczky [54] found stable equal sphere packings with zero density. The articles by Polák [171], Fejes-Tóth and Heppes [99], and Polák and Poláková [172] were devoted to problems of a different kind concerning the stability of packings.

§ 2. Coverings by Convex Bodies of Spaces E^n

1. Least Dense Coverings of a Space E^n by Equal Spheres. The problems of least dense, or sparsest, coverings are the duals of the problems of densest packings. However, in contrast to the latter, the problems of coverings have only comparatively recently attracted the attention of mathematicians. An important motive for the intensive development of this branch of the theory of distributions is the connection of the problems of coverings with the problems of computational mathematics (the tabulation of functions) and of information theory.

We shall take it that the letters C and R denote the same things as in § 1.1, but that the set $\{K + \bar{a}_i\}$ of closed spheres of unit radius covers the space E^n. The density of this covering is the lower bound

$$\rho_- = \liminf_{R\to\infty} \frac{N(R)}{R^n}, \tag{2.1}$$

where N(R) is the number of spheres $K + \bar{a}_i$ which lie in the sphere C. The problem of the sparsest coverings reduces to finding the coverings for which the density equals

$$\vartheta(n) = \inf \rho_-, \tag{2.2}$$

where the inf is taken over all coverings of E^n.

If the centers of the spheres form a point lattice Γ, then the covering is called a lattice covering. The density $\rho(\Gamma)$ of this covering can be found from the formula

$$\rho(\Gamma) = I_n \cdot \frac{R^n}{V}, \tag{2.3}$$

where I_n is the volume of the n-dimensional unit sphere, V is the volume of the basic parallelepiped (the fundamental domain) of the lattice, and R is the radius of the spheres of the covering, i.e., of the spheres forming the covering, with centers at the lattice points and such that the covering is destroyed by an arbitrarily small decrease in R. The density $\vartheta_L(n)$ of the sparsest lattice covering equals:

$$\vartheta_L(n) = \inf_{\{\Gamma\}} \rho(\Gamma) = I_n \cdot \inf_{\{\Gamma\}} \frac{R^n}{V}. \tag{2.4}$$

Lattices of least density are called optimal (in the sense of coverings).

A second problem, closely related with the seeking of optimal lattices, also is solved for lattice coverings: to find the local minima of the density $\rho(\Gamma)$ considered as a function of the parameters of the lattice Γ. Lattices on which these local minima are achieved are called extremal lattices.

Partitioning into Dirichlet domains and the partitioning dual to it, the so-called L-partitioning, have important applications in solving the problem of sparsest coverings. Let $\{X\}$ be a discrete system of points, for example, a point lattice. A polytope, called the Dirichlet domain of a given point X of the system, is the set of points distant from X by not more than some other point of system $\{X\}$. The collection of Dirichlet domains for all the points of the system forms a partitioning of the space. The polytope bounded by the convex hull of the collection of points of the system $\{X\}$, equidistant from Y and closest to it, is called the L-polytope for a given vertex Y of the partitioning into Dirichlet domains. The set of L-polytopes also forms a partitioning, called the L-partitioning.

As in the case of packings, the density $\vartheta(n)$ has been found only for $n = 2$. The following upper and lower bounds have been obtained for the quantities $\vartheta(n)$ and $\vartheta_L(n)$ for an arbitrary n:

$$\begin{gathered} cn(\ln n)^{\frac{1}{2}\log_2 2\pi e} \geqslant \vartheta_L(n) \geqslant \vartheta(n) \geqslant \tau_n, \\ \tau_n = \left(\frac{2n}{n+1}\right)^n \sigma_n \sim \frac{n}{e\sqrt{e}} (n \to \infty), \end{gathered} \tag{2.5}$$

where c is some constant; see §1.1 concerning σ_n.

The lower bound was obtained by Coxeter, Few, and Rogers in the article [65] we have already mentioned. It is based on the following geometrical property, analogous to that on which the upper bound of the packing density $\delta(n)$ was based: the density of a covering of space E^n by equal spheres cannot be less than the density τ_n of the covering of an n-dimensional simplex by the parts belonging to it of spheres of equal radii with centers at its vertices.

Upper bounds for the densities $\vartheta_L(n)$ have been obtained in a very complicated way by Rogers [178]. In Chapter 5 of his book

[180], Rogers gave a simpler and shorter proof of the upper bounds which were somewhat weaker in comparison with (2.5),

$$\vartheta(n) \leqslant \vartheta_L(n) \leqslant c\,(n \log_2 n)^{\frac{1}{2}\log_2 2\pi e} \tag{2.6}$$

However, the obtaining of estimate (2.6) also necessitated the construction of a very subtle method. How significant this result of Rogers is can be seen from the fact that all the previous upper bounds had the form C^n for large n, where C is a constant greater than one (Davenport [66], Watson [193]). The closeness between the upper and lower bounds (2.5) also attracted attention [compare with estimates (1.4) and (1.6) for packings].

Rogers' method for obtaining the upper bounds (2.5) and (2.6) is such that from it there follows only the existence of lattices of density no greater than that indicated in the estimates. The problem of an actual determination of even one such lattice in E^n for all n or for an n starting from a certain N remains open to this day.

In the Sixties, optimal and extremal lattices were known only for n = 2, 3. The problem presents no particular difficulty on a plane; however, the passage to a three-dimensional space already complicates it somewhat. In 1954, Bambah [39] found the optimal lattice in E^3, Few [100] repeated his results by a simpler and more elementary method, and in 1956 Barnes [41] solved the problem of extremal lattices in E^3. Its solution by another method is contained in the article by Baranovskii [2].

It turned out that the so-called Voronoi principal lattice of the first type Γ_1^n is the optimal and the unique extremal lattice in the spaces E^2 and E^3. It is constructed on a frame with the metric form

$$n\sum_{i=1}^{n} x_1^2 - 2\sum_{\substack{i,j=1\\(i<j)}}^{n} x_i x_j, \tag{2.7}$$

and the density of the covering for it equals

$$\rho(\Gamma_1^n) = I_n\left[\frac{n(n+2)}{12(n+1)}\right]^{n/2}\sqrt{n+1} \sim (1.19)^n \quad \text{as} \quad n \to \infty. \tag{2.8}$$

Thus,

$$\vartheta_L(2) = \rho(\Gamma_1^2) = \frac{2\pi}{\sqrt{27}} = 1.209\ldots, \quad \vartheta_L(3) = \rho(\Gamma_1^3) = \frac{5\sqrt{5}\pi}{24} = 1.464\ldots, \tag{2.9}$$

moreover,

$$\vartheta(2) = \vartheta_L(2).$$

In recent years notable progress has been made in solving the problem of the least dense lattice covering and, moreover, a significant part of this progress is due to a group of Soviet mathematicians (Delone, Ryshkov, Gametskii, and Baranovskii) headed by Boris Nikolaevich Delone, a Corresponding Member of the Academy of Sciences of the USSR. The impetus for their work on this problem was Academician Sobolev's problem on seeking the best lattice of interpolation points (Delone [18]).

In 1962-1963, Gametskii established first the extremality of the lattice Γ_1^n for all n [12] and then, what is more, its optimality on the set of lattices of first type [13]. (In spaces E^2 and E^3 all lattices are of this type; in space E^4 there are, besides it, two more types, the second and third; further, as n increases, the number of types grows rapidly, and the problem of the types of lattices has not been solved for $n \geq 5$.) By types we mean here the types of primitive lattices, i.e., lattices of a "general form," while special lattices are treated as limiting cases of primitive ones. Voronoi [11] and Delone [15, 70] have studied the types of four-dimensional lattices.

Bleicher [46] proved the extremality of the lattice Γ_1^n almost simultaneously with Gametskii.

In 1963, Delone and Ryshkov [20] solved the fundamental problem in the theory of lattice coverings in E^4, namely, they proved the optimality of the lattice Γ_1^4 on the set of all four-dimensional lattices. Thus, one further value of the density of sparsest lattice coverings was found:

$$\vartheta_L(4) = \frac{2\pi^2}{5\sqrt{5}} = 1.765\ldots. \tag{2.10}$$

The values of $\vartheta_L(n)$ for $n > 4$ have not been established as yet.

Gametskii, in [12, 13], and Delone and Ryshkov, in [20], investigated the expression for the sum $\sum_{i=1}^{(n+1)!} R_i^2$ of the square of the radii of the Dirichlet domains for a lattice, and its minimality was established for the lattice Γ_1^n on the set of lattices of type I for all n and on the set of all four-dimensional lattices, respectively. Since in every lattice the radius of the spheres of the covering $R = \max_{i=1,2,\ldots,(n+1)!} R_i$ and since in the case of lattice Γ_1^n all the radii are equal, the minimality of $\sum_{i=1}^{(n+1)!} R_i$ for this lattice allowed us to make relative to it the conclusion stated above.

In the majority of papers devoted to lattice coverings the investigations are based on the lattice-specified partitioning of spaces into Dirichlet domains (the Voronoi parallelepipeds) and on the dual to it partitioning into L-polytopes (see [15]). It is the topologically different types of these partitionings which generates the classification of the lattices into types.

The structure of the partitionings in E^n was known for $n \leq 4$ and, only in the case of type I lattices, for arbitrary n (Voronoi [11], Delone [15, 70]). Ryshkov [31] uncovered a number of new geometric properties of the n-dimensional Voronoi parallelepiped of type I, one of which, in particular, was utilized by Gametskii in [12].

It was in this same paper [31] that Ryshkov first introduced "snakelets" for describing L-simplexes (in the case of primitive lattices L-polytopes are simplexes). The closed polygonal line with links $\langle \bar{a}_{i_1}, \bar{a}_{i_2}, \ldots, \bar{a}_{i_{n+1}} \rangle$ is called the snakelet $\bar{a}_1, \bar{a}_2, \ldots, \bar{a}_{n+1}$ of the collection Z of vectors $\bar{a}_{i_1}, \bar{a}_{i_2}, \ldots, \bar{a}_{i_{n+1}}$ having zero sum. If any n vectors of collection Z are linearly independent, then the convex closure of the snakelet is an n-dimensional simplex.

The concept of a snakelet proved to be very convenient for describing the structure of the L-partitionings of lattices in spaces E^n with arbitrary values of n. With the aid of snakelets, Ryshkov [32] found and described the structures of the L-partitionings of lattices of type II and III and of a new type, which he named type IV, for arbitrary n (type IV exists for $n \geq 5$), and on the basis of all this he obtained the following result concerning coverings: the lattice Γ_1^n is optimal on the set of lattices of types I–III when $n \geq 4$ and on the set of lattices of types I–IV when $n \geq 6$.

From Rogers' estimate (2.6) when compared with (2.8) (and also from Davenport [66], Watson [193], and Ryshkov [33] which we shall discuss in detail below) it follows that beginning with some n the lattice Γ_1^n is already not optimal. Ryshkov's results are directed toward the conjecture that optimal lattices which are not Γ_1^n belong to types essentially different from the first, since types II–IV are close to the first in the geometric structure of the partitionings.

Ryshkov's snakelets have also been used by Gametskii [13] and Baranovskii [4–7].

In 1964, at the Second Geometric Conference in Khar'kov, Baranovskii and Ryshkov [8] reported on a second extremal lattice, besides Γ_1^4, in space E^4. A detailed exposition of the result has been published in [7]. The extremal lattice Γ_2^4 which was found was determined by the quadratic form

$$(1+\alpha)(x_1^2+x_2^2)+\frac{1}{2}(3-\beta)(x_3^2+x_4^2)-\alpha x_1x_2-(x_1x_3+x_2x_3+ \\ +x_1x_4+x_2x_4)+\beta\cdot x_3x_4 \quad (\alpha=0.5441\ldots,\ \beta=0.5001\ldots) \qquad (2.11)$$

and belongs to the set of lattices of type II. The density of the covering for it equals

$$\rho(\Gamma_2^4)=1.8836\ldots. \qquad (2.12)$$

In [4, 5], Baranovskii gave a complete solution of the problem of the extremal lattices of a four-dimensional space. It turned out that besides the previously known lattices Γ_1^4 and Γ_2^4, there is only one more lattice which is extremal, the lattice Γ_3^4 of third type with the quadratic form

$$\sum_{i=1}^{4} x_i^2+(1-\gamma)x_1x_2-(x_1x_3+x_2x_3+x_1x_4+x_2x_4)+ \\ +\gamma x_3x_4,\ \gamma=\frac{2}{3+\sqrt{3}} \qquad (2.13)$$

and with covering density

$$\rho(\Gamma_3^4)=1.9287\ldots$$

The following theorem plays a fundamental role in the solution of the problem of extremal lattices in space E^4 [4, 5]: if the star* of the lattice Γ is not symmetric relative to any of those hyperplanes relative to which it is affine-symmetric, then lattice Γ is not extremal.

Later, Baranovskii [6] proved this theorem for any n and with its help established that on the set of lattices of the first type the lattice Γ_1^n is the unique extremal lattice, that when $n \geq 13$ there are no extremal lattices among the lattices of type II in space E^n, and that when $4 \leq n \leq 12$ the lattice Γ_2^n can be a unique extremal lattice with the form

$$[1 + \alpha_n \frac{n-2}{2}] \sum_{i=1}^{n-2} x_i^2 + \frac{n-1-\beta_n}{2} (x_{n-1}^2 + x_n^2) - \\ - \alpha_n \sum_{i,j=1}^{n-2} x_i x_j - (x_{n-1} + x_n) \sum_{i=1}^{n-2} x_i + \beta_n x_{n-1} x_n, \tag{2.14}$$

where α_n and β_n are functions of n, $\alpha_n > 0$, $\beta_n \in [0, 1]$. It is most probable that the lattice Γ_2^n indeed is extremal; however, this has not been proved for $n > 4$.

The extremality of a lattice of the third type Γ_3^4 with form (2.13) was established by Dickson [73] by a method different than that in [4, 5]. He took Γ_3^4 as the given lattice and, using the Voronoi tables of the structure of the Dirichlet domain of four-dimensional lattices of type III, he proved that the density $\rho(\Gamma)$ has a local minimum at the point Γ_3^4 in the lattice space. Since in the Dirichlet domain of lattice Γ_3^4 of the 12 radii which may be distinct only three are equal to the radius R of the covering spheres, it is feasible to carry out the calculations involved in proving the extremality of this lattice.

Relying on one of Davenport's lemmas [66], Ryshkov [33] constructed the lattice Γ_*^{2m} whose covering density satisfies the inequality

$$\rho(\Gamma_*^{2m}) < I_{2m} \left[\frac{5(m+1)}{18\sqrt{3}}\right]^m (m+1). \tag{2.15}$$

* The collection of L-polytopes (L-simplexes in the case of a primitive lattice) having a common vertex is called the star of the L-partitioning of lattice Γ.

From (2.8) it follows that

$$\rho(\Gamma_*^{2m}) \leqslant \rho(\Gamma_1^{2m})$$

for m ≥ 57, i.e., the covering density of the lattice Γ_*^{2m} is less than the covering density of the lattice Γ_1^n for all even n ≥ 114. In the case of odd n, a lattice can be constructed in the same way which is better than Γ_1^n when n ≥ 201.

The method of constructing lattices used in [33] forces us to conjecture that lattices better than Γ_1^n in the sense of coverings exist and in a space of considerably lower dimensions than those found. It would be of interest to know the first such dimension.

2. Covering of Space E^n by Equal Spheres (Other Problems). Erdös, Few, and Rogers [80] considered a system of equal spheres in E^n, partially covering the space. The set of points of E^n each of which belongs to at least two spheres of the system, was called the overlapping of this system. Lower bounds were given when n ≥ 2^{20} for the covering density (it equals the difference between the density δ of the system and the density θ of the part of the space covered by it; the densities δ and θ were defined in the usual way), and specific conditions relative to θ were given.

In the space E^5, Woods [196] indicated a class of lattices with unit determinant, possessing the property that any sphere of radius $(\sqrt{5})/2$ contains at least one point of the lattice, i.e., the covering radius r of such lattices is not greater than $(\sqrt{5})/2$ [we recall that for the lattice Γ_1^5 with d = 1 the radius $r = \sqrt{(^{35}/_{72})}\sqrt{6}$ is considerably smaller than $(\sqrt{5})/2$].

Cleaver [58] proved the theorem: the equality $\inf\limits_{\{\Lambda\}} d(\Lambda) = 1$, is valid on the set $\{\Lambda\}$ of lattices of space E^4, for each of which there exists a hypersphere of unit radius not containing lattice points inside it and the minimum distance between the lattice points is not less than one and, moreover, the lower bound is achieved only for a cubic lattice and for a lattice with edges $\bar{a}_1 = (1, 0, 0, 0)$, $\bar{a}_2 = \left(\frac{1}{2}, \frac{\sqrt{3}}{2}, 0, 0\right)$, $\bar{a}_3 = \left(0, \frac{1}{\sqrt{3}}, \sqrt{\frac{2}{3}}, 0\right)$, $\bar{a}_4 = (0, 0, 0, \sqrt{2})$. From this theorem there was obtained a corollary concerning the lattice coverings of space E^4.

3. Covering by Convex Bodies. Together with the problem of the least dense covering of a space by equal spheres there was considered the problem of the same kind of covering of the space by a set of bodies $\{K + \overline{a}_i\}$ obtained by parallel displacements of a convex body K along vectors $\overline{a}_i$. In a number of cases the estimates obtained here, considered as a special case when K is a sphere, are the best also in the problem of the sparsest covering by spheres.

Rogers [178] found that for any convex body K the density $\vartheta_L(K)$ of the most sparse lattice covering of E^n satisfies the inequality

$$\vartheta_L(K) \leqslant n^{\log_2 \ln n + C} \quad (C = \text{const}). \tag{2.16}$$

This is still the best estimate to date. The upper bound (2.5) is a particular result of [178] for the case when K is a sphere.

Erdös and Rogers [78, 79] examined the covering of E^n by convex bodies under the additional condition that the points of the space do not belong to too large a number of covering bodies. They established [78] that for a sufficiently large n there exist coverings constructed for an arbitrary convex body K, with density less than

$$\nu = n \ln n + n \ln \ln n + 4n, \tag{2.17}$$

and, moreover, not one point of the space E^n belongs to more than $e\nu$ bodies. If the body K is symmetric and the covering is a lattice covering, then there exist coverings for which the maximal number N of bodies $K + \overline{a}_i$ to which points of space E^n belong satisfies the inequality

$$N < 2^n(\nu + 1), \tag{2.18}$$

while for any lattice covering $N > 2^{n+1} - 1$ [79].

The article by Bambah, Rogers, and Zassenhaus [40], and partially the article by Lekkerkerker [152] are devoted to problems on the coverings of a plane by convex domains.

Groemer [116] pointed out the essential difference of the problem of covering by asymmetric convex bodies as compared with symmetric ones.

Let K be a convex body in E^n, symmetric about the origin; let Λ be a lattice; and let $\{tK + \overline{a}_i\}$, $\overline{a}_i \in \Lambda$, be a lattice covering by the bodies tK. The quantity inf t is called the inhomogeneous minimum μ of body K relative to lattice Λ.

Chalk [56] proved an inequality concerning μ.

4. Covering of a Plane by Noncongruent Circles. The problem of the sparsest covering by noncongruent spheres has been considered only for the case of E^2, where we are required to find such a covering by circles whose radii fall in the segment [q, 1] (Molnar [158]).

Jucovic and Leso [143] indicated bounds for the covering density when $q \in [0.088, 0.169]$.

Florian [104] considered coverings by circles of two sorts with radii ratios 0.2482. . . and 0.502. . . and for them indicated the values of the least density.

5. Covering of a Convex Body by Bodies Homothetic to It. A certain convex body K is given in the space E^n. We are required to find the minimum number of bodies homothetic to it with homothety coefficient $k < 1$ which can cover K. This covering problem falls in the area of combinatorial geometry. Concerning this and other problems of combinatorial geometry we refer the reader to the recent book by Boltyanskii and Gokhberg [9] and to the book by Hadwiger and Debrunner [38]. After 1961 the problem of covering by homothetic bodies has been treated by Soltan [35-37] wherein this problem is related to other problems of combinatorial geometry, namely, those on the illumination of convex bodies.

6. Other Covering Problems. Firey and Groemer [103] proved a number of theorems on the coverings of convex sets in E^n by polytopes of minimal sizes.

Eggleston [75] found the smallest regular triangle covering each of the sets of a given class. In another article [76] he obtained interesting results concerning minimal universal coverings.

The articles [125, 126, 168, 171, 172], already mentioned in §1, are also devoted in part to the topic of coverings. They consider covering problems dual to the packing problems of which we spoke when we surveyed them in §1. Thus, [171, 172] contain results on the stability of coverings. Moon and Moser [168] found an estimate for the sizes of the largest parallelepipeds covered by a set of convex bodies whose diameters and total volume are bounded by a number D and, respectively, a number v.

§ 3. Partitionings in Spaces E^n

1. Regular Partitionings of a Space E^n. Regular partitionings are the most important ones in applications (in crystallography, physics, etc.). A partitioning of a space is called regular if in the group G (the group is discrete and with a bounded fundamental domain) of motions of this space joining the partitioning to itself there is at least one motion joining any one of the bodies of the partitioning with any other. When the bodies of the partitioning are convex, then they are polytopes. If these polytopes adjoin along whole faces of dimension $n - 1$ (the space of the partitioning is n-dimensional), then the partitioning is called normal. For example, the Dirichlet partitioning (§ 2.1) by an arbitrary discrete system of points is normal.

The basic papers in the theory of regular partitionings of space E^n are by Federov, Minkowski, Voronom, and Delone. Modern crystallography is based on the works of Fedorov and Delone.

In this short survey we shall not dwell on the description of the general state of the theory of regular partitionings, referring the reader to the article by Venkov and Delone [10] and to the book [90], but we shall examine only the state of the art in certain individual branches of this theory in connection with the listing of the achievements in recent years (starting with 1961-1962; article [10] contains the plenary report of the Fourth All-Union Mathematical Congress and reflects the state of the theory of regular partitionings up to 1961).

Delone [17] proved the fundamental theorem in the theory of stereohedra (the bodies of a regular partitioning): the number of topologically different normal regular partitionings of a space E^n is bounded only as a function of n. In the proof of this theorem it was shown that the number of $(n - 1)$-dimensional faces of a stereohedron is not greater than $2(2^n - 1) + 2^n(h - 1)$, where h is the number of collections of stereohedra whose centers of gravity form a lattice; it was also shown that the number of stereohedra meeting at one vertex is not greater than $2^n \cdot g!$, where g is the index relative to G of the subgroup of parallel displacements of group G.

Delone and Sandakova [19] have given a finite algorithm which for a given n allows us to compute with a finite number of operations the topology of regular partitionings of E^n into Dirichlet regions (Voronoi had obtained such an algorithm in a special case).

Sandakova [34] considered general normal isogonal partitionings in E^3 (i.e., those in which the group of combinations is transitive relative to the vertices of the partitioning but not of the bodies, as in the case of stereohedra) connected with the group of parallel displacements. The vertices of such a partitioning $\{Q\}$ form a lattice Γ. She proved a theorem on the fact that the partitioning $\{Q\}$ is symmetric about any point of lattice Γ and about the midpoint of any segment connecting two points of Γ, and a theorem on the fact that every partitioning $\{Q\}$ can be translated by an affine transformation into an L-partitioning of the lattice.

Zamorzaev [22] was the first to find examples of regular partitioning of space E^3 which were not normal. In [22] it was shown how to construct n-dimensional generalizations of these partitionings. The existence of nonnormal regular partitionings in E^n shows that when $n \geq 3$ the requirement of normality in the fundamental theorem in the theory of stereohedra (Delone [17]) cannot be weakened.

In this case if the group G of the partitioning is the group of parallel displacements, the bodies of the partitioning are called parallelohedra. Veronoi [11] and Delone [15, 70] have studied the parallelohedra of a normal partitioning of spaces E^3 and E^4. Voronoi proved that any primitive parallelohedron (primitive means that $n + 1$ parallelohedra meet at each vertex of the partitioning) is the affine image of a Dirichlet partitioning for some lattice. In the articles [31, 32], already mentioned in connection with coverings, Ryshkov investigated the geometrical peculiarities of an n-dimensional parallelohedron of type I, gave a description more detailed than that by Voronoi [11] of the affine structure of L-partitionings of lattices in E^n of type II, and described the structure of L-partitionings of n-dimensional lattices of type III and of a new type which he called the fourth (which appears beginning with $n = 5$). The descriptions of these structures was in terms of the concept of snakelets newly introduced by Ryshkov (see §2.1).

Gametskii's note [14] was devoted to certain geometric characteristics of the partitionings of lattices of type I, connected with the reduced Zelling parameters of these lattices.

The results of Delone, Ryshkov, and Shtogrin [21] are connected with Sandakova's theorem (for example, see [19]); if a frame E is a frame in E^n, reduced in the Hermite — Minkowski sense, then the coordinates with respect to it of the vertices of the parallepiped $\Pi(E)$ are in absolute value less than some number which depends only on n. This theorem has many applications in the geometry of lattices. One of the particular results in [21] is the proof of Jordan's theorem on the number of nonequivalent finite groups of n-dimensional numerical matrices, as an immediate corollary of Sandakova's theorem.

A distinctive isoperimetric problem concerning a regular tessellation on a plane (see § 7.2) and on a hyperbolic plane was considered by Fejes-Tóth [89]. Heppes' results [129] also are connected with regular tessellations.

Groemer [120] established one property of the polytopes of a regular partitioning of E^n.

Regular partitionings on a pseudo-Euclidean plane were examined by Baltag and Zamorzaev [1].

A polygon is called a planigon if equal or congruent and symmetric polygons form a partitioning of the plane and, moreover, is such that the motion leading one of the polygons into another leads the partitioning into itself. Delone [16] proved that 46 different planigons exist in E^2. A system of circles is called homogeneous if for any pair of circles of the system we can find a motion leading one of the circles into the other and the system of circles into itself. Consequently, the Dirichlet domain of a homogeneous system of circles is a planigon. Sinogovitz [186] established that 131 homogeneous systems exist in E^2 and he enumerated them beginning with the characteristics of the signs. Horváth [140] pointed out the correspondence between homogeneous systems of circles, enumerated by Sinogovitz, and the planigons which are Dirichlet domains for the systems of circles corresponding to them.

Delone's survey article [72] was devoted to the theory of regular Dirichlet partitionings in E^n.

Zeitler [199] obtained the finite-difference equation for parqueting, i.e., a partitioning of a plane by equal and similarly arranged polygons.

2. Irregular Partitioning of a Plane. Partitionings of Planar Figures. The problem of the partitioning of a plane into congruent convex hexagons was considered by Bollobas [53]. He established that the hexagons which can be taken as the basis of such a partitioning have at least two equal sides and in it there are three angles having a sum of 2π; if there are only two equal sides then it can be taken as the basis of the partitioning if and only if they are equal and parallel.

The partitioning of a strip into congruent convex polygons is called a honeycomb if in every polygon (cell) there is one and only one side, termed the fundamental, lying on one of the straight lines bounding the strip, and in congruent cells the fundamentals correspond to each other. Bleicher and Fejes-Tóth [48] showed that among the cells of a given area, forming a honeycomb of a given width, the smallest perimeter obtains either for an isoceles triangle with a vertex angle $\geq 120°$ or for a pentagon composed of a rectangle and an isoceles triangle with a vertex angle of 120°.

Conway [59] and Trustrum [190] considered the problem of the partitioning of a square with side N (N is a positive integer) by the least possible number $f(N)$ [$f(N) > 1$] of squares with integral sides where, moreover, the collection of different side lengths should not have a common factor other than unity. Conway computed the values of $f(N)$ for $N \leq 100$ and found the estimates

$$\log_2 N \leqslant f(N) \leqslant 6\sqrt[3]{N} + 1. \qquad (3.1)$$

Trustrum improved the upper bound ($N > 2$),

$$f(N) \leqslant 6 \log_2 (3N - 1) - 10 < 6 \log_2 N. \qquad (3.2)$$

Vit'tek [191] proposed a method for the partitioning of a square into 25 pairwise incongruent squares.

Goldberg and Stewart's article [111] was devoted to the problem of the dissection of each of n equal polygons P in like manner into $k = k(n, P)$ parts such that from the nk pieces obtained there could be put together a polygon similar to the original P.

Walkup [192] found the necessary and sufficient conditions for the partitioning of a rectangle into T-tetraminoes each of which is a collection of four unit squares put together in the form of the letter T.

Lindgren [155] examined problems concerning the partitioning of a regular decagon; for example, how to partition it into two and three parts which serve as the faces of a tessellation on a plane, etc. Lindgren's book [156] is devoted to the elementary problems of dissecting plane figures.

An equilateral closed triangle with a side of unit length can be partitioned into n subsets and for a given n the lower bound d_n of the maximum diameter of the subsets can be sought. Graham [113] found the values of d_n for $n \leq 15$ (except $n = 14$). He conjectured that $d_n \sim 2/\sqrt{6n}$ as $n \to \infty$.

Golomb [112] concerned himself with the possible partitioning of a figure by parts which are equal among themselves and similar to the figure itself.

Moon and Moser [167] obtained estimates of the number of methods of dissecting a regular m-gon into triangles whose sides are the sides of the m-gon and of its whole diagonals.

Groemer's articles [118, 119] were devoted to the partitionings of the space E^n and of the convex body $M \subset E^n$ by bodies homothetic to M.

§4. Distributions on a Sphere, Cylinder, and Other Surfaces of Constant Curvature

1. Densest Packing on a Sphere. Let a unit sphere S_n be given in the space E^n. By a ball of radius r on the sphere S_n we mean the collection of points on the surface of sphere S_n, whose geodesic distances from a fixed point (the center) do not exceed r. The problem of the densest packing of balls of equal radius on the sphere is usually formulated thus: how to distribute the m points P_i $(i = 1, 2, \ldots, m)$ on S_n so that $\theta_m = \inf_{i \neq j} \theta(P_i, P_j)$ [$\theta(P_i, P_j)$ is the angular distance between the points P_i and P_j] would be smallest?

In the case of a sphere in three-dimensional space the solution of the densest packing problem is known for $m \leq 12$ (see [90]) and for $m = 24$ (Robinson [174]); for other not very large values of m there have been found only specific favorable configurations on the basis of which lower bounds have been obtained for the density. The upper bounds for the density d_m are found from the inequality [82]

$$d_m \leqslant \frac{m}{2}\left(1 - \frac{1}{2}\operatorname{cosec}\frac{m}{m-2}\frac{\pi}{6}\right). \tag{4.1}$$

Strohmajer [189], for certain m between 18 and 52, proposed distributions of largest possible density. Goldberg [109, 110] indicated the packing of 18 circles with diameters $\theta_{18} = 48°38'$ and $\theta'_{18} = 49°33'$ and of 33 circles with diameters $\theta_{33} = 35°22'$ and $\theta'_{33} \doteq 35°25'$. The survey article [61] by Coxeter is devoted to the packing of circles on a sphere.

Of great interest is another survey paper by Coxeter [62] devoted to equal ball packings on the surfaces of spheres S_n. Apart from a historical survey of results, beginning with Kepler, Gregory (the controversy with Newton as to whether a ball can be touched by 13 balls equal to it), and a consideration of the Schlafli function, this article proposed a new upper bound for the density of the packing. This estimate is based on the same assumption as was Rogers' estimate (1.4) (see §1.1). A rigorous proof of this has been given thus far only for the case of the sphere S_3. The question of the packings of balls with angular diameter $\theta = \pi/3$ was discussed in detail, i.e., how many, N_n, balls of S_n, equal to a given one, can touch it.

Leech [150, 151] pointed out the very large values of the number N_n for $n = 2^m$ and $n \leq 24$. When the values of N_n obtained by him were compared with the upper bounds of Coxeter [62], it was discovered that they were close for $n \leq 8$ (for example, $N_8 = 240$, Coxeter's estimate is 244) and for $n = 24$ (196,500 and 263,285, respectively).

Blachman and Few [45] examined multiple packings of equal balls on the surface of sphere S_n. They have obtained upper bounds for the number $N_n^k(\theta)$ of balls with angular diameter θ in the case of a k-fold packing. Fejes-Tóth [95] found the densest 3-fold packing (and the sparsest 3-fold covering) of 12 equal circles on the sphere S_3.

A survey of the earlier results on packings on spheres S_n can be found in the Appendix by Yaglom to the Russian translation of the book [82].

2. Other Distributions on a Sphere. Rogers [179] considered the problem of the least dense covering of the surface of sphere S_n by equal spherical balls. He found the following

upper bound for the density $\rho_n(k)$ of the covering of the unit sphere S_n $(n \geq 9)$ by balls each subtended by a chord of length $k < 2$,

$$\rho_n(k) \geqslant \left[n \ln n + n \ln \ln n + n \ln \frac{2}{k} + \frac{1}{2} \ln 16n\right]\left(1 - \frac{2}{\ln n}\right)^{-1}. \quad (4.2)$$

Article [179] contains a number of other results concerning coverings of S_n.

By a spherical net on sphere S_3 we mean a connected system of arcs of large circles on the surface of the sphere, which divides the surface into a finite number of spherical polygons and, moreover, in such a way that at least three polygons meet at every vertex. The net is called isogonal if all the angles at the vertices of the net are equal. In particular, all nets composed of regular polygons are isogonal. Heppes [130] found all the isogonal nets. It turned out that besides the five regular nets and the degenerate net of lunes there exist only five more isogonal nets.

Davies [68] examined the partitioning of the surfaces of the spheres S_3 and S_4, respectively, by triangles and by tetrahedra which are congruent to each other or which are images of each other under a mirror mapping, and which are contiguous, respectively, along whole sides and faces. The partitioning of the surface of S_3 into isosceles triangles is investigated in detail in [68].

Chapters II and VII of [90] are devoted to distributions on a sphere.

3. Distributions on a Cylinder. In the space E^3 we consider a cylinder with perimeter (of an orthogonal section) U. The set of points whose geodesic distances from a fixed point (the center) do not exceed r is called a circle of radius r on the cylinder. Articles [47, 86] are devoted to packings and coverings by circles of equal radii on a cylinder. Fejes-Tóth [86] established that the most dense packing of equal unit circles on a cylinder has the density $\pi/2\sqrt{3}$ and is achieved on cylinders with the perimeter $U = \sqrt{p^2 + pq + q^2}$; $p, q = 0, 1, \ldots$; $p + q > 0$. Bleicher and Fejes-Tóth [47] proved the theorems: 1) the smallest density of the densest packings of unit circles equals $\pi/4$ and is achieved only on a cylinder with perimeter $2\sqrt{2}$; 2) the largest density of the sparsest covering of a cylinder by circles equals $\pi/2$ and is achieved only in the case $U = 2$.

4. Packings and Coverings on Arbitrary Surfaces of Constant Curvature in Space E^3. For the packing of circles of radius r on a surface F with curvature $\varkappa =$ const , Molnár [163] gave the following estimate of density δ:

$$\delta \leqslant d(a) = \frac{3 \operatorname{cosec} \frac{\pi}{a} - 6}{[a] - 3 - \frac{6}{\pi} \tan^{-1}\left\{\sqrt{3} \cdot \tan\frac{\pi}{a} \cot\left(1 - \frac{[a]}{a}\right)\pi\right\}}, \tag{4.3}$$

$$d(6) = \lim_{a \to 6} d(a) = \frac{\pi}{\sqrt{12}}, \quad \csc\frac{\pi}{a} = 2\cos\sqrt{\varkappa}r.$$

It was assumed that the packing consists of not less than three circles. Estimate (4.3) is an improvement over that by Fejes-Tóth ([82], p. 242 of Russian translation).

In the case of the packings and coverings of surface F by unequal circles whose radii fall in the segment [r, R], Molnár conjectured the following estimates of the covering density ϑ and of the packing density δ:

$$\vartheta \geqslant \inf \frac{1}{2\pi\Delta} \sum_{i=1}^{3} \alpha_i K_i, \quad \delta \leqslant \sup \frac{1}{2\pi\Delta} \sum_{i=1}^{3} \alpha_i k_i, \tag{4.4}$$

where K_1, K_2, K_3 and k_1, k_2, k_3 are the areas of two triples of circles on surface F; α_1, α_2, α_3 are the angles of the triangle formed by the centers of the triple of circles and Δ is its area. The first triple of circles is made up of three arbitrary circles having only one common point, the second triple is a packing in which the smaller circle touches two others and not one of the circles intersects the opposite side of the triangle formed by their centers. The radii of the circles of these triples fall in the segment [r, R].

For the coverings of surface F by unequal circles, see Imre [141].

§ 5. Minkowskian Distributions

A centrally symmetric convex body with interior points will be called a disk. A set of disks in the space E^n forms a Minkowskian distribution (in short: M-distribution) if not one of the disks of the set contains inside it the center of symmetry of some other disk. The problem of the most dense M-distributions of disks of various forms has been examined.

On a plane Fejes-Tóth [92] considered the M-distribution of circles with radii $r_1, r_2, \ldots, r_i, \ldots$ ($r = \inf r_i > 0$, $R = \sup r_i < \infty$) and showed that the density of such a distribution $\delta \leq 2\pi/\sqrt{3} = 3.627\ldots$ and, moreover, equality is achieved only in the case when the distribution has been formed by circles of equal radius and has been distributed on the whole plane. In [97] he investigated the case of a finite number of circles, and in this distribution always $\delta < 2\pi/\sqrt{3}$. The results obtained by Fejes-Tóth in [97] for a plane were extended by Molnár [164] to the case of an arbitrary surface of constant curvature.

Fejes-Tóth [92] showed also that there exists an M-distribution of unequal disks, having an arbitrarily high density, although for congruent disks of like orientation, as follows from the well-known Minkowski theorem, the density does not exceed four. He posed the question [91]: does there exist an upper bound for the densities of any distributions of equal disks (not necessarily of like orientation) ?

Continuing the investigations of Fejes-Tóth, Bleicher and Osborn [49] established that on a plane for any positive number m there exists an M-distribution of equal rectangles having precisely m orientations, with a density not less than $m + 3 - \log_2 m$. By the same token it was shown that there does not exist an upper bound for the density of M-distributions of congruent disks. The following theorem also was proven in [49]: in the space E^n there exists an M-distribution of congruent disks having at most m different orientations, with a density as close to $m \cdot 2^n$ as desired.

Furthermore, Bleicher and Osborn investigated homogeneous M-distributions when every point of the space except for a set of measure zero is contained in one and the same number of discs.

Fejes-Tóth's survey article [96] contains an interesting generalization of M-distributions and problems relating to them.

§6. Other Distributions in Spaces E^n

On the plane E^2 we consider a finite system Q of closed circles (where by circles we understand also circles with limit radii, i.e., points and closed halfplanes) such that any two circles have a nonempty intersection. The inclusion number s (internal inclusion number s') is the smallest number of points such that each circle of the system contains (contains inside) at least one of these points. Ungar and Sekers (see [82]) showed that $s \leq 7$ for any Q. Stachó [188] improved their results by showing that always $s \leq s' \leq 5$. On the other hand, a system of circles is known (Grünbaum [123], Dantser) for which $s = 4$.

Eggleston [77] considered a saturated system of curves on a plane: a system of circles with radii $r, r_1, r_2, \ldots, r_i, \ldots, r = \inf r_i > 0$, such that any circle of radius r has at least one point in common with the system. He proved the theorem: the least density of a saturated system of circles equals $\pi/6\sqrt{3}$ and is achieved for a system of circles of radius r whose centers form a lattice constructed on an equilateral triangle with side $2\sqrt{3}r$.

The system $\{K + \overline{a}_i\}$ of convex bodies is called connected if the set $\bigcup_i (K+\overline{a}_i)$ is connected. Groemer [122] obtained a number of results concerning least dense connected lattice distributions of convex bodies K. Another of Groemer's articles [115] is devoted to asymptotic distributions. For given numbers $\alpha > 0$ and $\beta > 0$ he constructed a certain system $\{X\}$ of sets X such that the Lebesgue measure $\mu(X) \geq \alpha$ and the diameter $d(X) \leq \beta$. Using parallel displacements along the vectors $\overline{a}_i$ he constructed the distributions $\{X + \overline{a}_i\}$ for $X \in \{X\}$. These distributions are called asymptotic if $\mu\left(E^n - \bigcup_i (X + \overline{a}_i)\right) = 0$, i.e., each distribution is a covering to within a point set of measure zero.

Groemer's article [117] contains a detailed examination of the general questions of the distributions of convex and star bodies and of the definition of densest distributions.

Besicovitch [43] established that the minimum length of the arc of a curve not covered by a regular triangle with unit side does not exceed 0.982.

§ 7. Distributions in Non-Euclidean Spaces

1. Packings and Coverings. Fejes-Tóth's article [83], containing a survey of the results up to 1958-1959, was devoted to the questions of the densest packings and sparsest coverings of constant-curvature spaces, mainly of two and three dimensions, by equal spheres (of finite or infinite radius). By that time a solution had been obtained (by Fejes-Tóth) for the problem of the densities δ_2 of

the densest packing and ϑ_2 of the sparsest covering of the Lobachevskii plane H^2 by not less than three equal circles: $\delta_2 = 3/\pi$, $\vartheta_2 = \sqrt{12}/\pi$, and, moreover, these densities are achieved in the cases, respectively, of a packing of horocycles and of a covering by horocycles. Article [83] gives a detailed discussion of the proposition (a): the density δ of the densest packing (ϑ of the sparsest covering) of constant-curvature spaces by equal spheres of radius r does not exceed (is not less than) the density d(r) of the packing [D(r) of the covering] of a regular simplex with side 2r (with a radius r of the circumscribed sphere) by spheres with centers at its vertices. Proposition (a) has been proven only in the case of two-dimensional spaces.

In 1964, Böröczky and Florian [55] proved proposition (a) with respect to packings in the Lobachevskii space H^3 and on the basis of (a) established that the densest packing in this space is achieved by means of horospheres and has the density

$$\delta_3 = \lim_{r\to\infty} d(r) = 0.853\ldots. \tag{7.1}$$

Zeitler [198] indicated a one-parameter family of regular coverings of the space H^3 by horospheres. One of the coverings of this family has the minimum density

$$\left\{\sum_{k=0}^{\infty}\left[\frac{1}{(1+3k)^2} - \frac{1}{(2+3k)^2}\right]\right\}^{-1} = 1.280\ldots. \tag{7.2}$$

Density (7.2) coincides with the lower bound of the density of the covering obtained on the basis of proposition (a).

The estimates (4.3) and (4.4) found by Molnár [163, 166], as well as Dominyák's results [74] concerning stable coverings of two-dimensional constant-curvature spaces, have been extended to the case of the plane H^2.

Fejes-Tóth [93] found a distribution of $k \leq 6$ points on an elliptical plane such that the least distance between them is maximal for a given k, i.e., he solved the problem of the densest packing of $k \leq 6$ equal circles on this plane.

2. Partitionings of Lobachevskii Spaces H^n. A partitioning of a plane by polygons, under which two polygons hav-

ing more than one point in common adjoin along a whole side, is called a tessellation. If the midpoints of the edges, meeting at a given vertex, of the polygons of a tessellation are joined in cyclic order, we obtain a polygon called the vertex figure of the given vertex. If the polygons (the faces of the tessellation) and the vertex figures are regular, the tessellation is called regular and is denoted by the Schlafli symbol $\{p, q\}$, where p is the number of sides of a face and q, of the vertex figure.

The articles by Coxeter [63], Horváth [139], and Zeitler [200] are devoted to regular tessellations on the plane H^2. Coxeter considered the collection of two tessellations $\{p, q\}$ and $\{m, n\}$ when the vertices of the tessellation $\{p, q\}$ are a part of the vertices of the tessellation $\{m, n\}$. Such a collection of tessellations was called a compound tessellation. Vertex-regular and face-regular compound tessellations were investigated in [63].

If in a regular tessellation we choose some face P_0, then the collection of faces of the tessellation contiguous with P_0 (along sides or vertices) is called the first belt, while the k-th belt is the collection of faces of the tessellation contiguous with the outer face of the (k − 1)-st belt but not occurring in this belt. Horváth [139], using a finite-difference equation, described an arrangement of belts of the tessellation $\{p, q\}$ and obtained a formula for the function

$$G(p, q) = \lim_{k\to\infty} \frac{U_k}{\sum_{l=1}^{k} U_l}, \qquad (7.3)$$

where U_k is the area of the k-th belt. Zeitler [200] mainly reiterates the results of [139].

Pogorelov [30] found the necessary and sufficient conditions that among the set $\{P\}$ of convex polytopes of space H^3 with a given structure (in any two polytopes of set $\{P\}$ the faces, edges, and vertices may be associated while retaining the membership relation) there exists a polytope P_0 with right dihedral angles such that the mirror images of the faces of P_0 can partition H^3 into polytopes congruent to P_0.

Garner [106] considered, in spaces H^n with $n \geq 3$, partitionings into polytopes of a special form, in which the permutation of the vertices of any face and the permutation of the faces adjacent to one and the same vertex, leads to reflections. It was established that in H^5 there exists only one such partitioning, and that they do not exist when $n > 5$. Another paper by Garner [107] is devoted to the partitionings of H^n into regular polytopes.

An important result concerning normal regular partitionings (see §3.1) of space H^3 was obtained by Makarov [27]. He indicated a simple geometric method for constructing a countable infinite set of nonisomorphic discrete groups with finite fundamental domain (the Fedorov groups) in H^3, from which ensues at least the countability of the number of different topological types of polytopes (prisms), each of which partitions H^3 normally and regularly. In another paper [28], Makarov constructed analogous partitionings of H^3 into unbounded polytopes of finite volume.

LITERATURE CITED

1. I. A. Baltag and A. M. Zamorzaev, "On the theory of Fedorov groups and of regular partitionings on a pseudo-Euclidean plane," Uch. Zap. Kishinevsk. Univ., 82: 82-96 (1965).
2. E. P. Baranovskii, "The minima of the density of a lattice covering of a space by equal spheres," Uch. Zap. Ivanovsk. Gos. Ped. Inst., 34: 71-76 (1963).
3. E. P. Baranovskii, "The filling of n-dimensional Euclidean spaces by equal spheres. I," Izv. Vysshikh. Uchebn. Zavedenii, Matematika, No. 2, pp. 14-24 (1964).
4. E. P. Baranovskii, "Local minima of the density of the lattice coverings of a four-dimensional Euclidean space by equal spheres," Dokl. Akad. Nauk, SSSR, 164(1): 13-15 (1965).
5. E. P. Baranovskii, "Local minima of the density of the lattice covering of a four-dimensional Euclidean space by equal spheres," Sibirsk. Mat. Zh., 7(5): 974-1001 (1966).
6. E. P. Baranovskii, "Extremal lattice coverings," Uch. Zap. Ivanovsk. Gos. Ped. Inst., p. 44 (1967).
7. E. P. Baranovskii, and S. S. Ryshkov, "A second local minimum of the density of a lattice covering of a four-dimensional Euclidean space by equal spheres," Sibirsk. Mat. Zh., 7(4): 731-739 (1966).
8. E. P. Baranovskii and S. S. Ryshkov, "A second local minimum of the density of a lattice covering of a four-dimensional Euclidean space by equal spheres," in: Reports Abstracts Second All-Union Geom. Conf., Sept. 17-23, 1964, Kharkov University, Kharkov (1964), pp. 10-11.
9. V. G. Boltyanskii and I. Ts. Gokhberg, Theorems and Problems of Combinatorial Geometry, "Nauka," Moscow (1965), 108 pp.

191. L. Vit'tek, "Dokonaly rozklad stvorca n 25 stvorcov," Math.-Fyz. Casop., 14(3): 234-235 (1964).
192. D. W. Walkup, "Covering a rectangle with T-tetrominoes," Am. Math. Monthly, 72(9): 986-988 (1965).
193. G. L. Watson, "The covering of space by spheres," Rend. Cirkolo Mat. Palermo, 5(1): 93-100 (1956).
194. M. R. von Wolff, "A star domain with densest admissible point set not a lattice," Acta Math., 108: 53-60 (1962).
195. A. C. Woods, "The densest double lattice packing of four-spheres," Mathematika, 12(2): 138-142 (1965).
196. A. C. Woods, "Lattice coverings of five space by spheres," Mathematika, 12(2): 143-150 (1965).
197. H. Zeitler, "Überdeckungsprobleme, Übung der Raumanschauung in der Arbeitsgemeinschaft," Prax. Math., 7(3): 69-72 (1965).
198. H. Zeitler, "Eine reguläre Horosphärenüberdeckung des hyperbolischen Raumes," Elem. Math., 20(4): 73-79 (1965).
199. H. Zeitler, "Differenzengleichungen bei Parkettierungen," Prax. Math., 8(3): 61-64 (1966).
200. H. Zeitler, "Über Netze aus regulären Polygonen in der hyperbolischen Geometrie," Elem. Math., 22(3): 56-62 (1967).

10. B. A. Venkov and B. N. Delone, "The regular partitionings of spaces," in: Proc. Fourth All-Union Math. Congr., 1961, Vol. 1, Akad. Nauk SSSR, Leningrad (1963), pp. 49-55.
11. G. F. Voronoi, "Investigations on primitive parallelepipeds," in: Collected Works, Vol. II, Izd. Akad. Nauk Ukr. SSR, Kiev (1952), pp. 239-268.
12. A. F. Gametskii, "On the theory of coverings of an n-dimensional Euclidean space by equal spheres," Dokl. Akad. Nauk SSSR, 146(5): 991-994 (1962).
13. A. F. Gametskii, "The optimality of Voronoi's lattice of the first type among lattices of the first type of any number of dimensions," Dokl. Akad. Nauk SSSR, 151(3): 482-484 (1963).
14. A. F. Gametskii, "Geometric meaning of the reduced Zelling parameters of n-dimensional Voronoi lattices of the first type," in: Investigations on Algebra and Math. Analysis, Karta Moldovenyaske, Kishinev (1965), pp. 122-127.
15. B. N. Delone, "The geometry of positive quadratic forms," Usp. Mat. Nauk, 3: 16-22 (1937); 4: 102-164 (1938).
16. B. N. Delone, "Theory of planigons," Izv. Akad. Nauk SSSR, Ser. Mat., 23(3): 365-386 (1959).
17. B. N. Delone, "Proof of the fundamental theorem in the theory of stereohedra," Dokl. Akad. Nauk SSSR, 138(6): 1270-1272 (1961).
18. B. N. Delone, "A problem of S. L. Sobolev," Usp. Mat. Nauk, 17(3): 205-207 (1962).
19. B. N. Delone and N. N. Sandakova, "Theory of stereohedra," Tr. Mat. Inst. Akad. Nauk SSSR, 64: 28-51 (1961).
20. B. N. Delone and S. S. Ryshkov, "Solution of the problem of the least dense lattice covering of a four-dimensional space by equal spheres," Dokl. Akad. Nauk SSSR, 152(3): 523-524 (1963).
21. B. N. Delone, S. S. Ryshkov, and M. I. Shtogrin, "On a theorem of Sandakova from the theory of positive quadratic forms," Mat. Zametki, 1(3): 253-262 (1967).
22. A. M. Zamorzaev, "Nonnormal regular partitionings of a Euclidean space," Dokl. Akad. Nauk SSSR, 161(1): 30-32 (1965).
23. N. K. Ignat'ev, "Geometric descriptions of optimal coding," Sb. Tr. Gos. NII Min. Svyazi SSSR, No. 8(12), pp. 85-145 (1958).
24. N. K. Ignat'ev, "A practical method for finding dense packings of n-dimensional spheres," Sibirsk. Mat. Zh., 5(4): 815-819 (1964).
25. N. K. Ignat'ev, "Finding dense packings of n-dimensional spheres," Sibirsk. Mat. Zh., 7(4): 820-825 (1966).
26. J. W. S. Cassels, Introduction to the Geometry of Numbers [Russian translation], "Mir," Moscow (1965), 421 pp. [English edition: Springer-Verlag, New York.]
27. V. S. Makarov, "One class of partitionings of a Lobachevskii space," Dokl. Akad. Nauk SSSR, 161(2): 277-278 (1965).
28. V. S. Makarov, "One class of discrete groups of a Lobachevskii space, having an infinite fundamental domain of finite measure," Dokl. Akad. Nauk SSSR, 167(1): 30-33 (1966).
29. C. N. Mergelyan, "Uniform approximations of functions of a complex variable," Usp. Mat. Nauk, 7(2): 31-122 (1952).

30. A. V. Pogorelov, "Regular partitioning of a Lobachevskii space," Mat. Zametki, 1(1): 3-8 (1967).

31. S. S. Ryshkov, "The structure of an n-dimensional parallelepiped of the first type," Dokl. Akad. Nauk SSSR, 146(5): 1027-1030 (1962).

32. S. S. Ryshkov, "Some remarks on the types of n-dimensional parallelepipeds and on the density of lattice coverings of space E^n by equal spheres," Dokl. Akad. Nauk SSSR, 162(2): 277-280 (1965).

33. S. S. Ryshkov, "Effectuation of a method of Davenport in the theory of coverings," Dokl. Akad. Nauk SSSR, 175(2): 303-305 (1967).

34. N. N. Sandakova, "Proof of the analogs of the Minkowski and Voronoi theorems for three-dimensional isogonal partitionings connected with the group of parallel displacements," Dokl. Akad. Nauk SSSR, 154(2): 271-273 (1964).

35. P. S. Soltan, "The problems of covering and of illuminating convex bodies," Bul. Akad. Shtiintse RSS Mold., Ser. Estestv. i Tekhn. Nauk, No. 1, pp. 49-57 (1963).

36. P. S. Soltan, "Concerning the problem of covering and illuminating convex bodies," Uch. Zap. Kishinevsk. Univ., 82: 69-74 (1965).

37. P. S. Soltan, "Relations between the problems of covering and illumination of convex bodies," Bul. Akad. Shtiintse RSS Mold., Ser. Fiz.-Tekhn. i Mat. Nauk, No. 4, pp. 91-93 (1966).

38. H. Hadwiger and H. Debrunner, Combinatorial Geometry in the Plane [Russian translation], "Nauka," Moscow (1965), 171 pp. [English edition: Holt, Rinehart and Winston, New York.]

39. R. P. Bambah, "On lattice covering by spheres," Proc. Nat. Inst. Sci. India, 20(1): 25-52 (1954).

40. R. P. Bambah, C. A. Rogers, and H. Zassenhaus, "On coverings with convex domains," Acta Arithm., 9(2): 191-207 (1964).

41. E. S. Barnes, "The covering of space by spheres," Can. J. Math., 8(2): 293-304 (1956).

42. E. S. Barnes, "The construction of perfect and extreme forms. II," Acta Arithm., 5(2): 205-222 (1959).

43. A. S. Besicovitch, "On arcs that cannot be covered by an open equilateral triangle of side 1," Math. Gaz., 49(369): 286-288 (1965).

44. N. M. Blachman, "The closest packing of equal spheres in a larger sphere," Am. Math. Monthly, 70(5): 526-529 (1963).

45. N. M. Blachman and L. Few, "Multiple packing of spherical caps," Mathematika, 10(19): 84-88 (1963).

46. M. N. Bleicher, "Lattice coverings of n-space by spheres," Can. J Math., 14(4): 632-650 (1962).

47. M. N. Bleicher and L. Fejes-Tóth, "Circle packings and circle-coverings on a cylinder," Michigan Math. J., 11(4): 337-341 (1964).

48. M. N. Bleicher and L. Fejes-Tóth, "Two-dimensional honeycombs," Am. Math. Monthly, 72(9): 969-973 (1965).

49. M. N. Bleicher and J. M. Osborn, "Minkowskian distributions of congruent discs," Acta Math. Acad. Sci. Hung., 18(1-2): 5-17 (1967).

50. H. F. Blichfeldt, "A new principle in the geometry of numbers with some applications," Trans. Am. Math. Soc., 15: 227-239 (1914).

51. W. J. Blundon, "Multiple packing of circles in the plane," J. London Math. Soc., 38(2): 176-182 (1963).

52. W. J. Blundon, "Note on a paper of A. Heppes," Acta Math. Acad. Sci. Hung., 14(3-4): 317 (1963).

53. B. Bollobás, "Filling the plane with congruent convex hexagons without overlapping," Ann. Univ. Sci. Budapest. Sec. Math., 6: 117-123 (1963).

54. K. Böröczky, "Über stabile Kreis- und Kugelsysteme," Ann. Univ. Sci. Budapest. Sec. Math., 7: 79-82 (1964).

55. K. Böröczky and A. Florian, "Über die dichteste Kugelpackung im hyperbolischen Raum," Acta Mat. Acad. Sci. Hung., 15(1-2): 237-245 (1964).

56. J. H. H. Chalk, "A local criterion for the covering of space by convex bodies," Acta Arithm., 9(3): 237-243 (1964).

57. T. W. Chaundy, "The arithmetic minima of positive quadratic forms," Quart. J. Math., 17: 166-192 (1946).

58. F. L. Cleaver, "On coverings of four-space by spheres," Trans. Am. Math. Soc., 120(3): 401-416 (1965).

59. J. H. Conway, "Mrs. Perkins's quilt," Proc. Cambridge Phil. Soc., 60(3): 363-368 (1964).

60. R. Courant, "The least dense lattice packing of two-dimensional convex bodies," Commun. Pure Appl. Math., 18(1-2): 339-343 (1965).

61. H. S. M. Coxeter, "The problem of packing a number of equal nonoverlapping circles on a sphere," Trans. N. Y. Acad. Sci., 24(3): 320-331 (1962).

62. H. S. M. Coxeter, "An upper bound for the number of equal nonoverlapping spheres that can touch another of the same size," in: Convexity, Am. Math. Soc., Providence, R. I. (1963), pp. 53-71.

63. H. S. M. Coxeter, "Regular compound tessellations of the hyperbolic plane," Proc. Roy. Soc., A278(1373): 147-167 (1964).

64. H. S. M. Coxeter and J. A. Todd, "An extreme duodenary form," Can. J. Math., 5: 384-392 (1951).

65. H. S. M. Coxeter, L. Few, and C. A. Rogers, "Covering space with equal spheres," Mathematika, 6(12): 147-157 (1959).

66. H. Davenport, "The covering of space by spheres," Rend. Cirkolo Mat. Palermo, 1: 92-107 (1952).

67. H. Davenport, "Problems of packing and covering," Rend. Semin. Mat. Univ. e Politecn. Torino, 24: 41-48 (1965).

68. H. L. Davies, "Packings of spherical triangles and tetrahedra," Proc. Colloq. Convexity, Copenhagen, 1965, Copenhagen (1967), pp. 42-51.

69. B. N. Delone (Delaunay), "Sur la sphere vide" (in the Trans. Congr. of Mathematicians at Toronto in 1924).

70. B. N. Delone (Delaunay), "Sur la partition regulière de l'espace à 4 dimensions," Izv. Akad. Nauk SSSR, Ser. Mat., No. 1, pp. 79-110 (1929).

71. B. N. Delone (Delaunay), "Sur la sphere vide. II," Izv. Akad. Nauk SSSR, Ser. Mat., 7: 793-800 (1934).

72. B. N. Delone (Delaunay), "Theorie der regulären Dirichletschen Zerlegungen des n-dimensionalen euklidischen Raumes," Schriftenreiche Inst. Math. Dtsch. Akad. Wiss. Berlin, No. 13, pp. 27-31 (1963).

73. T. J. Dickson, "An extreme covering of 4-space by spheres," J. Austral. Math. Soc., 6(2): 179–192 (1966).
74. I. Dominyak, "Stabilis kürrendszerek sürüsegeröl," Magyar Tud. Akad. Mat. Fiz. Tud. Oszt. Közl., 14(4): 401-413 (1964).
75. H. G. Eggleston, "On covering a regular polygon with a triangle," Proc. Cambridge Phil. Soc., 58(1): 8-11 (1962).
76. H. G. Eggleston, "Minimal universal covers in E^n," Israel J. Math., 1(3): 149–155 (1963).
77. H. G. Eggleston, "A minimal density plane covering problem," Mathematika, 12(2): 226-234 (1965).
78. P. Erdös and C. A. Rogers, "Covering space with convex bodies," Acta Arithm. 7(3): 281-285 (1962).
79. P. Erdös and C. A. Rogers, "The star number of coverings of space with convex bodies," Acta Arithm., 9(1): 41-45 (1964).
80. P. Erdös, L. Few, and C. A. Rogers, "The amount of overlapping in partial coverings of space by equal spheres," Mathematatika, 11(2): 171-184 (1964).
81. G. Fejes-Tóth, "Über die Blockierungszahl einer Kreispackung," Elem. Math., 19(3): 49-53 (1964).
82. L. Fejes-Tóth, Lagerungen in der Ebene, auf der Kugel und im Raum, Springer-Verlag, Berlin (1953). [Russian translation: Distributions on a Plane, on a Sphere, and in Space, Gos. Izd. Fiz.-Mat. Lit., Moscow (1958), 363 pp.]
83. L. Fejes-Tóth, "Kugelunterdeckungen und Kugelüberdeckungen in Räumen konstanter Krümmung," Arch. Math., 10(4): 307-313 (1959).
84. L. Fejes-Tóth, "Verdeckungen einer Kugel durch Kugeln," Publ. Math. Debrecen, 6: 234-240 (1959).
85. L. Fejes-Tóth, "Neuere Ergebnisse in der diskreten Geometrie," Elem. Math., 15(2): 25-36 (1960).
86. L. Fejes-Tóth, "Dichteste Kreispackungen auf einem Zylinder," Elem. Math., 17(2): 30-33 (1962).
87. L. Fejes-Tóth, "Mi a 'Diszkrét geometria?'" Magyar Tud. Akad. Mat. Fiz. Tud. Oszt. Közl., 13(3): 229-238 (1963).
88. L. Fejes-Tóth, "Üjabb eredmények a diszkrét geometriában," Magyar Tud. Akad. Mat. Fiz. Tud. Oszt. Közl., 13(4): 341-354 (1963).
89. L. Fejes-Tóth, "Isoperimetric problems concerning tessellations," Acta Math. Acad. Sci. Hung., 14(3-4): 343-351 (1963).
90. L. Fejes-Tóth, Regular Figures, Pergamon Press, Oxford (1964), Chap. XI.
91. L. Fejes-Tóth, "Ungelöste Probleme," Elem. Math., 20(3): 61-63 (1965).
92. L. Fejes-Tóth, "Minkowskian distribution of discs," Proc. Am. Math. Soc., 16(5): 999-1004 (1965).
93. L. Fejes-Tóth, "Distribution of points in the elliptic plane," Acta Math. Acad. Sci. Hung., 16(3-4): 437-440 (1965).
94. L. Fejes-Tóth, "On the permeability of a circle-layer," Studia Sci. Math. Hung. (ex Akad. Mat. Kutato Int. Közl.), 1(1-2): 5-10 (1966).
95. L. Fejes-Tóth, "Mehrfache Kreisunterdeckungen und Kreisüberdeckungen auf der Kugel," Elem. Math., 21: 34-35 (1966).
96. L. Fejes-Tóth, "Packings and coverings in the plane," in: Proc. Colloq. Convexity, Copenhagen, 1965, Copenhagen (1967), pp. 78-87.

97. L. Fejes-Tóth, "Minkowskian circle-aggregates," Math. Ann., 171(2): 97-103 (1967).
98. L. Fejes-Tóth, "On the arrangement of houses in a housing estate," Studia Sci. Math. Hung., 2(1-2): 37-42 (1967).
99. L. Fejes-Tóth, and A. Heppes, "Über stabile Körpersysteme," Compositio Math., 15(2): 119-126 (1963).
100. L. Few, "Covering of space by spheres," Mathematika, 3(6): 136-139 (1956).
101. L. Few, "Multiple packing of spheres," J. London Math. Soc., 39(1): 51-54 (1964).
102. L. Few, "Multiple packing of spheres: a survey," in: Proc. Colloq. Convexity, Copenhagen, 1965, Copenhagen (1967), pp. 88-93.
103. W. J. Firey and H. Groemer, "Convex polyhedra which cover a convex set," J. London Math. Soc., 39(2): 261-266 (1964).
104. A. Florian, "Zum Problem der dünnsten Kreisüberdeckung der Ebene," Acta Math. Acad. Sci. Hung., 13(3-4): 397-400 (1962).
105. A. Florian, "Dichteste Packung inkongruenter Kreise," Monatsh. Math., 67(3): 229-242 (1963).
106. C. W. L. Garner, "Polyhedra and honeycombs in hyperbolic space," Doctoral Dissertation, University of Toronto, 121 pp, 1964. Dissertation Abstr., 26(5): 2775 (1965).
107. C. W. L. Garner, "Coordinates for vertices of regular honeycombs in hyperbolic space," Proc. Roy. Soc., A293(1432): 94-107 (1966).
108. E. N. Gilbert, "Randomly packed and solidly packed spheres," Can. J. Math., 16(2): 286-298 (1964).
109. M. Goldberg, "Packing of 33 equal circles on a sphere," Elem. Math., 18(5): 99-100 (1963).
110. M. Goldber, "Packing of 18 equal circles on a sphere," Elem. Math., 20(3): 59-61 (1965).
111. M. Goldberg and B. M. Stewart, "A dissection problem for sets of polygons," Am. Math. Monthly, 71(10): 1077-1095 (1965).
112. S. W. Golomb, "Replicating figures in the plane," Math. Gaz., 48(366): 403-412 (1964).
113. R. L. Graham, "On partitions of an equilateral triangle," Can. J. Math., 19(2): 394-409 (1967).
114. H. Groemer, "Über die dichteste gitterförmige Lagerung kongruenter Tetraeder," Monatsh. Math., 55(1): 12-15 (1962).
115. H. Groemer, "Über Zerlegungen des Euklidischen Raumes," Math. Z., 79(4): 364-375 (1962).
116. H. Groemer, "Eine Bemerkung über die Überdeckung des Raumes mit unsymmetrischen konvexen Körpern," Monatsch. Math., 66(5): 410-416 (1962).
117. H. Groemer, "Existenzsätze für Lagerungen im Euklidischen Raum," Math. Z., 81(3): 260-276 (1963).
118. H. Groemer, "Über die Zerlegung eines konvexen Körpers in homothetische konvexe Körper," Math. Ann., 154(1): 88-102 (1964).
119. H. Groemer, "Über die Zerlegung des Raumes in homothetische konvexe Körper," Monatsh. Math., 68(1): 21-32 (1964).

120. H. Groemer, "Über Würfel- und Raumzerlegungen," Elem. Math., 19(3): 25-27 (1964).
121. H. Groemer, "Über gewiss dichteste Anordnungen von Punkten in der Ebene," Arch. Math., 15(4-5): 385-387 (1964).
122. H. Groemer, "Zusammenhängende Lagerungen konvexer Körper," Math. Z., 94(1): 66-78 (1966).
123. B. Grünbaum, "On intersections of similar sets," Portug. Math., 18(3-4): 155-164 (1959).
124. G. Hajós, "Über Kreiswolken,"Ann. Univ. Sci. Budapest. Sec. Math., 7: 55-57 (1964).
125. R. J. Hans, "Some results on lattice packings and coverings," Dotoral Dissertation, Ohio State University, 1965. Dissertation Abstr., B27(1): 229 (1966).
126. R. J. Hans, "Extremal packing and covering sets," Monatsh. Math., 71(3): 203-213 (1967).
127. A. Heppes, "Mehrfache gitterförmige Kreislagerungen in der Ebene," Acta Math. Acad. Sci. Hung., 10(1-2): 141-148 (1959).
128. A. Heppes, "Über Kreis- und Kugelwolken," Acta Math. Acad. Sci. Hung., 12(1-2): 209-214 (1961).
129. A. Heppes, "Filling of a domain by discs," Magyar Tud. Akad. Mat. Kutato Inf. Közl., 8(3): 363-371 (1964).
130. A. Hepper, "Isogonale sphärische Netze," Ann. Univ. Sci. Budapest. Sec. Math., 7: 41-48 [1963 (1964)].
131. A. Heppes, "On the number of spheres which can hide a given sphere," Can. J. Math., 19(2): 413-418 (1967).
132. A. Heppes, "On the densest packing of circles not blocking each other," Studia Sci. Math. Hung., 2(1-2): 257-263 (1967).
133. A. Heppes, "On the densest packing of circles not blocking each other," in: Proc. Colloq. Convexity, Copenhagen, 1965, Copenhagen (1967), p. 142.
134. A. Heppes, and Jozsef Molnar, "Ujabb eredmények a diszkret geometriaban. II," Mat. Lapok, 13(1-2): 39-72 (1962).
135. A. Heppes and Jozsef Molnar, "Ujabb eredmenyek a diszkret geometriaban. III," Mat. Lapok, 16(1-2): 19-41 (1965).
136. K. E. Hirst, "The Apollonian packing of circles," J. London Math. Soc., 42(2): 281-291 (1967).
137. E. Hlawka, "Überdeckung durch konvexe Scheiben," Sitzber. Berliner Math. Ges., pp. 28-36 (1961-1964).
138. M. J. Hogendijk, "Random dense packing of spheres with a discrete distribution of the radii," Philips Res. Rept., 18(2): 109-126 (1963).
139. J. Horváth, "Über die regulären Mosaike der hyperbolischen Ebene," Ann. Univ. Sci. Budapest. Sec. Math., 7: 49-53 (1964).
140. J. Horvath, "Bemerkungen zur Theorie der Planigone," Ann. Univ. Sci. Budapest. Sec. Math., 8: 147-153 (1965).
141. M. Imre, "Kreislagerungen auf Flächen konstanter Krümmung," Acta Math. Acad. Sci. Hung., 15(1-2): 115-121 (1964).
142. E. Jucovic, "Über die minimale Dicke einer k-fachen Kreiswolke," Ann. Univ. Sci. Budapest. Sec. Math., 9: 143-146 (1966).

143. E. Jucovič and J. Lešo "Eine Bemerkung Überdeckung der Ebene durch inkongruente Kreise," Mat.-Fyz. Časop., 16(4): 324-328 (1966).
144. S. Kravitz, "Packing cylinders into cylindrical containers," Math. Mag., 40(2): 65-71 (1967).
145. D. G. Larman, "A note on the Besicovitch dimension of the closest packing of spheres in R_n," Proc. Cambridge Phil. Soc., 62(2): 193-195 (1966).
146. D. G. Larman, "On the exponent of convergence of a packing of spheres," Mathematika, 13(1): 57-59 (1966).
147. D. G. Larman, "An asymptotic bound for the residual area of a packing of discs," Proc. Cambridge Phil. Soc., 62(4): 699-704 (1966).
148. D. G. Larman, "On the Besicovitch dimension of the residual set of arbitrarily packed discs in the plane," J. London Math. Soc., 42(2): 292-302 (1967).
149. Z. Laszlo, "Az egységgömböt kitöltö körrendszerek kerület-és sugarösszegének vizsgálata," Magyar Tud. Akad. Mat. Fiz. Tud. Oszt. Közl., 16(1): 17-32 (1966).
150. J. Leech, "Some sphere packing in higher space," Can. J. Math., 16(4): 657 682 (1964).
151. J. Leech, "Notes on sphere packings," Can. J. Math., 19(2) : 251-267 (1967).
152. C. G. Lekkerkerker, "Stapelen en overdekken van figuren in net platte vlak," Rept. Math. Centrum, No. 8 (1964).
153. A. Lelek, "O pokryciach i rozkladach figur geometrycznych," Matematyka 16(3): 81-85 (1963).
154. N. Lindecrantz, "Method for optimum cutting of rectangular sheets," Nord. Tidskr. Informationsbehandl., 4(1): 30-35 (1964).
155. H. Lindgren, "Dissecting the decagon," Mat. Gaz., 46(358): 305-306 (1962).
156. H. Lindgren, Geometric Dissections, D. Van Nostrand Co., Princeton, N. J., (1964).
157. Z. A. Melzak, "Infinite packings of discs," Can. J Math., 18(4): 838-852 (1966).
158. J. Molnar, "Unterdeckung und Überdeckung der Ebene durch Kreise," Ann. Univ. Sci. Budapest. Sec. Math., 2: 33-40 (1959).
159. J. Molnar, "Körelhelyezések allando görbületü felüteken," Magyar Tud. Akad. Mat. Fiz. Tud. Oszt. Közl., 12(3): 223-263 (1962).
160. J. Molnar, "Estensione del teorema di Segre-Maler allo spazio," Atti Accad. Naz. Lincei. Rend. Cl. Sci. Fis. Mat. e Natur., 35(3-4): 166-168 (1963).
161. J. Molnar, "Sui sistemi di punti con esigenza di spazio," Atti Accad. Naz. Lincei. Rend. Cl. Sci. Fis. Mat. e Natur, 36(3): 336-339 (1964).
162. J. Molnar, "Térigényes körelhelyezésekröl," Magyar Tud. Akad. Mat. Fiz. Tud. Oszt. Közl., 14(2): 113-136 (1964).
163. J. Molnar, "Kreislagerungen auf Flächen konstanter Krümmung," Math. Ann., 158(5): 365-376 (1965).
164. J. Molnar, "Aggregati di cerchi di Minkowski," Ann. Mat. Pura ed Appl., 71: 103-108 (1966).
165. J. Molnar, "Collocazioni di cerchi con esigenza di spazio," Ann. Univ. Sci. Budapest. Sec. Math., 9: 71-86 (1966).
166. J. Molnar, "Kreispackungen und Kreisüberdeckungen auf Flächen konstanter Krümmung," Acta Math. Acad. Sci. Hung., 18(1-2): 243-251 (1967).

167. J. W. Moon and L. Moser, "Triangular dissections of N-gons," Can. Math. Bull., 6(2): 175-178 (1963).
168. J. W. Moon and L. Moser, "Some packing and covering theorems," Colloq. Math, 17(1): 103-110 (1967).
169. N. Oler, "The slackness of finite packing in E_2," Am. Math. Monthly, 69(6): 511-514 (1962).
170. N. Oler, "Packings with lacunae," Duke Math. J., 33(1):141-144 (1966).
171. V. Polak, "On one L. Fejes-Tóth's problem of stable packing," Spisy Prirodovéd. Fak. Univ. Brne, No. 9, pp. 433-447 (1962).
172. V. Polak and N. Polakova, "Remarks to some problems of discrete geometry. Part I-II," Spisy Přirodovéd. Fak. Univ. Brne, No. 6, pp. 293-323 (1964).
173. H. Rademacher, R. Dickson, and M. Plotkin, "A packing problem for parallelepipeds," J. Combin. Theory, 1(1): 3-14 (1966).
174. R. M. Robinson, "Arrangement of 24 points on a sphere," Math. Ann., 144(1): 17-48 (1961).
175. R. M. Robinson, "Packing closed sets," Fundam. Math., 56(2): 189-198 (1964).
176. C. A. Rogers, "Existence theorems in the geometry of numbers," Ann. Math., 48: 994-1002 (1947).
177. C. A. Rogers, "The packing of the equal spheres," Proc. London Math. Soc., 8(32): 609-620 (1958).
178. C. A. Rogers, "Lattice covering of space," Mathematika, 6(1): 33-39 (1959).
179. C. A. Rogers, "Covering a sphere with spheres," Mathematika, 10(20): 157-164 (1963).
180. C. A. Rogers, Packing and Covering, Cambridge University Press, Cambridge (1964), Chap. VIII.
181. J. Schaer, "The densest packing of 9 circles in a square," Can. Math. Bull., 8(3): 273-277 (1965).
182. J. Schaer, "On the densest packing of spheres in a cube," Can. Math. Bull., 9(3): 265-270 (1966).
183. J. Schaer, "The densest packing of five spheres in a cube," Can. Math. Bull. 9(3): 271-274 (1966).
184. J. Schaer, "The densest packing of six spheres in a cube," Can. Math. Bull., 9(3): 275-280 (1966).
185. J. Schaer and A. Meir, "On a geometric extremum problem," Can. Math. Bull., 8(1): 21-27 (1965).
186. U. Sinogovitz, "Die Kreislagen und Packungen kongruenter Kreise in der Ebene," Z. Krist., 100: 461-508 (1939).
187. J. Smalley, "Simple regular sphere packing in three dimensions," Math. Mag., 36(5): 295-299 (1963).
188. L. Stacho, "Über ein Problem für Kreisscheibenfamilien," Acta Sci. Math., 26(3-4): 273-282 (1965).
189. J. Strohmajer, "Über die Verteilungen von Punkten auf der Kugel," Ann. Univ. Sci. Budapest. Sec. Math., 6: 49-53 (1963).
190. G. B. Trustrum, "Mrs. Perkins's quilt," Proc. Cambridge Phil. Soc., 61(1): 7-11 (1965).